Wireless for the Newton®

· SOFTWARE DEVELOPMENT FOR MOBILE COMMUNICATIONS ·

LIMITED WARRANTY AND DISCLAIMER OF LIABILITY

ACADEMIC PRESS, INC. ("AP") AND ANYONE ELSE WHO HAS BEEN INVOLVED IN THE CREATION OR PRODUCTION OF THE ACCOMPANYING CODE ("THE PRODUCT") CANNOT AND DO NOT WARRANT THE PERFORMANCE OR RESULTS THAT MAY BE OBTAINED BY USING THE PRODUCT. THE PRODUCT IS SOLD "AS IS" WITHOUT WARRANTY OF ANY KIND (EXCEPT AS HEREAFTER DESCRIBED), EITHER EXPRESSED OR IMPLIED, INCLUDING, BUT NOT LIMITED TO, ANY WARRANTY OF PERFORMANCE OR ANY IMPLIED WARRANTY OF MERCHANTABILITY OR FITNESS FOR ANY PARTICULAR PURPOSE. AP WARRANTS ONLY THAT THE MAGNETIC DISKETTE(S) ON WHICH THE CODE IS RECORDED IS FREE FROM DEFECTS IN MATERIAL AND FAULTY WORKMANSHIP UNDER THE NORMAL USE AND SERVICE FOR A PERIOD OF NINETY (90) DAYS FROM THE DATE THE PRODUCT IS DELIVERED. THE PURCHASER'S SOLE AND EXCLUSIVE REMEDY IN THE EVENT OF A DEFECT IS EXPRESSLY LIMITED TO EITHER REPLACEMENT OF THE DISKETTE(S) OR REFUND OF THE PURCHASE PRICE, AT AP'S SOLE DISCRETION.

IN NO EVENT, WHETHER AS A RESULT OF BREACH OF CONTRACT, WARRANTY OR TORT (INCLUDING NEGLIGENCE), WILL AP OR ANYONE WHO HAS BEEN INVOLVED IN THE CREATION OR PRODUCTION OF THE PRODUCT BE LIABLE TO PURCHASER FOR ANY DAMAGES, INCLUDING ANY LOST PROFITS, LOST SAVINGS OR OTHER INCIDENTAL OR CONSEQUENTIAL DAMAGES ARISING OUT OF THE USE OR INABILITY TO USE THE PRODUCT OR ANY MODIFICATIONS THEREOF, OR DUE TO THE CONTENTS OF THE CODE, EVEN IF AP HAS BEEN ADVISED OF THE POSSIBILITY OF SUCH DAMAGES, OR FOR ANY CLAIM BY ANY OTHER PARTY.

Any request for replacement of a defective diskette must be postage prepaid and must be accompanied by the original defective diskette, your mailing address and telephone number, and proof of date of purchase and purchase price. Send such requests, stating the nature of the problem, to Academic Press Customer Service, 6277 Sea Harbor Drive, Orlando, FL 32887, 1-800-321-5068. APP shall have no obligation to refund the purchase price or to replace a diskette based on claims of defects in the nature or operation of the Product.

Some states do not allow limitation on how long an implied warranty lasts, nor exclusions or limitations of incidental or consequential damage, so the above limitations and exclusions may not apply to you. This Warranty gives you specific legal rights, and you may also have other rights which vary from jurisdiction to jurisdiction.

THE RE-EXPORT OF UNITED STATES ORIGIN SOFTWARE IS SUBJECT TO THE UNITED STATES LAWS UNDER THE EXPORT ADMINISTRATION ACT OF 1969 AS AMENDED. ANY FURTHER SALE OF THE PRODUCT SHALL BE IN COMPLIANCE WITH THE UNITED STATES DEPARTMENT OF COMMERCE ADMINISTRATION REGULATIONS. COMPLIANCE WITH SUCH REGULATIONS IS YOUR RESPONSIBILITY AND NOT THE RESPONSIBILITY OF AP.

Wireless for the Newton®

· SOFTWARE DEVELOPMENT FOR MOBILE COMMUNICATIONS ·

Julie McKeehan · Neil Rhodes

AP PROFESSIONAL

Boston San Diego New York
London Sydney Tokyo Toronto

This book is printed on acid-free paper. ♾

AP PROFESSIONAL
1300 Boylston Street, Chestnut Hill, MA 02167

An imprint of ACADEMIC PRESS, INC.
A Division of HARCOURT BRACE & COMPANY

United Kingdom Edition published by
ACADEMIC PRESS LIMITED
24–28 Oval Road, London NW1 7DX

Library of Congress Cataloging-in-Publication Data

McKeehan, Julie.
Wireless for the Newton: Software Development for Mobile Communications/Julie McKeehan and Neil Rhodes.
p. cm.
Includes bibliographical references and index.
ISBN 0-12-484801-X
1. Newton (Computer)—Programming. 2. Computer software—Development. 3. Mobile communication systems—Computer programs.
I. Rhodes, Neil, 1962- . II. Title.
QA76.8.N48M34 1995
005.265—dc20 95-3658
CIP

Printed in the United States of America
95 96 97 98 IP 9 8 7 6 5 4 3 2 1

For Fred O. Francis, beloved teacher.
For helping me learn that what you
don't know may be the most fun yet.
—JM

To my father, Fen.
—NR

Contents

Acknowledgments

It has been an exciting year. It was just 12 months ago that we finished the first book on this subject. As was the case with that book, here as well, we have many people we would like to acknowledge for their contributions to this work.

First of all we would like to thank the people on the Newton team—everyone from the people in Developer Support, to the programmers and managers on the Newton and NTK teams. To Gabriel Acosta-Lopez, a very good friend, funny guy, and innovative leader in PIE developer training. He is more responsible than most for seeing to it that this book actually got written. First, by making us (against our will) write articles on Newton programming that formed the basis of the book, and then for giving us the time to turn those articles into a book.

We also want to acknowledge Mike Engber on the Newton team and Christopher Bell, Bob Ebert, Jim Schram, and Maurice Sharp of PIE Developer Technical Support. They have gone over our training materials for errors more times that we can remember. They also reviewed this manuscript for errors that might have crept in—catching quite a few we might add. It is great to have the generous assistance of such fine people. There is also the important contribution of people like James Joaquin and Tony Espinoza, without whose efforts there would have been no demonstration version of NTK in either this book or the last. If you find the demo version of NTK useful, you have them specifically to thank for its presence.

The contributions of our Newton programming students at Apple Developer University are also littered throughout this book. From them, we learned what subjects we explained well and which subjects we needed to work on more. The folks at Apple may have seen to it that we understood Newton correctly; but our students patiently helped us refine the training materials, and in the long run made this book all the better for it.

Thanks to Jay Galvin and Rob Sirota of Motorola for their help in the Marco programming portion of the Communications chapter. They generously loaned us not only a Marco Newton, but technical aid, and documentation at crucial points in the creation of this manuscript

On the book end of things, there are many people to whom we owe a debt of gratitude. We particularly thank our editor, Charles Glaser, at AP PROFESSIONAL.

He is the kind of editor who understands that we have lives that, at times, involve neither writing nor teaching. We thank him immensely for enriching both our personal and business endeavors and look forward to many years of friendship together. Authors could *still* have no better editor. We want to thank all the other people of AP PROFESSIONAL for their work in copy editing and promotion, particularly Mary Treseler for her ongoing management of our books. There is also our agent, Carole McClendon, whose skill and professionalism found us this publisher.

Last of all, we have some personal acknowledgments. This book could not have happened without the unending support of people in our day-to-day lives. Their support is all the more precious because they put up with us while we write. They do so, not because they are enamored with either the Newton, or new technologies, or technical books, but rather in spite of those things. To our children, who most especially dislike our frequent disappearances to write, we give our thanks and love. To our friends and family, thank you once again to putting up with people who occasionally drop off the face of the earth. If they did not love us and thereby grant us enormous amounts of slack—there would be no books, *ever.* Our heartfelt and loving gratitude to you all.

—Julie McKeehan and Neil Rhodes
March 1995

Preface

No pessimist ever discovered the secrets of the stars, or sailed to an uncharted land, or opened a new heaven to the human spirit.

—*Helen Keller*

Here is what we have learned after two years—Newton programming is fun. While it may be true that the press has, at times, been unkind to the Newton we believe it is because they do not know where this product is going. The Macintosh in 1984 was not at all what it has become today. Just so, the Newton will evolve along just as amazing a path if people have the vision and patience to see it through. What was the case, however, two years ago, is still the case now: any good programmer can sit down and write a Newton program in months (not years) of time. The Newton development environment makes customizing software a snap, and a joy as well. Newtons and NewtonScript really are fun and we think worth your time.

So, "Why Newton?" you might ask. Because Newton is part of our future, and the direction in which computing will move. Wireless technology will reshape the world, just as the personal computer did before it. So the question you should ask yourself is simple, "Do you want to help or watch from the sidelines?" Try and imagine what your software and a Newton could accomplish: for that is what the Newton needs, your imaginative software. The hardware may be coming along nicely, but it will only end up as so much landfill without really good software.

This book picks up where we left off in the last one (*Programming for the Newton® Software Development with NewtonScript*™). Advanced aspects of Newton programming and wireless technology are all covered in one fell swoop. Plus, with the help of Apple Computer, we have provided you with a full-featured demonstration version of NTK. Thus, we hope that the two books give you everything you need to write the best Newton program in the universe.

There is a floppy disk at the back of the book. It contains many things that will be of interest to you. Among these things are:

- A full-featured demonstration version of Newton Toolkit, the development environment for the Newton.
- Several sample applications that we create within the book.
- Demo versions of the ViewFrame debugger and Newt-RTFM electronic documentation.

How to Read this Book

You should read this preface first. Next, you should read Appendix F on page 393 if you have not used NTK before. This will familiarize you with the development environment before you begin using it. It is also there that you can learn about the demonstration version of NTK included in this book (see "Differences between Demo NTK and NTK" on page 396).

After that, what material you read is up to you. For the most part, each subject is self-contained in a chapter and does not require knowledge from other chapters (exceptions are the Routing, Communications, Filing, and Find chapters which build on one another). This book does assume, however, that you already have a basic understanding of NewtonScript and programming on the Newton. Without that material, you will quickly become lost. This book is meant to be a supplemental volume and not a comprehensive introduction to Newton programming.

The Structure of the Book

The chapters are loosely ordered. Roughly the first third of the book deals with communications on the Newton. First we discuss routing features which are standard on the Newton and then discuss wireless technology and programming. The last two-thirds of the book deals with a potpourri of subjects (basically everything there wasn't room for in the first book). You can, of course, read straight through the book, but should feel free to skip a topic if you already understand it.

The Structure of Each Chapter

Most chapters have been divided into sections:

- First, we discuss a topic and give you a general overview of the material.
- Next, we cover any relevant Newton User Interface issues.
- Third, we give you a step-by-step implementation guide for the feature we are covering.
- In most cases, we also give you a review checklist at the end of each chapter for later reference.

What You Need to Know to Program the Newton

Programming Experience

Here are our expectations concerning you as a reader:

- You already know how to program.
- You already posses some basic NewtonScript knowledge.
- You have done a little bit of Newton programming.

Our book makes no effort to explain standard programming concepts such as parameters, functions, or variables or other keys concepts such as NewtonScript inheritance.

If you don't have experience programming the Newton, a good introduction is *Programming for the Newton: Software Development with NewtonScript*™ by Julie McKeehan and Neil Rhodes (ISBN 0-12-484800-1).

Newton Experience

This book does expect you to be reasonably familiar with a Newton as a user and have at least tried some Newton applications.

About Errors in This Book

Even though many people gave time and energy to see to it that this book is without mistake, some no doubt still remain. While the presence of such errors is solely our responsibility, we would be grateful if you report any that you find in the text or code. If you find an error we are unaware of, we will also gratefully acknowledge you in the next edition. We read and keep every suggestion that readers send to us or the publisher (both email and letters work). So you know, we already have the list made of people who sent us material on the first book. Their names will go into that book's new edition.

You can send suggestions or corrections to:

NeilRhodes@eworld.com
JMcKeehan@eworld.com

or:

Calliope Enterprises, Inc.
700 E. Redlands Blvd., Suite 154
Redlands, CA 92373

Chapter 1

"Beam me up, Scotty!"

—— *James T. Kirk*

An Overview of Routing
The Routing Frame and Required Slots
Implementing the Routing User Interface
Implementing the Routing Frame
Duplicate
Delete
Card Actions
Undo
Beaming
Printing
Faxing
Mailing
A Quick Guide to Routing
Summary

Sending something somewhere is one of the things that the Newton does best. This sending—or routing as it is called on the Newton—can be done by faxing, printing, beaming or mailing an item. Along with these routing capabilities, we will also cover how to duplicate and delete an item, as well as how to "Undo" the duplicate or delete. But to begin our whole discussion, let's get a bird's-eye view of routing and its user interface elements.

An Overview of Routing

Fully implementing routing within an application involves a variety of tasks. Most of these tasks, however, are handled within the same piece of code, the routing frame. Once you have created a routing frame and added some appropriate slots (containing both variables and methods) to your application you will have implemented 90% of routing. The implementation process is cumulative; less is required for each consecutive feature. Indeed, when you get to mailing, you will only be writing a few lines of code.

FIGURE 1.1 shows a typical routing process in action. In each case, the user picks a routing action, the routing frame gets accessed, and then the appropriate subframe (like print, fax, or beam) gets accessed in turn. The subframe contains a symbol which refers to a slip to be displayed (like `'zapSlip`), or a message to be sent to your application base view (like `'DuplicateActionScript`).

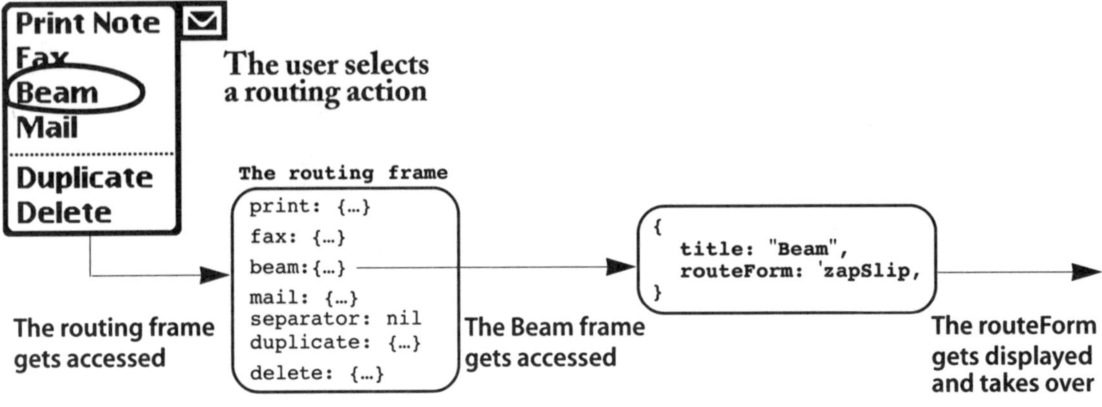

FIGURE 1.1 Parts in the routing process.

The Routing Frame and Required Slots

Your application's routing frame indicates to the Newton operating system and to the user which routing functions it supports. The items that you find listed in a typical routing button usually include Duplicate, Delete, Print, Fax, Mail, and Beam. Each of these items in turn corresponds to a slot in the routing frame:

```
                   routingFrame := {
                      print: …,
                      fax: …,
                      beam: …,
                      mail: …,
A dotted line ——— separator: …,
                      duplicate: …,
                      delete: …
                   }
```

Each of these routing frame slots is itself a frame containing two or three slots of its own. Here is a fully filled-out routing frame to give you a better idea of what it looks like:

```
routingFrame := {
   print: {
      title: "Print Note",
      routeForm: 'printSlip,
      formats: [kFormatSymbol],
   },
   fax: {
      title: "Fax",
      routeForm: 'faxSlip,
      formats: [kFormatSymbol],
   },
   beam: {
      title: "Beam",
      routeForm: 'zapSlip,
    },
   mail: {
      title: "Mail",
      routeForm: 'mailSlip,
      formats: [kFormatSymbol],
   },
   separator: nil, // dotted line in picker
   duplicate: {
      title: "Duplicate",
      routeScript: 'DuplicateActionScript,
   },
```

```
        delete: {
            title: "Delete",
            routeScript: 'DeleteActionScript,
        },
        card: ROM_cardAction,
    };
```

The Newton operating system fills in the items in the action picker in precisely the same order that it finds them in this routing frame (making this the singular case in Newton programming where the order, not the names, of the elements in a frame matters). A slot with a `nil` value specifies a separator line in the picker list.

In the preceding example, the first slot found is `print`. The operating system then looks in the `title` slot or calls the function in the `GetTitle` slot of the print frame to determine what text string to put in the Action button list. Given the above title slots of each action in the routing frame, we would end up with an Action button as shown in FIGURE 1.2.

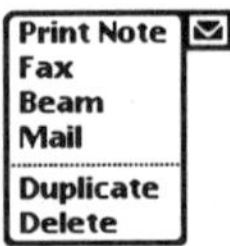

FIGURE 1.2 The Action picker.

A Quick Look at One Action Item

Now, starting with the `print` slot, let's summarize each of the print action's slots and their functions:

```
print: {
    title: "Print Note",
    routeForm: 'printSlip,
    formats: [kFormatSymbol],
  },
```

When the user picks the "Print Note" item in the action button, the symbol in the `routeForm` slot is used to determine what slip to open. In this case, the `routeForm` slot contains a reference to the system symbol `printSlip`.

The `formats` Slot

The last slot in the print frame is a `formats` slot. This contains an array of symbols that specify the formatting styles available to the user. In our example, there is only one style, `kFormatSymbol`. These styles are displayed in the format picker that the user sees when selecting an appropriate routing function (*e.g.*, mailing, printing, faxing, etc.). Format pickers for different routing slips may have the same types of format styles (see FIGURE 1.3) or ones that vary.

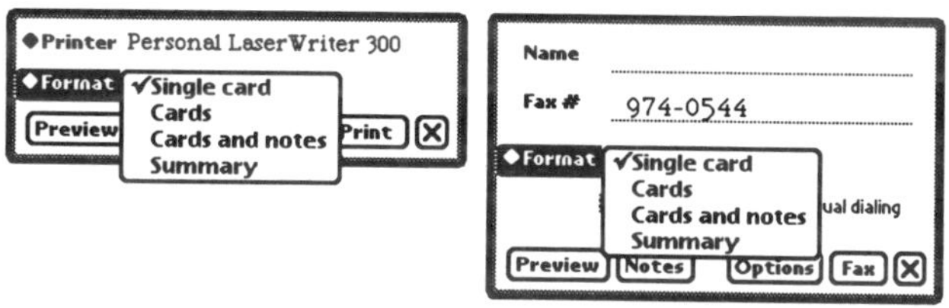

FIGURE 1.3 Different slips with the same format styles.

The Other Slots

With the exception of the `card` slot, the action slot in each action frame is quite similar in its construction. TABLE 1.1 shows an overview of the contents of each action slot.

ACTION FRAME	SLOT IN FRAME	SYMBOL IN SLOT	DESCRIPTION OF SYMBOL
fax	`routeForm`	`'faxSlip`	*System symbol*
beam	`routeForm`	`'zapSlip`	*System symbol*
mail	`routeForm`	`'mailSlip`	*System symbol*
delete	`routeScript`	*`'DeleteActionScript`*	*Your method*
duplicate	`routeScript`	*`'DuplicateActionScript`*	*Your method*

TABLE 1.1 Action slots in the routing frame and their contents.

Required Application Slots

You will have to ensure that the system knows about your routing capabilities. Thus, when your application is installed, its routing frame (typically stored as a slot in the base template) needs to be installed into the global routing frame. The following lines in your application's `InstallScript` will accomplish this:

```
route := partFrame.theForm.routingFrame;
routing.(kAppSymbol) := route;
```

Routing also requires the addition of certain slots to your application base view. These slots are `appSymbol`, `target`, `targetView`. Here is a description of each.

`appSymbol`	This slot contains your application symbol.
`target`	This slot identifies the frame that is being acted on by the routing action. It is up to you to maintain the contents of this slot.
`targetView`	This slot points to the view that is being routed. It is passed as a parameter to the `routeScript` methods. Generally, the value of this slot will be your application base view. It is up to you to maintain the contents of this slot.

Implementing the Routing User Interface

All routing actions are implemented using a `protoActionButton`. You can attach these buttons to either the application as a whole, or to any particular part of it. As you can see in FIGURE 1.4, `protoActionButtons` can contain the entire set of routing actions or a subset of choices.

FIGURE 1.4 The button from which routing actions are chosen.

Where to Put **protoActionButtons**

Generally, `protoActionButtons` are attached to some type of status bar in an application. If the button is for the entire application, then it resides in the same location regardless of application. It is the second to last button on the right-hand side of the status bar (on the left of the close box). Looking in FIGURE I.5 at one built-in application and two custom applications, you can see the placement of action buttons for an entire application.

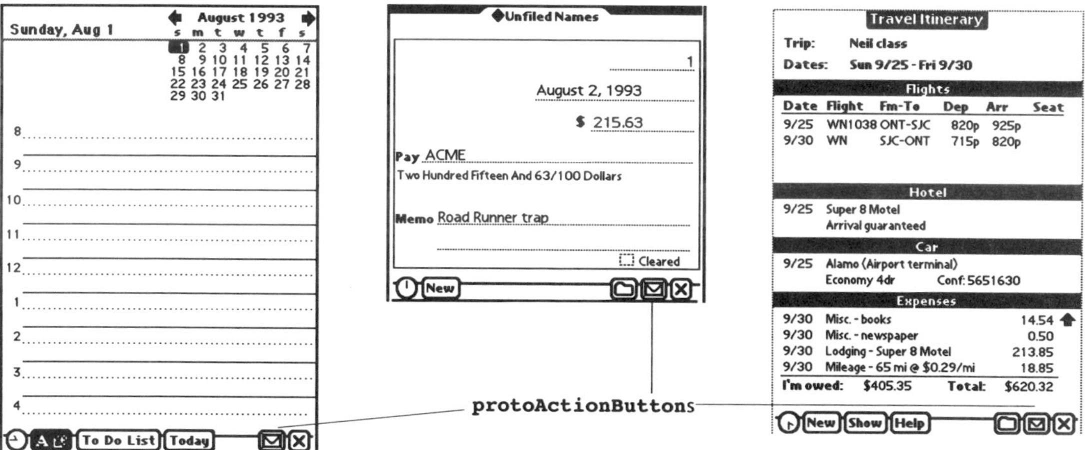

FIGURE I.5 Application level `protoActionButtons`.

If you are adding an action button to some smaller part of an application, then its placement is less straightforward. Certainly one useful approach is to leave the action button in the same place but have it affect whatever view is displayed upon the screen; if you are in the overview it affects the whole application, if you are in a display view then it acts on whatever the view contains.

If you are using borders with embedded action buttons similar to the built-in Notes application, then the placement of your action buttons should be identical to it. In this case, at the right-hand side of each displayed bar. FIGURE I.6 shows a set of action buttons within an application.

Without a status bar present, placing the action button is not as straightforward. Basically, the rule of thumb you should follow is placing it in some logical relation to the element that it is affecting. You should also show a very strong preference for the right-hand side of the element—thereby maintaining consistency with the placement of other `protoActionButtons`.

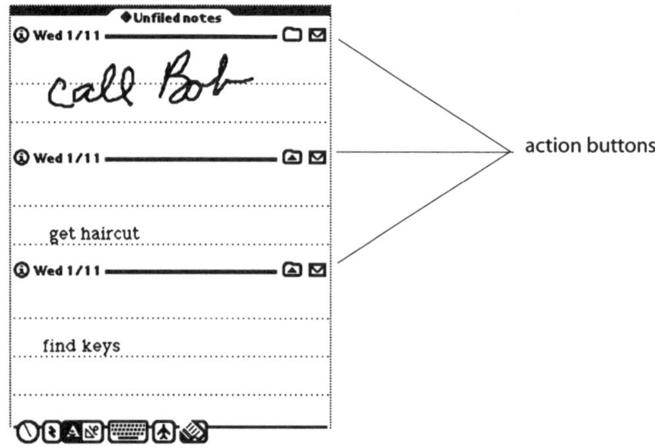

FIGURE 1.6 Action buttons within applications.

Using a `protoActionButton` *in an Application*

If you are using a `protoApp` for your application, you'll need to switch your base template to a `clView` instead. This is because of the `protoActionButton`. Unfortunately, you can't make the `protoActionButton` a child of the status bar in a `protoApp`. Changing to a `clView` provides you with this needed access to the status bar.

Here are the steps for turning a `protoApp` application into a `clView` one:

1. First, create `viewJustify` and `viewFlags` slots in the base template (this will create slots with the default `protoApp` values).

2. Change the `_proto` slot of the base template to a `clView` (the slot will change to a `viewClass` slot automatically).

3. Draw a `protoTitle` at the top of the base template. Set its `title` to the title of your application.

4. Draw a `protoStatus` (not a `protoStatusBar`, which has no close box) in the base template. It should automatically draw itself at the bottom of the base template.

5. The final modification to the base template is to add a `declareSelf` slot. Set it to `'base` (if it is not already done). This is similar to

declaring it; at run time this adds a `base` slot to the base that points to itself. This is used by the close box, whose action when tapped is to call `base:Close()`.

Implementing the Routing Frame

Now that the routing user interface issues have been covered, it is time to turn to the most general level of routing implementation: creating the routing frame, adding required slots, and installing the routing frame in the routing global.

The Checklist

Here are the first four steps to routing:

1. Add the `protoActionButton` to the application.
2. Add the application slots: `appSymbol`, `target`, and `targetView`.
3. Hide the action button in the overview.
4. Install and deinstall the routing frame.

1. Adding the `protoActionButton` *to the Application*

Create your `protoActionButton` as a child of the `protoStatus`.

Note: If you already have buttons in the status bar, make the `protoAction-Button` the first child. The justification of the `protoActionButton` assumes that it will be the first child.

---

2. Preparing the Application Slots

The `appSymbol`, `target`, and `targetView` slots must now be set up.

Setting the `appSymbol`

The `appSymbol` can simply be set to the `kAppSymbol` constant you should have already defined in your Project Data file (see FIGURE 1.7).

Setting the `targetView`

The `targetView` slot will normally refer to your application base view; thus you can set it up in the application's `viewSetupFormScript` method with this code:

```
self.targetView := self;
```

Note: The `targetView` might point at some view other than the application base view if your application supports multiple action buttons. As an example, the Notepad application has one `protoActionButton` for each note; in this case, the `targetView` specifies which view will be acted on.

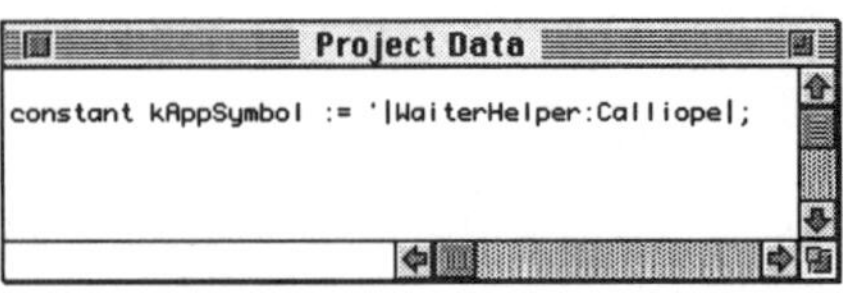

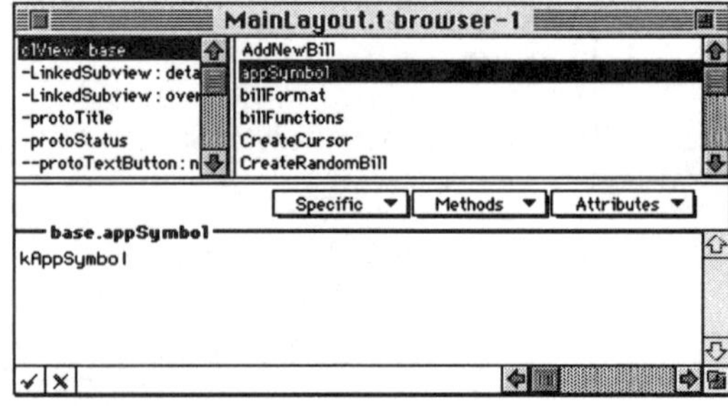

FIGURE 1.7 Setting the `appSymbol` slot.

Setting the `target`

Last of all, the `target` slot needs to be set to the frame that will be acted on by the action button—the frame that will be deleted/duplicated/printed, etc. In most applications, the target points to the currently displayed soup entry.

In the Words sample application, the target will be set in the detail view whenever a new entry is displayed. Here is the detailView's `DisplayWord` method in which the target gets set:

```
func(entry)
begin
   :SaveCurrentEntry();
   self.currentEntry := entry;
   SetValue(wordEdit, 'text,
      Clone(currentEntry.word));
   self.entryHasChanged := nil;
   target := entry;
end
```

Just as you have to set the slot when there is something to act on, so you have to nil it out when there is no available target. For example, if the overview is being displayed, usually there is no target. In such cases, the `target` should be set to `nil`. Here is the `viewShowScript` of the Words application overview that handles that task:

```
func()
begin
   :DisplayWordsFromSoup();

   target := nil;
end
```

3. Hide the Action Button in the Overview

In many applications, it does not make sense to let the user use the action button from the overview. For those applications, the action button should not appear.

In the Words application, when the overview is shown, the action button is hidden, and when the detail view is shown the action button is revealed. To send these hiding and showing messages to the action button, however, it must be named and declared to the application (see FIGURE I.8).

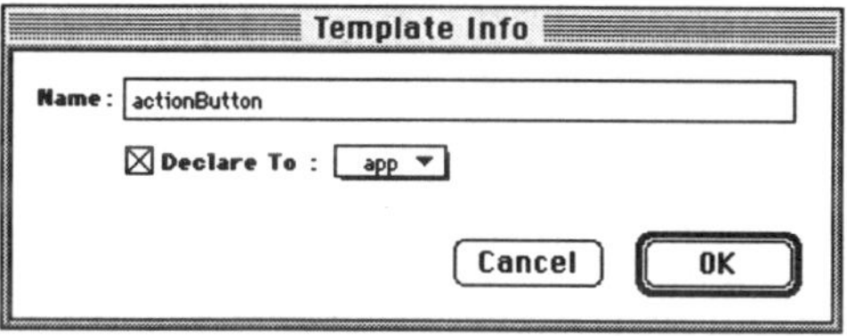

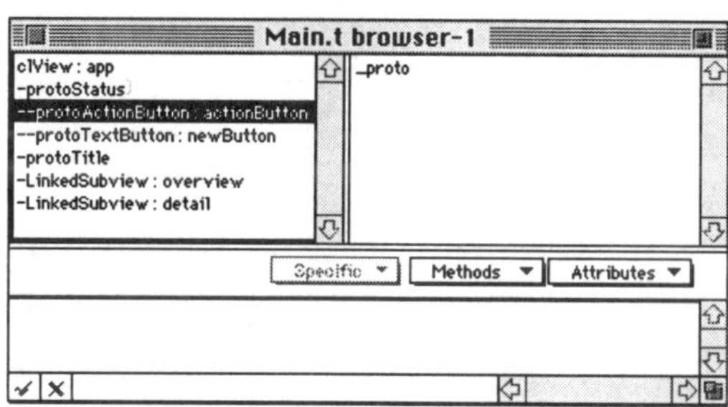

FIGURE I.8 Naming and declaring the action button.

Here is the overview's `viewShowScript` where the hiding occurs:

```
func()
begin
   :DisplayWordsFromSoup();

   actionButton:Hide();
   target := nil;
end
```

Here is the detail's `viewShowScript` where the showing happens:

```
func()
begin
   actionButton:Show();
   :DisplayWord(theCursor:Entry());
end
```

4. Installing and Deinstalling the Routing Frame

As we said earlier, all the items in the action picker are specified as slots in the routing frame. When an application is opened, it must install this routing frame; and when the application is closed, it must deinstall it. This is accomplished by installing your routing frame as a slot in the `routing` global variable. The slot name is the application symbol.

Typically this routing frame is created in NTK as a slot within the application's base template. For now, simply add an empty frame to this slot as shown in FIGURE 1.9. None of the routing actions will work—or show up for that matter—in the button yet.

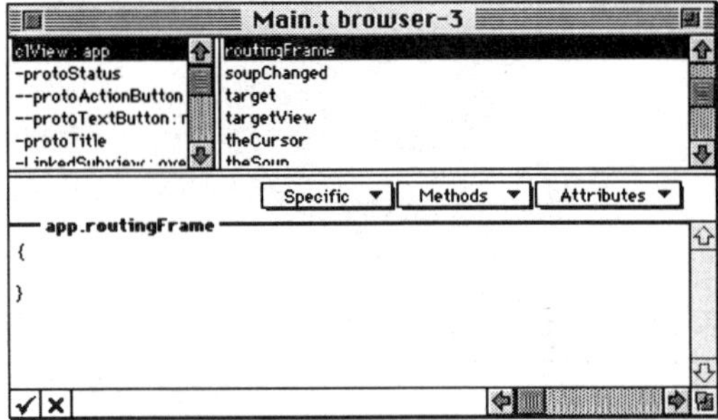

FIGURE 1.9 An empty routing frame in an application.

Then, within your application's `InstallScript`, you write code that installs the `routingFrame` in the `routing` global:

```
route := partFrame.theForm.routingFrame;
routing.(kAppSymbol) := route;
```

To clean up after itself, the application's `RemoveScript` should remove this slot from the `routing` global:

```
RemoveSlot(routing, 'kAppSymbol);
```

At this point, you should now have an action button that appears in your application and have all of the necessary slots added to your application. Each of the necessary slots is noted in FIGURE 1.10. Once these slots are in place you will be in a position to implement individual actions.

We will now discuss these individual routing actions. Notice that while you will be implementing them in the following order, we will be taking a brief detour in the middle to talk about how to "Undo" an action:

- Duplicate
- Delete
- Card Actions
- Beaming
- Printing
- Faxing
- Mailing

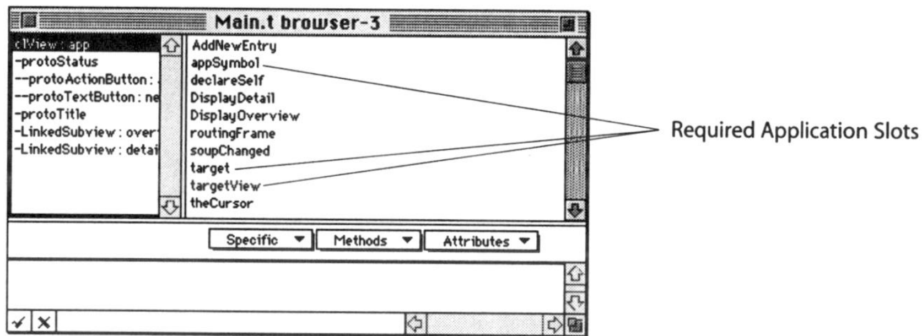

FIGURE 1.10 Necessary slots for routing in the application base template.

Duplicate

To support any item in the action button, you must add a slot for that item to the routing frame. This slot, which is itself a frame, will contain the title of the item in the picker, along with information about what to do if this item is chosen from the picker.

The Duplicate Checklist

There are two tasks here:

1. Add Duplicate to the Routing Frame.
2. Add a `DuplicateActionScript` method as a slot in the base template.

1. Add a Duplicate Slot to the Routing Frame

To implement Duplicate, we will add the following slot in the routing frame:

```
routingFrame:= {
   duplicate:{
      title: "Duplicate item",
      routeScript: 'DuplicateActionScript,
   },
}
```

The italicized portions are those elements that you might wish to name differently in your application. In particular, "*item*" should correctly name whatever item is being acted on. For example, if we had a checkbook application the title slot would contain "Duplicate Check", while an expense application would use, "Duplicate Expense" and so on.

Note: Only the first item in the routing frame names what type of item (check, expense, etc.) is being acted on. Subsequent routing actions should not use the *item* word.

The `routeScript` slot in our duplicate frame contains the message that is sent to your application base view when the user selects this item from the picker. Although you need not name the method in this slot `DuplicateActionScript`, it is common to do so.

The name of the slot, `duplicate`, is unused; although it is common to give it this name.

2. *Add a* `DuplicateActionScript` *to Your Application*

When the user picks the "Duplicate Item" in the action button this script will be executed by the system. What it should do is, of course, duplicate whatever item is being displayed. This method will be in your application base template and will contain the necessary information to accomplish this task.

As an example, here is the `DuplicateActionScript` from the Words application:

```
func(entryToDuplicate, targetView)
begin
   local newEntry := DeepClone(entryToDuplicate);

   theSoup:AddToDefaultStore(newEntry);
   detail:DisplayWord(newEntry);
end;
```

The code makes a complete copy of the current entry, adds it to the soup, and then displays it.

Delete

Implementing Delete in your application is very similar to what you just did for Duplicate. Here is the checklist of the two tasks it requires. There is also an optional third task.

The Checklist

1. Add Delete to the routing frame.
2. Add a `DeleteActionScript` method to your base template.
3. Support the Delete animation in your method.

1. Add a Delete Slot to the Routing Frame

To implement Delete, we will put this code into the routing frame immediately following the duplicate slot:

```
routingFrame:= {

   duplicate: … //previously added

   delete:{
      title: "Delete",
      routeScript: 'DeleteActionScript,
   },
}
```

The `routeScript` slot in our delete frame similarly contains the message that is sent to your application base view when then user selects this item. As was true before, you need not name the method in this slot, `DeleteActionScript`, though it is common to do so. Last of all, `delete`, while not necessary as a name for this slot, is the standard terminology.

2. Add a `DeleteActionScript` *to your Application*

When the user picks "Delete" in the action button this script will be executed by the system. The script is then responsible for deleting whatever item is being displayed. This method is also in your application base template.

Here is the `DeleteActionScript` from the Words application to look at as a guide:

```
func(entryToDelete, targetView)
begin
   local nextEntry;

   EntryRemoveFromSoup(entryToDelete);

   nextEntry := theCursor:Next();
   if not nextEntry then
      nextEntry := theCursor:Prev();

   if not nextEntry then
      :DisplayOverview();
   else
      detail:DisplayWord(nextEntry);
end;
```

The code deletes the specified entry, and then displays the next entry. If there is no next entry, it displays the previous entry. If there is no previous entry, it switches to the overview.

3. Support the Delete Animation

You may have noticed on the Newton that many applications animate the deleting of an item. Adding this animation to your application is very straightforward. In your `DeleteActionScript`, replace the last line of code:

```
detail:DisplayWord(nextEntry);
```

with the following line:

```
detail:Delete('DisplayWord, [nextEntry]);
```

`Delete` is a system method that performs the animation. The effect it provides makes it look like the deleted entry is crumpled up and tossed into a trash can as the new entry slowly appears behind it. This method takes two parameters:

`methodSymbol`	This specifies a message which will draw the new contents of the screen.
`[`*`method parameters`*`]`	Parameters required by *`methodSymbol`*.

The `Delete` method saves a copy of the current screen contents, and then sends the message specifed by the `methodSymbol` to itself. After the new method draws (offscreen), the `Delete` method uses the old screen buffer and the new screen buffer to do a crumpling animation.

Card Actions

Card actions are ways the user can maintain where an item is stored. FIGURE I.11 shows the difference in the action popup depending upon whether a PCMCIA card is present or not.

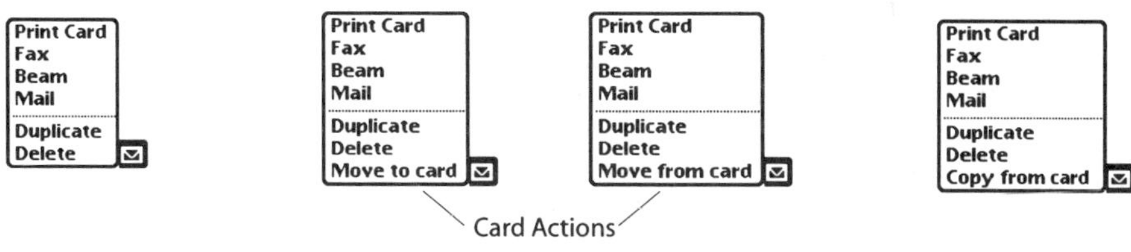

FIGURE I.11 The different forms of the card item in the action menu.

The Checklist

There is only one task necessary for implementing Card actions:

1. Add a card slot to the routing frame.

Add a Card Slot to the Routing Frame

The slot will contain the title of the item in the picker, along with information about what to do if this item is chosen from the picker. Because the title of the picker for card actions varies (Move to Card/Move from Card), the card action uses a function to specify the title. Happily this is one of the easiest items to support in the routing frame. All you have to add is:

```
card: ROM_cardAction,
```

Since each application will have the same title for the card (one of "Move to card", "Move from card", "Copy from card", or nothing) and the same action, there is a frame in the ROM which contains the appropriate slots. One of these slots is a `GetTitle` method which returns the correct title, and an `actionScript` method containing the symbol `'cardActionScript` which carries out the necessary action when the user selects this picker item.

Undo

Now that you have added Duplicate and Delete to your application, this is the logical point to add Undo capabilities as well. This task will allow the user to undo a duplicate or delete request. As you might have expected, your code will do the opposite of what the user had originally requested. Undoing a Delete adds the deleted entry back into the soup and displays it on the screen. Undoing a Duplicate removes the newly duplicated entry from the soup and displays a different entry on the screen.

Adding Undo for Duplicate

It is necessary to complete two tasks to add Undo capabilities to Duplicate.

The Checklist

1. Modify the `DuplicateActionScript` slightly.
2. Write a new method called `UndoDuplicate` and add it to your base template.

1. Modify your DuplicateActionScript

Let us look at the first task. Your new `DuplicateActionScript` will change by only one additional line of code:

```
func(entryToDuplicate, targetView)
begin
  local newEntry := DeepClone(entryToDuplicate);

  theSoup:AddToDefaultStore(newEntry);
  detail:DisplayWord(newEntry);
  AddUndoAction('UndoDuplicate, [newEntry]);
end;
```

New Undo Action

2. Add an UndoDuplicate Method to your Application

`AddUndoAction` is a global function that registers with the system a method that will be called when the user taps the Newton's built-in Undo button. The system keeps track of the last two undoable items in this stack, replacing the oldest item with each new incoming item.

Now, of course, you need to write the method, `UndoDuplicate`, that you just sent to the `AddUndoAction` queue. This method will be a slot in your base template. Here is Words' `UndoDuplicate` to use as your guide:

```
func(newEntry)
begin
  EntryRemoveFromSoup(newEntry);

  if Visible(overview) then
    overview:DisplayFromSoup()
  else begin
    if theCursor:Entry() = 'deleted then begin
      // the user was looking at the just-deleted entry

      local nextEntry := theCursor:Next();
```

```
        if not nextEntry then
            nextEntry := theCursor:Prev();
        if not nextEntry then
            :DisplayOverview()// removed the last entry
        else
            detail:DisplayWord(nextEntry);
      end;
   end;
end;
```

Note: There is a global function, `ClearUndoStacks`, which clears the stack of currently undoable actions. After making this call, an Undo will yield the message shown in FIGURE 1.12. Most applications will have no need to call this function.

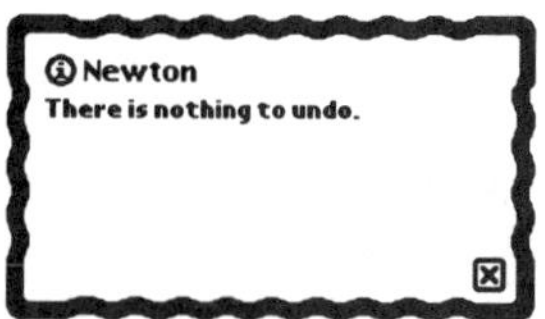

FIGURE 1.12 Notification when the user taps Undo and there is nothing to undo.

Adding Undo for Delete

Adding delete's Undo also requires two tasks.

The Checklist

1. Modify the `DeleteActionScript` slightly.
2. Write a new method called `UndoDelete` and add it to your base template.

1. Modify the DeleteActionScript

Let us look at the first task. Your new `DeleteActionScript` will have one additional line of code that looks like this:

```
func(entryToDelete, targetView)
begin
   local nextEntry;

   EntryRemoveFromSoup(entryToDelete);
   nextEntry := theCursor:Next();
   if not nextEntry then
      nextEntry := theCursor:Prev();
   if not nextEntry then
      :DisplayOverview();
   else
      detail:Delete('DisplayWord, [nextEntry]);
   AddUndoAction('UndoDelete, [entryToDelete]);
end;
```

New Undo Action

2. Add an UndoDelete Method

Now, of course, you need to write the method, `UndoDelete`, that you just sent to the `AddUndoAction` queue. Here is Words' `UndoDelete` to use as your guide:

```
func(deletedEntry)
begin
   theSoup:AddToDefaultStore(deletedEntry);
   :DisplayDetail(deletedEntry);
end
```

That is all there is to adding undo capabilities to Duplicate and Delete. As you can see it is a fairly straightforward process.

Beaming

You beam an item from one Newton to another using the infared port. FIGURE 1.13 shows the various steps that the user takes when sending a beam.

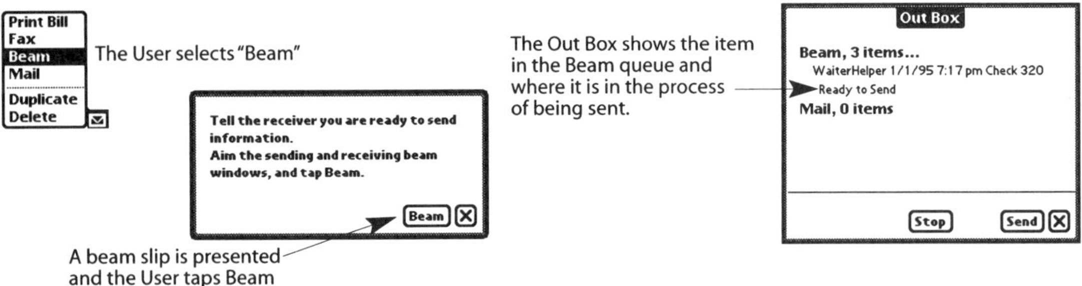

FIGURE 1.13 Beaming an item.

Depending on the settings, a receiving Newton may automatically receive a beam, or the user may have to explicitly choose "Receive" from the In Box. FIGURE 1.14 shows what the user may see (depending on the settings) on the receiving Newton. As shown, the user may first wish to confirm an incoming beam; after doing so he or she can view the item in the In Box and then put it away with a simple tap.

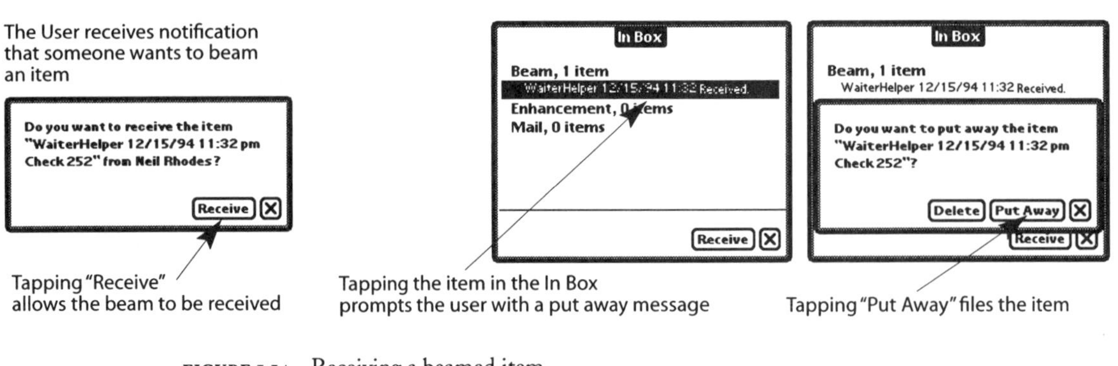

FIGURE 1.14 Receiving a beamed item.

As you can see from FIGURE 1.13 and FIGURE 1.14 there are some things that you need to support and some system-provided material. First, you are responsible for adding a beam frame to the routing action button. When the user taps "Beam", the system in turn presents the beam slip. You are next responsible for providing the one line of title information that describes the item in the Out/In Box and in the beam slips (see FIGURE 1.13). This same title will again be used to describe the item on the receiving Newton. The last thing you need to provide is a method to handle putting away the item away once it has been received.

Now, let us look at the implementation of beaming and how you would do this as a series of steps:

The Checklist

1. Add "Beam" to the action picker.
2. Add a title for items that are routed to the In/Out Box.
3. Test sending and receiving an item.
4. Allow a receiving Newton to put away an item.

1. Add Beam to the Action Picker

To the routing frame of your application, add a new beam slot:

```
beam: {
   title: "Beam",
   routeForm: 'zapSlip,
 },
```

2. Add a Title for Items that Are Routed to the In/Out Box

To support titles for items in the In/Out Box you need to add a `SetupRoutingSlip` method to the application base template. Here is an example of this method:

```
func(fields)
begin
   fields.title := "Words:" && target.word &&
      DateNTime(Time());
end
```

This method should create a `title` slot with a title that is descriptive to the user. For instance, providing the date and information which identifies the target will make it easier for the user to recognize a particular entry in the In/Out Box. The title should be no more than 44 characters long, and is used in the Out Box of the sender and in the In Box of the receiver. If necessary, append an ellipsis (…) character to handle the display of longer strings.

3. Testing Sending and Receiving an Item

After implementing beaming in the action picker and adding your `SetupRoutingSlip`, you are able to beam an item. To test, set the Beam preferences on the receiving Newton to those found in FIGURE 1.15.

Beam
- [x] Confirm before receiving
- [ ] Put away beams immediately
- [ ] Receive beams automatically
- [x] Open In Box while receiving
- [ ] Close In Box after receiving

FIGURE 1.15 Beam preferences.

Now, select Beam from the action menu, then send the beam to another Newton (make sure to tap Receive from the In Box on the other Newton). The item should appear in the In Box on the other Newton.

Caution: Make sure the Inspector is not connected while the beaming occurs. The current Newton system allows only one communication channel to be open at a time; thus, beaming and serial communication cannot occur at the same time.

4. Allow a Receiving Newton to Put Away an Item

To put away an item on a receiving Newton, you must take several conditions into account in your code:

- A beam can be received when your application is closed.
- A beam can be received before your application has ever been run, and thus your soup may not exist on the receiving Newton.
- A received item may come from a nonexistent folder.

Each of these conditions must be handled properly in your `PutAway` method. When a Newton receives a beam and the user selects "Put Away" (or has `PutAway` handle the beams immediately), the `PutAway` message is sent to the application with a matching application symbol. Or, if you want to be tricky, a sending application can modify the `appSymbol` slot in the `fields` frame (in the `SetupRoutingSlip` method) and beam to a different application. For instance, our application could beam a name to the Names application.

Now that we have an idea of what type of conditions we need to account for in our method, let us look at a sample `PutAway`. Remember that `PutAway` is a method in your application base view:

```
func(item)
begin
   local newEntry := item.body;
   local theSoup := nil;
   local appIsOpen := call kViewIsOpenFunc
      with (self);

   CheckThatFolderExists(newEntry);
   if appIsOpen then
      theSoup := self.appSoup
   else begin
      theSoup := call kRegisterCardSoupFunc with
         (kSoupName, kSoupIndexes,
         kAppSymbol, kAppObject);
   end;

   theSoup:AddToDefaultStore(newEntry);
   if not appIsOpen then
      call kUnRegisterCardSoupFunc
         with (kSoupName);
   BroadcastSoupChange(kSoupName);
end
```

The above method serves as a good example for applications which put away soup entries. It is worth noting in this method that the `item` frame is roughly the same as the `fields` frame from `SetupRoutingSlip`. Also, `item.body` contains the target from the sending Newton.

`CheckThatFolderExists` should only be used by those applications that support folders. It is used to handle a particular type of condition: If Jean sends an item from "Jean's Folder" to Fred's Newton, what folder should the item appear in if Fred doesn't have a "Jean's Folder?" Answer: "Unfiled." `CheckThatFolderExists` compares the `labels` slot of its argument to the list of folders on this machine. If the folder is found, `CheckThatFolderExists` does nothing, otherwise the function sets the `labels` slot value to `nil`.

Now, look at how the preceding code deals with the situation in which the application is not open on the receiving Newton. This application stores the open soup in a slot called `appSoup` which it nils out in the application's `viewQuit-`

Script. If the app is running, PutAway uses that soup. If the app is not running, PutAway calls kRegisterCardSoupFunc to create the soup (if necessary) on all writable stores (including the default store). Notice that if we called kRegisterCardSoupFunc, we must call kUnregisterCardSoupFunc.

BroadcastSoupChange will call our application's SoupChanged method which will redisplay the view if the application is open. If we wanted to be robust in our programming, it wouldn't hurt to have an exception handler in this code as well:

```
try
   theSoup:AddToDefaultStore(...)
   onException |evt.ex| do begin
      if not appIsOpen then
         call kUnRegister...
      Rethrow();
   end;|
if not appIsOpen then
   call kUnRegister...
```

Unfortunately, AddToDefaultStore doesn't currently throw an exception, so the exception handler will not execute. In the future, though, AddToDefaultStore might throw exceptions and then this code would correctly handle it.

To test Put Away, make sure that the application with the PutAway method has been installed on the receiving Newton. Otherwise, there won't be an application that can put the beamed item away. Set your Beam preferences to those shown in FIGURE 1.16 and test receiving a beam with your application closed, and with your application open (if your application is open and a beamed item is autojatically put away, it should appear automatically in your application).

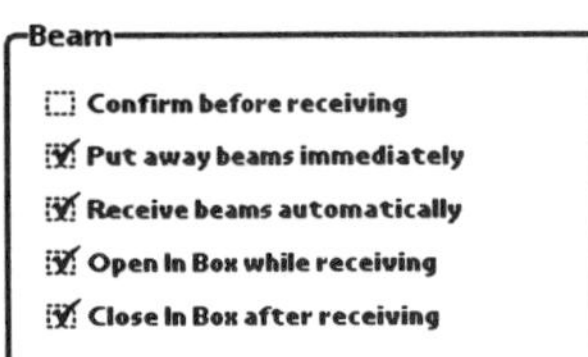

FIGURE 1.16 Beam preferences when testing PutAway.

Printing

Printing on the Newton uses a number of different frames located in different parts of the system. Remember that within the `routing` global our application's routing frame is stored in a slot whose symbol is our application signature. In the print subframe, there is a `formats` slot that is an array. This array contains an entry for each item in the format picker of the print slip. Also, each entry in this array is a symbol. Each of these symbols is a slot in the root view, that points to another view. The view protos from a format frame which you normally store in your base template. FIGURE I.17 shows this graphically.

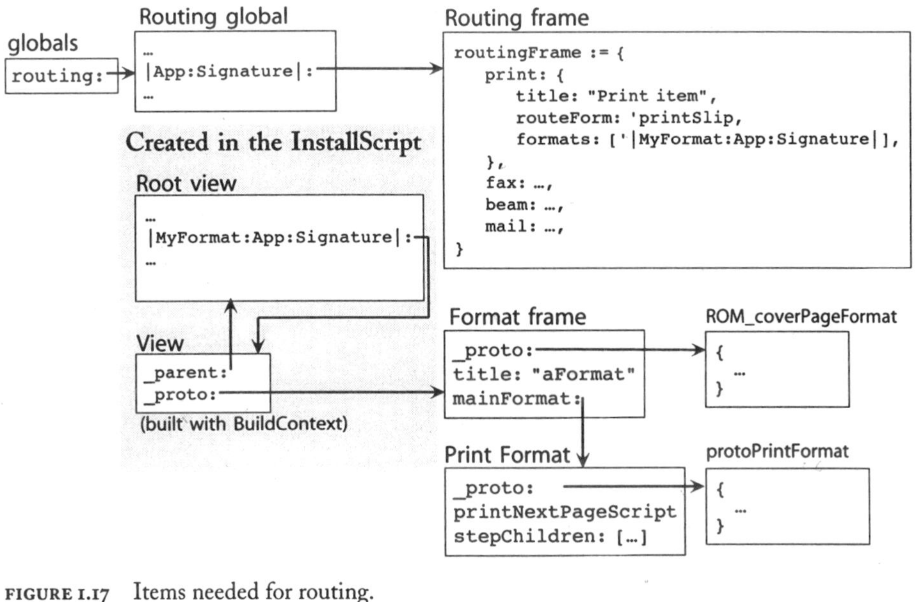

FIGURE I.17 Items needed for routing.

The Checklist

Here is the entire set of steps needed to support printing:

1. Create a print format and add it to your project.

2. Create a format frame in your base template.

3. In your `InstallScript`, call `BuildContext` on the format frame from step 2 and save the result in a slot in the root view.

4. In your `RemoveScript`, remove the slot from the root view.

5. Add Print to your routing frame.

6. Modify the `SetupRoutingSlip` to save data.

7. Modify the `printNextPageScript` in the print format.

1. Create a Print Format and Add It to Your Project

You create a print format by using the special "New Print Format" layout provided by NTK. The topmost template in this layout needs to proto from a `protoPrintFormat`. This template not only has a proto slot but a `printNextPageScript` slot as well. By default, `printNextPageScript` only handles one page but you can add code to support multiple pages (see step 7).

Once you compile the application, the Print format layout is saved into the package and is accessible by name (e.g., `printFormat_`*`nameOfLayoutFile`*) at compile time (in an evaluate slot, for instance).

Create any children you wish to have on the printed page. The `protoPrintFormat` view will be sized so that it is the size of the printable area on the page (the page size inset by the margins). You can use justification with child views to place them in appropriate places on the page. For instance, you might want a header at the bottom-right of each page. This could be done with a `protoStaticText` with parent relative bottom and parent relative right justification.

2. Create a Format Frame in Your Base Template

The next step is to add a format frame to the application base template. Here is a typical format frame with italics showing the parts you might want to customize:

```
{
   _proto: ROM_coverPageFormat,
   title: "Bill",   //text in the format picker
   mainFormat: printFormat_billFormat,
}
```

The `mainFormat` slot points to the `printFormat` you created in the last step. You'll normally create this format frame as a slot in the application base template named `formatFrame`.

3. *Call* `BuildContext` *and Save the Result in a Slot in the Root View*

First, create a new constant that will refer to the application's new print format:

```
constant kFormatSymbol :=
      '|MyPrintFormat:App:Signature|;
```

It is possible for an application to have multiple print formats. They should each have unique symbols. Then the `InstallScript` will have code similar to the following, which uses `BuildContext` to create a view based on the format frame:

```
InstallScript(partFrame)
begin
   local appBaseTemplate :=
      partFrame.theForm;
   local formatFrame :=
      appBaseTemplate.formatFrame;
   GetRoot().(kFormatSymbol) :=
      BuildContext(formatFrame);
   ...
end;
```

Remember that `BuildContext` takes a template and creates a view protoing from that template (its parent is the root view). The view is stored in the root view in a slot with a unique symbol.

4. *In Your* `RemoveScript`, *Remove the Slot from the Root View*

Here is where you clean up and remove what is no longer needed.

```
RemoveScript(partFrame)
begin
   ...
   RemoveSlot(GetRoot(), kFormatSymbol);
end
```

5. Add Print to the Routing Frame

Now that you have created the necessary layouts and instantiated them, you need to add print to the routing frame. After this step, print will be a choice in the action picker:

```
print: { title: "Print item",
         routeForm: 'printSlip,
         formats: [kFormatSymbol],
       },
```

If print is the first routing action in the frame, then the `title` slot string should contain "Print *Object*" (where *Object* stands for whatever is appropriate for your application) instead of just "Print."

6. Modify `SetupRoutingSlip` to Save Data

For beaming, the target is automatically stored in the `body` slot of the `fields` frame after the call to `SetupRoutingSlip`. For printing, the target is not automatically saved. You need to save whatever information you'll need when you actually print in the `fields.body` slot. This information may be the target, it may be some information from the target, or something else entirely. Save this information in the `SetupRoutingSlip` method:

```
func(fields)
begin
   fields.title := kAppName &&
      DateNTime(Time()) &&
      distinctive info from target;
   fields.body := information needed for printing;
end
```

Don't write to any slots in the `fields` frame other than the `body` slot.

7. Modify `printNextPageScript` in the Print Format

The `printNextPageScript` message will be sent to your `protoPrintFormat` at the end of each page. If there are no more pages, it should return `nil`. If there are more pages, it should prepare for the next page and return `true`:

```
printNextPageScript : func()
begin
   if moreToPrint then begin
      send messages to children to update their data
      create new children possibly
      remove children possibly
   end else
      return nil;
end
```

When View Messages are Sent During the Printing Process

The system implements printing differently depending on the type of printing (the differences are usually transparent to the application programmer). For PostScript printers, each view on the page is converted to a PostScript equivalent, and then the PostScript representation of the printed page is sent to the printer. For bitmap printers, the Newton images the page into an offscreen bitmap and then sends those bits to the printer. Since there is not enough memory to hold the entire page in an offscreen bitmap, the Newton images the page in horizontal bands from top to bottom. All views that intersect this band are then drawn, and the bits for that band are sent to the printer. The offscreen buffer is then used for the next band.

The views involved in printing receive a number of different system messages. At the beginning of printing, the `protoPrintFormat` view and all of its descendants receive these messages:

- `viewSetupFormScript`
- `viewSetupChildrenScript`
- `viewSetupDoneScript`

Thus, these are good times for the programmer to do any one-time initializations. At the end of each page, the `protoPrintFormat` receives a `printNextPageScript` message. During each band, each view that intersects the band is also sent the `viewDrawScript` message.

From any of these methods, you can access the `fields` frame (via parent inheritance). Thus, you can access any information you saved in the `fields.body` slot in the `SetupRoutingSlip` method.

Supporting More Than One Print Format

If you want to support more that one type of print format, simply create multiple print formats and add them to your project. Then, create multiple format frames in your application base template. Build all of these multiple format views in your `InstallScript` instead of just the original one. In the `RemoveScript`, remove the slots you created in the `InstallScript`. Your last duty is to create special format symbols for each of the format frames and then add the multiple symbols to the formats slot in the routing frame.

Faxing

Here is the good news—once you've implemented printing you have already gone a long way to adding fax capabilities. To support faxing, you need a separate entry in the routing frame, and you need to make sure your imaging is quick enough to avoid timing out. Lets take a look the checklist for faxing:

Faxing Checklist

1. Add fax to the routing frame.
2. Verify that your page images quickly.

1. Add Fax to the Routing Frame

```
fax: {
        title: "Fax item",
        routeForm: 'faxSlip,
        formats: [kFormatSymbol],
    },
```

Any format frame that supports printing will also support faxing. For both faxing and printing, a format frame must proto from `ROM_coverPageFormat`. The name of this proto is somewhat misleading since this proto is also necessary for printing.

2. Verify that Your Page Images Quickly

Faxing requires careful attention because of the very real possibility of timing out during the middle of a fax. Thus, it is imperative that the actual sending of a fax be speedy. Fax machines will time-out when about 5 seconds elapse without receiving information. Thus, once the connection has been made everything must proceed at a fairly brisk pace. Further, each band must image quickly because all imaging occurs while the fax connection exists. As a result, you will need to do the following:

- Make `printNextPageScript` and `viewDrawScript` quick.
- Consider using multiple views whose height is less than the entire page (only views that intersect a band need to be drawn).

These considerations are especially necessary if you have complicated pages to fax. For many applications, however, all that is required is some testing that demonstrates that faxing does not time-out.

Mailing

Mailing involves some additions to the format frame to support the creating of text that will be sent.

The Checklist

1. Add mail to the routing frame.
2. Add a method to the application that will create text to send.
3. Add a `textScript` slot to the format frame that references this new method.
4. Support enclosures by adding an `attachment` slot to the format frame.
5. Add a `PutAway` method to put away enclosures.

1. Add Mail to the Routing Frame

```
mail: {
        title: "Mail",
        routeForm: 'mailSlip,
        formats: [kFormatSymbol],
    },
```

2. Add a Method to the Application that Will Create Text to Send

When mail is sent from a Newton, there is always text associated with it. For non-Newton receivers, this is the only thing that is mailed. Newton mail receivers, however, can also receive frames associated with the mail message (which can be put away just like a beam). You must write a method that will create this text. This method is a slot in the application base template.

```
appBase.GetTextVersionOfTarget:=
func(fields, target)
begin
        // for a checkbook program, perhaps
    return "Pay" && target.payee;
end
```

3. Add a `textScript` Slot to the Format Frame that References This New Method

```
myFormatFrame : {
    _proto: ROM_coverPageFormat,
    title: "Check",
    mainFormat: printFormat_checkFormat,
    textScript: 'GetTextVersionOfTarget,
}
```

4. Support Enclosures by Adding an `attachment` Slot to the Existing Format Frame

If this format supports an enclosure, add the `attachment` slot. Here is the same format frame you just reviewed with this added:

```
myFormatFrame : {
    _proto: ROM_coverPageFormat,
    title: "Check",
```

```
        mainFormat: printFormat_checkFormat,
        textScript: 'GetTextVersionOfTarget,
        attachment: true,
    }
```

Now, the Newton operating system will clone the `target` and enclose it with the mail message, just as happens in beaming. Some formats shouldn't support enclosures (for example, imagine mailing a list of all names in your Names application). For these types of formats, set the `attachment` slot to `nil`.

The `PutAway` method you added for beaming will be used to put away mail enclosures too.

Notice how interconnected each routing process is within the format frame. This format frame can be used for faxing and printing as long as it has the correct proto. It can also be used for mailing as long as it has a `textScript` slot (and optionally an `attachment` slot). Thus, one format frame can be used for all three routing actions.

Summary

In this chapter, we stepped through the routing process—everything from the user interface, to required application slots, to adding each feature available in a typical routing action button. A quick review of each step we took is available in TABLE 1.2 for your reference. Remember that if you handle routing incrementally, it is a fairly easy aspect of your Newton application.

Routing Checklist

You will find a full checklist of each routing action combined in one convenient location in TABLE 1.2. Each of the routing actions is covered as well as the elements that you need to add to the routing frame.

Routing Checklist

The Routing Frame and Application Slots

1. Add the `protoActionButton` to the application.
2. Add the application slots: `appSymbol`, `target`, and `targetView`.
3. Hide the action button in the overview.
4. Install and deinstall the Routing Frame.

Supporting Duplicate

1. Add Duplicate to the routing frame.
2. Add a `DuplicateActionScript` method to your base template.

Supporting Delete

1. Add Delete to the routing frame.
2. Add a `DeleteActionScript` method to your base template.
3. Support the `Delete` animation in your method.

Supporting Card Actions

1. Add Card to the routing frame.

Supporting Undo for Duplicate

1. Modify the `DuplicateActionScript` slightly.
2. Write a new method called `UndoDuplicate` and add it to your base template.

Supporting Undo for Delete

1. Modify the `DeleteActionScript` slightly.
2. Write a new method called `UndoDelete` and add it to your base template.

Adding Beam

1. Add Beam to the action picker.
2. Add a title for items that are routed to the In/Out Box.
3. Test sending and receiving an item.
4. Allow a receiving Newton to put away an item.

TABLE 1.2 A quick guide to routing.

Routing Checklist

Adding Print

1. Create a print format and add it to your project.
2. Create a format frame in your base template.
3. In your `InstallScript`, call `BuildContext` and create a slot in the root view.
4. In your `RemoveScript`, remove the slot in the root view.
5. Add Print to the routing frame.
6. Modify `SetupRoutingSlip` to save data.
7. Modify `printNextPageScript` in the print format.

Adding Fax

1. Add Fax to the routing frame.
2. Verify that your page images quickly.

Adding Mail

1. Add Mail to the routing frame.
2. Add a method to the application that will create text to send.
3. Add a text slot to the format frame that references this new method.
4. Support enclosures by adding an `attachment` slot to the format frame.

TABLE 1.2 A quick guide to routing.

Chapter 2

Communications

Good communication is as stimulating as black coffee, and just as hard to sleep after.

—Ann Morrow Lindbergh

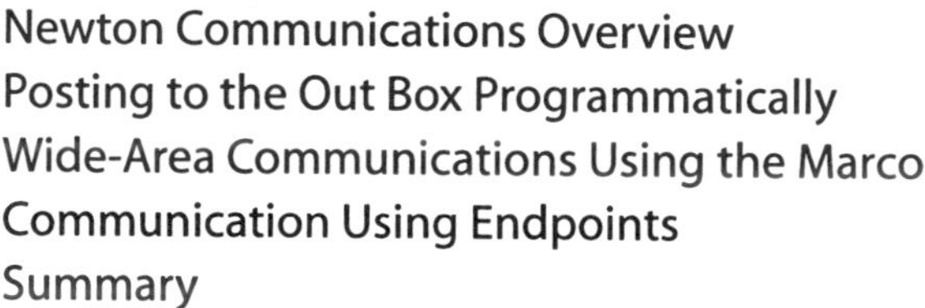

"Anytime, anywhere communications" is one of the mottos of the Newton. Indeed, communications is one of the hottest areas of development within the computer industry, in general, and on the Newton, in particular.

In this chapter, we talk about many of the advances that have occurred in both wired and wireless communications and how the Newton fits into this picture. Perhaps, most importantly, we will also develop two communications applications and describe the Newton communication architecture. From this material, you will be able to gain a clear understanding of what is involved in this aspect of Newton programming.

Newton Communications Overview

The Newton communications architecture was designed to provide a number of different communication alternatives. All current Newtons ship with a serial port, a PCMCIA slot, and an infrared port. Further, the first two of these hardware ports provide a fair measure of expandability. In this section, we'll discuss the various communications options available for the Newton now, as well as options that might be available in the future.

Wired

There are a handful of wired communications options on the Newton. These include serial, modem, AppleTalk, and various PCMCIA possibilities.

Serial

The serial communication operates over the serial port (an RS-422 port that is compatible with the older RS-232 standard). Nominally, speeds of up to 230,400 bps can be specified, although any speeds over 57,600 bps will require flow control and an error-correcting protocol due to buffer and latency issues.

Software compression (MNP5 or V42.bis) and/or error-correction (MNP) can be added to a serial connection, and this can certainly increase the effective throughput.

Modem

There are two Apple modems, an external modem that connects to the serial port, and a PCMCIA modem. Other external or PCMCIA modems can also be used, although they require the Modem Enabler software package.

Modem communication can use built-in hardware compression and error correction. Another alternative is to use software MNP5 or V42.bis compression and/or MNP error-correction.

AppleTalk

A LocalTalk connector can be plugged into the serial port to connect to an AppleTalk network. This provides the ability to print to AppleTalk printers, to see zones and entities on the network, and to connect using ADSP (AppleTalk Data Session Protocol).

A Newton application can connect to ADSP servers on the network, but a Newton cannot become a server. The reason for this restriction is that the NBP (Name Binding Protocol) requires that packets be sent to the network fairly often; a requirement that is fairly impractical for a battery-powered device.

Other

The PCMCIA slot provides the potential for many other wired communications options. Some of these include:

Ethernet	An RJ45 connector to a 10Base-T network could easily be connected to a PCMCIA card. The protocol might be EtherTalk, or could be TCP/IP. A TCP/IP stack is under development, although no details have been revealed about when it might be available.
Fiber-optic	We know of no plans for fiber-optic PCMCIA cards for Newton, but it's certainly within the realm of possibility.

Wireless

Wireless communication options are certainly attractive for portable devices. Many wireless options exist today and others will be coming in the future.

IR/Beaming

Currently, every Newton has an infrared port. This port can be used at high speeds (19200 bps) or low speeds (9600 bps) to communicate from Newton to Newton. The Newton can also operate as an infrared remote control for many consumer devices such as TVs and stereos.

Local Area

Local-area networking can be done using wireless LocalTalk alternatives. Currently, three LocalTalk alternatives exist, with a fourth to be available in mid-95:

Photonics Cooperative	This unit uses diffuse infrared that bounces off ceilings and walls to other units. It connects to the serial port.
Digital Ocean Grouper	This unit uses spread-spectrum radio that transmits to other units. It connects to the serial port.
Dayna Serial Roamer	This unit uses spread-spectrum radio that transmits to other units or to an access point connected to an Ethernet network. It connects to the serial port.
Dayna PCMCIA Roamer	This unit uses spread-spectrum radio that transmits to other units or to an access point connected to an Ethernet network. It is a PCMCIA card. This unit is projected to ship in mid-95.

All of these units provide standard AppleTalk functionality. Further, the ones that connect to the serial port appear to the Newton as standard LocalTalk connectors; the fact that they are wireless is transparent to the Newton. The PCMCIA card redirects all LocalTalk communication from the serial port to the card.

The Photonics Cooperative consists of a small box with an infrared transmitter/receiver which is connected to another box that contains 3 AA batteries. This infrared transceiver uses diffuse IR, rather than the focused IR used by the built-in IR port.The IR signals bounce off of ceilings and walls, where they then hit other Photonics Cooperative units. These units are meant to work with a range of about 25 feet of each other. To connect to a wired LocalTalk network, you can connect a wired LocalTalk connector to one of these units.

We've used some of these units to set up a long-term small wireless LocalTalk network in the office; other than occasional setup problems in aiming them (at an angled ceiling no less), they worked quite well. For more information about this product, you can contact the company:

Photonics Corporation
2940 North First Street
San Jose, CA 95134
(408) 955-7930

The Digital Ocean Grouper is a unit that fits under the bottom of a MessagePad 100. A small connector plugs into the serial port. The transceiver uses spread-spectrum technology; rather than sending a strong radio signal on one frequency, it sends weak radio signals on many different frequencies. It does not require line of sight (one Grouper can be in one room and another in a different room—radio waves penetrate through walls). These units are claimed to transmit up to 800 feet. For more information about this product, you can contact the company:

Digital Ocean
11206 Thompson Avenue
Lenexa, KS 66219-2303
(913) 888-3380

The Dayna Serial Roamer is a small unit with a connector that plugs into the serial port. Nicely enough, it uses standard cellular phone rechargeable battery packs. This unit also uses spread-spectrum technology which allows these devices to communicate up to a distance of 150 feet indoors. Longer ranges are achievable for line of sight. A separate device, the Network Access Point, has a transceiver along with Ethernet 10Base-T and 10Base-2 connections. This provides a one-step mechanism for connecting to an existing Ethernet network.

The Dayna PCMCIA Roamer is a version of the Dayna Serial Roamer on a PCMCIA card. It can communicate with the other Dayna products. The great advantage of this product is its small size. It appears that battery life will not be as great as with the external units, however, due to a lack of separate power supply.

We've tested a prerelease version of the Serial Roamer along with a shipping version of the Network Access Point. The Serial Roamer works quite well. We achieved a range of about 125 feet indoors with no problems. We like the standard battery which makes replacement and recharging simple and allows a choice of different battery technologies that allow longer battery life.

Dayna Communications
Sorenson Research Park
849 West Levoy Drive
Salt Lake City, UT 84123
(801) 269-7200

Wide-Area Communications

As well as local-area wireless communication, wide-area communication is available as well. We will look at services that exist both today and in the future.

ARDIS

The ARDIS network is a two-way packet radio network in the United States, Puerto Rico, and the Virgin Islands. It covers hundreds of metropolitan areas. At the time this was written, coverage was claimed to more than 90% of U.S. businesses and more than 80% of the U.S. population.

While the ARDIS network started as an internal IBM field service network, it is now used by many commercial and individual customers. Because of its initial target audience (field engineers servicing equipment inside offices), the ARDIS network provides coverage not just on the city streets, but inside buildings. To begin with, service was only provided at 4800 bps. Now, all new equipment supports 19200 bps as well.

Newton access to ARDIS is provided by the Marco, a Motorola Newton. We discuss this in great detail later in this chapter (see "Wide-Area Communications Using the Marco" on page 50).

In addition, access to the ARDIS network can be made through an external radio-modem, the Motorola InfoTac, or through a forthcoming Motorola PCMCIA card.

RAM Mobile Data

RAM Mobile Data is a newer competitor to ARDIS. At the time this was written, coverage was claimed for more than 90% of the U.S. urban population. The Mobidem external radio-modem provides access to this network and can be connected to the Newton using a serial port.

Cellular Modems

A cellular modem transmits data over a voice connection, just as it would over a wired voice connection. One problem that must be solved to support this technology involves losing data as a call is switched from a channel in one cell to a channel in another. To handle this loss, special error-correction protocols are used for modems (for instance MNP 10).

Another problem is that cellular calls are charged for an entire session. Pricing is for the time spent, not the amount of data sent. Cellular modems can be used through the serial port of the Newton.

CDPD

Cellular Digital Packet Data (CDPD) is a technology to transmit data over the existing cellular wireless network. It sends data packets on channels that are not currently being used for voice. Thus, when a voice call starts on a channel, the packets hop to another channel. Pricing is per packet; charges are for the amount of data sent, not the time spent.

At the time this was written, CDPD was available only in very limited areas in the U.S. There is no CDPD service currently available for the Newton.

GSM

GSM is another technology that transmits data over cellular wireless networks. It is being used in Europe and Australia, and was available in many areas at the time this was written.

The GSM Connection Kit for Newton is available in Germany. It allows the German MessagePad 120 to do wireless communications, including wireless transmission of email and fax. It works in conjunction with the Nokia Cellular PCMCIA Card and Nokia 2110 or 2140 cellular phones.

ARDIS Personal Messaging

This service runs on top of the ARDIS network and provides peer to peer mail services. Using this service, users of the ARDIS network can send mail to other Personal Messaging users on the network.

ARDIS Personal Messaging has an interesting pricing feature. When a user sends mail to another user, the receiver doesn't pay for the packets. This contrasts with more traditional pricing where the sender pays to send the packets and the receiver pays to receive the packets.

Radiomail

Radiomail is not a wireless provider, but rather a higher level information and gateway service. It has hosts on both the ARDIS and RAM mobile data networks which gateway to the Internet. Some of the services it provides are:

Individual Internet address	Email sent to that Internet address is forwarded over either the ARDIS or RAM Mobile Data network.

Internet Gateway	Email can be sent from the ARDIS or RAM Mobile Data network via Radiomail to the Internet.
News Service	Headline updates can be sent from Radiomail to your mobile device.
Outgoing Fax	An email message sent with a phone number will be converted to a fax and sent. This allows sending text-based faxes from a mobile device that has no phone hookup.
Messages	People can call Radiomail and leave a message for you. The message will be sent to your mobile device just like an alphanumeric page.

MobileVision

This service from CE Software was just about to ship as this book was being written. It provides Marco users access to LAN mail servers.

It operates by setting up an InfoTac external radio-modem to a desktop machine on the LAN. The Marco then communicates with the desktop machine via the ARDIS network. Mail can be sent to and from the LAN mail server.

One-way Communications

The Newton also provides for one-way communication. PCMCIA pager cards provide a way for the outside world to communicate to an individual Newton. Two such pager cards are available:

Newton PCMCIA Messaging Card	This card is manufactured by Motorola; service is by MobilComm. Messages can be read only on the Newton or on Powerbooks with a PCMCIA slot.
Apple Mobile Messaging Card	This more recent card contains an alphanumeric display so that messages can be read without inserting the card into the Newton. The card is manufactured by Socket Services; service is by PageNet.

Both devices share a number of similarities. Each can be used in Apple Powerbooks with PCMCIA slots. Each product contains its own battery and pages can be received while the card is outside of the Newton. When either card is inserted, pages will appear in the In Box.

Advantages of the pager cards are their long battery life; since they don't need to transmit and receive strong signals, not much power is needed. Like standard pagers, battery life is about a month.

Posting to the Out Box Programmatically

Note: Before reading this section, it would be well worth your while to read Chapter 1 which describes the Newton implementation of Routing. It will make the following discussion much easier to follow.

In most cases, users create items in the Out Box by using the action button. Some applications, however, might want to post items in the Out Box without requiring this type of user intervention.

For example, FIGURE 2.1 shows an application that supports product registration programmatically; thus, when the user fills out a product registration slip, the information in the slip is sent to the application vendor via email.

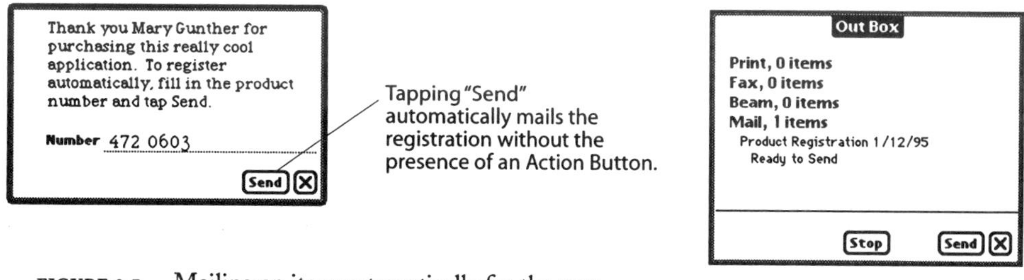

FIGURE 2.1 Mailing an item automatically for the user.

Using the `Send` *Method to Automatically Route Items*

To post programmatically, an application must first create the frame that will be routed via some particular transport. This frame, the `item` frame, will have a number of slots. Some slots will be common to all transports and some slots will be specific to a particular transport.

`Item` Slots Required for Most Routing Actions

These are the `item` slots that are common across a variety of transports (that is, printing, faxing, beaming, and mailing). Note that some of the slots are required while others are optional:

`appSymbol`	The symbol of the application that sent the item. It is used for beaming and mailing to determine which application's `PutAway` method to call. *Required.*
`body`	This is the object that will be beamed or mailed to a receiving Newton. It is required for beaming, but optional for mailing. For faxing and printing, this is the slot which contains information that will be used by the print format at imaging time.
`connect`	This boolean, if true, is a suggestion that the connection should occur immediately.
`hidden`	This boolean, if true, is a suggestion that the connection should occur quietly, without displaying the Out Box or other slips.
`printerFormat`	This is the symbol of the print format view which is in the root view. This slot is only used by faxing and printing.
`title`	This is a string used as the title for the item in the Out Box. *Required.*

Additional `Item` Slot for Printing

For printing, there is one additional required slot:

`printer`	This string specifies the printer to which the Newton will print. Note that the last chosen printer can be obtained with this code: `call kGetUserConfig with ('currentPrinter)`. *Required.*

Additional `Item` Slots for Mailing

For mailing, there are two additional slots:

`email`	This string specifies the mail address to which the item will be sent. *Required.*
`text`	This string contains the text of the mail message. *Required.*

Additional `Item` Slots for Faxing

For faxing, there are several additional slots, one of which is required:

`hasCoverPage`	A boolean specifying whether or not a cover page should be generated.
`name`	A string containing the receiving person's name. This is printed on the cover page.
`phoneNumber`	A string containing the fax phone number to dial. *Required.*
`toPhone`	A string containing the receiving person's voice phone number. This is printed on the cover page.

A Sample `Item` Frame

Now that we have discussed the various slots found in this frame here is a sample that might be used for mailing:

```
registrationItemFrame := {
   appsymbol:  kAppSymbol,
   body:       target, //frame filled w/user info
   connect:    true,
   hidden:     true,
   title:   "Registration slip from" && kAppName,
   email:   "joe@host.com",
   text:    "text of mail message",
}
```

Sending the `Item` Frame

Once you've created the item frame, you are now ready to send it. To do this, you use the Platforms file function `Send` to add it to the Out Box.

Here is the syntax of the function:

```
call kSendFunc with (transportSym, item)
```

The two parameters of `Send` can be briefly described as:

`transportSym`	A symbol for the type of transport being used on the item. For example, `'fax`, `'beam`, `'mail`, `'print`.
`item`	The `item` frame discussed above.

An example call is:

```
call kSendFunc with ('mail, registrationItemFrame)
```

Wide-Area Communications Using the Marco

The Motorola Newton, the "Marco," provides Newton users with an important feature previously unavailable in a Newton: built-in two-way wireless radio. This new feature gives the Newton user wireless access to a nationwide electronic network that includes a gateway to the Internet. This allows a wide range of new capabilities, including the ability to send and receive wireless mail to and from the Internet.

Marco Specialized Transports for Supporting Wireless Mail

The Marco Newton comes with a dual protocol (MDC4800 at 4.8 kbps, RDLAP at 19.2 kbps) radio modem that connects to the ARDIS network. A user can have an ARDIS Personal Messaging (ARDIS PM) service agreement with ARDIS which enables sending messages to and from other ARDIS PM users or to hosts on the ARDIS network. A user can also establish a service agreement with Radiomail, which provides a mail gateway to the Internet.

To enable this feat of wireless magic, Motorola has provided two specialized transports which handle the interaction with the two services:

ARDIS Personal Messaging	This transport provides the ability to send and receive messages from other ARDIS PM users.
Radiomail	This transport provides the ability to send and receive messages from the Internet.

Programming the Marco

Any application that has provided mail support (see "Mailing" on page 33) will automatically support wireless email. Once the user has chosen how to send mail (for example, ARDIS PM to another ARDIS PM user, Radiomail to send via the Internet, NewtonMail using a wired modem, or some custom transport), it will be automatically sent from any application via the specified transport.

New Motorola APIs

Motorola has also offered the programmer other ways of making a Newton application interact specifically with the Marco. There are three APIs that a Newton application can use, depending on the type of application it is:

Mail-based	Applications that send mail directly (without using the action menu) and receive mail directly (without requiring the user to use the In Box). These applications should use the Universal Mail Services (UMS) API.
Transport-specific	Applications that require a specific transport and communicate with that transport's API. Built-in transports on Marco are Radiomail and ARDIS PM. Other transports could be provided in the future.
Dedicated host	These applications write directly to the Wireless Manager endpoint. They send packets to a specific host on the ARDIS network.

Use of any of these API's makes your application Marco-specific.

Of these three types of transports, we will be dealing with the first, mail-based applications. We will cover it in detail since the vast majority of Marco-specific applications will be of this type.

> *Note:* It is possible, using the Wireless Manager endpoint, to write your own transport (a service which is IO Box-based, and which is available to all applications). For example, you could provide a transport that accesses your enterprise-wide email system. By simply connecting to a host on the ARDIS network and writing transport software that would connect to the email system. Incoming mail would appear in the In Box. Users could use the action button to mail to a user on the enterprise email system. The mail would go to the Out Box, and then out via the ARDIS network to the host on that network.

For information on the transport-specific APIs, on the Wireless Manager API, or on writing your own transport, you should contact Motorola at MotoNewt.Dev@applelink.apple.com.

Writing to the Mail-based API

There are two parts to mail-based applications: sending mail without going through the action button and mail slip, and receiving mail without requiring the user to go through the In Box. Sending mail without going through the action button has already been covered (see "Posting to the Out Box Programmatically" on page 47). The second part, receiving mail automatically, is done by registering your application.

Registering an Application and Sending Mail

To register an application to receive mail, you use the `RegisterMailPackage` method (a method of the `motorola` global frame). This method takes one parameter, your application symbol.

When a piece of mail arrives, the mail system goes through the registered applications, sending each a message along with the piece of incoming mail. Each application checks to see whether the incoming mail belongs to it. If it does, it handles it and returns `nil`. Otherwise, it returns the mail which is passed to the next registered application.

If no registered application handles the mail, it is placed in the In Box (as always). FIGURE 2.2 shows a graphic representation of this process.

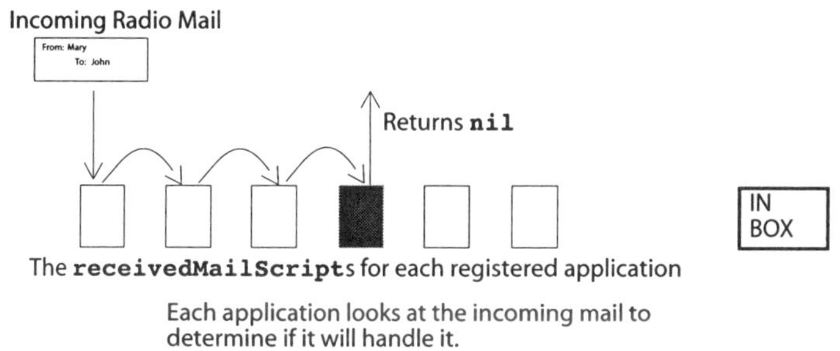

FIGURE 2.2 How the Marco handles incoming mail.

Receiving Incoming Mail

The message sent when incoming mail arrives is `receivedMailScript`. The parameter this method takes is a frame containing at least the following slots:

`dateStr`	The date of the mail as a string.
`fromEmailAddress`	The email address of the sender.
`fromName`	The user-readable name of the sender.
`toEmailAddress`	The email address the item was sent to.
`name`	The user-readable name at the mail address the item was sent to.
`title`	The title of the mail message.
`text`	The body of the mail message.

Unregistering an Application

To unregister your application, use the `UnregisterMailPackage` method. This is also a slot in the `motorola` global frame. You pass your application symbol as a parameter with code similar to the following:

```
UnregisterMailPackage(kApplicationSymbol);
```

The time at which you register and unregister your application will depend quite a bit on your particular application. Certainly, you'll want to register when you expect to receive mail. Different applications, however, might want to do this at different times. Here are some examples of when:

- When your application is opened.
- When your application is installed.
- When you send a message that requires a reply.
- At specific times. Perhaps your application receives mail at 10 AM each day. In that case, you could register an alarm for 9:55 AM that in turn registers your application for mail.

You'll deregister when you no longer expect to receive mail, or when your application can no longer receive it. The latter occurs when your application is removed. Therefore, you should minimally unregister in your `RemoveScript`. As to the former situation, any of the following answers might be appropriate for different applications:

- When your application is closed.
- When the mail you're expecting has finished arriving.
- At specific times. Perhaps your application listens for mail at 10 AM, but if no mail arrives by 10:15 AM, then it won't be arriving at all. In that case, you could unregister at 10:15 AM.

In general, you should try to minimize the amount of time your application is registered. Otherwise, receiving mail can be quite time consuming for the user. Imagine the time involved if every application always looks at the incoming mail in turn and then passes it on to the next application.

Implementing a Marco Mail-Based Application

Now that you have had an opportunity to get some sense of what is involved in creating a mail-based application for the Marco, let us look at a sample application that does this.

A Stock Quote Application

The application we'll be creating obtains stock quotes from an Internet stock information service. The user selects a stock symbol and taps Lookup (see FIGURE 2.3). This then sends a mail message to an Internet address, which in turn sends back information about that stock symbol. This mail message is received by the application, which then displays it for the user in a view (see FIGURE 2.4).

Note: The Internet service we are using for this application, QuoteCom, provides stock quotes and other information. They have many levels of service, one of which is free and allows up to 5 stock quotes per day. In order to receive free quotes, you must have registered for service. For further information, including information on registering for free quotes, send email to info@quote.com.

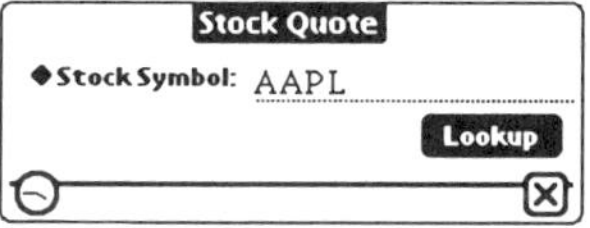

FIGURE 2.3 Looking up a stock quote.

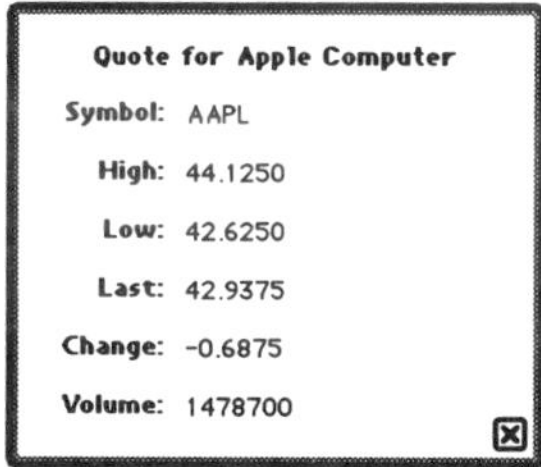

FIGURE 2.4 Receiving a stock quote.

Sending a Stock Quote

The first thing that needs to be done is to send the request. To receive a quote, we must send mail to `services@quote.com` with the subject "`Quote` *`symbol`*". First we create the frame that we will send out:

```
local mailFrame := {
   appSymbol: kAppSymbol,
   connect:true,
   email: services@quote.com,
   hidden: true,
   text: "",   // The text of the message is blank
   title: "Quote" && stockSymbol,
};
```

Now, we call the `Send` function specifying our newly created `mailFrame` and the transport we want to use:

```
call kSendFunc with ('mail, mailFrame);
```

Note: The Marco does allow sending frames in mail messages over its transports (just as NewtonMail does). Of course, the receiver of the mail must be a Newton in order to do anything with the frame.

Receiving a Stock Quote

Next, we must install our application so that it is notified when mail arrives. Although we could do this in our `InstallScript`, that would really be overkill. Rather, it is better to install only when we are expecting to receive mail. This way unasked-for mail is never sent to us; instead, mail comes only in response to an outgoing item. Therefore, we will install our application to receive mail only after sending a piece of mail. Likewise, when our response comes we will want to deinstall. This is a little trickier than is sounds, however. Here are the situations we need to make sure we take into account:

- *What if the user sends another stock lookup before the first has returned?*
- *What if the user removes the application while there is an outstanding request that hasn't been received yet?*

Thus, we'll also need to keep a count of how many outstanding requests we have. When we send a stock lookup request, we'll increment that count. If the count just went from 0 to 1, we'll install ourselves for mail notification:

```
numberOutstandingRequests:=
   numberOutstandingRequests + 1;

if numberOutstandingRequests = 1 then begin
   motorola:RegisterMailPackage(kAppSymbol);
end;
```

At this point, the mail has been sent, and our application has been registered to receive notification whenever mail arrives. If mail arrives, the `receivedMailScript` message is sent to our application's base view. Thus, we will create this method as a slot in our application base template, and within it check to see if the incoming mail belongs to our application or not. If the message does not belong to us, we also need to pass it along. Here is the `receivedMailScript`:

```
func(item)
begin
   if :IsQuoteResponse(item) then begin
      :RemoveRequestFromQueue();
      local error := :GetError(item);
      if error then
         :Notify(kNotifyAlert,
            EnsureInternal(kAppName),
            EnsureInternal(error))
      else begin
         local quote := :GetQuote(item);
            // EnsureInternal because floater may
            // be open when app is removed
         floater := BuildContext(EnsureInternal(
            pt_quoteResponseFloater));
         floater:Open();
         floater:SetResponse(quote);
      end;

      // let's delete the entry since we saw it...
      If IsSoupEntry(item) then
         EntryRemoveFromSoup(item);

      return nil;
   end else// it's not our mail
      return item;
end
```

In this code, we check to see whether the mail is a response to our quote request (with `IsQuoteResponse`). If it doesn't belong to us, we return the item, signifying that it's not ours. Otherwise, we need to take care of all of the following in this code:

- Reduce the count of outstanding requests (using `RemoveRequestFromQueue`),
- Deinstall our application from mail notification, if necessary, if the count is 0 (handled in `RemoveRequestFromQueue`).
- Read the information from the message. If the information is a valid quote, we need to display it in a floater (in `GetQuote`).
- If the information is not a valid quote, we put up a notify slip showing the error (handled in `GetError`).

Checking for a Valid Quote

Note that a response to a quote request will have a return address of "services@quote.com" and a subject of "QuoteCom Email Response". We look for both of those in `IsQuoteResponse`:

```
func(item)
begin
   constant kQuoteFromEMailAddress :=
      "services@quote.com";
   constant kQuoteTitle :=
      "QuoteCom Email Response";

   return StrEqual(item.fromEmailAddress,
      kQuoteFromEMailAddress) and
      StrPos(item.title, kQuoteTitle, 0) <> nil;
end
```

The text of a valid quote response will look like:

```
Tue Feb  7 14:24 EST                 Quotes may be delayed by exchanges
 Symbol           High        Low        Last       Change      Volume
 ---------------------------------------------------------------------
 BRK        24600.0000  24500.0000  24500.0000    +50.0000          60

 Name                   Earn/Shr    P/E  Div/Shr  52-wk Price Range
 ------------------------------------------------------------------
 Berkshire Hathaway          N/A   30.0     0.10  15150.00 - 25200.00
```

Handling an Error

We'll assume that if the text starts with "Error:" or if the "---" can't be found that the response is some sort of an error (perhaps we aren't registered with QuoteCom or we've exceeded our number of quotes for the day). Here is the `GetError` method which returns an error message, or `nil` for no error:

```
GetError := func(item)
begin
   constant kCharactersBeforeQuote := "----\n";

   if BeginsWith(item.text, "Error:") then
      return item.text
   else if not StrPos(item.text,
      kCharactersBeforeQuote, 0) then
      return "Can't find the quote"
   else
      return nil;
end
```

Getting the Text for a Quote Frame

We need to extract the information from the quote text and put it into a frame containing that information. An example of a desired quote frame is:

```
{
   name:     "Berkshire Hathaway",
   symbol:   "BRK",
   high:     "24600.0000",
   low:      "24500.0000",
   last:     "24500.0000",
   change:   "+50.0000",
   volume:   "60"
}
```

Here is the `GetQuote` method, which searches through the text using brute force alone. The method then, in turn, creates a quote frame:

```
func(item)
begin
   constant kCharactersBeforeQuote := "--\n";
   local startOfQuote := StrPos(
      item.text, kCharactersBeforeQuote, 0);
   if not startOfQuote then
      return nil;
```

```
local position := startOfQuote +
   StrLen(kCharactersBeforeQuote);
local stuff :=  {name: "", symbol: nil,
   high: nil, low: nil, last: nil,
   change: nil, volume: nil};

local GetString := func(s, sLength)
begin
   local endPos := position;
   // skip white spaces at the beginning
   while endPos < sLength and
      IsWhiteSpace(s[endPos]) do begin
      endPos := endPos + 1;
   end;
   position := min(endPos, sLength - 1);
   // now skip over
   while endPos < sLength and
      not IsWhiteSpace(s[endPos]) do begin
      endPos := endPos + 1;
   end;
   endPos := min(endPos, sLength - 1);
   local s := SubStr(s, position,
      endPos - position);
   position := endPos;
   return s;
end;

locallength := StrLen(item.text);
stuff.symbol := call GetString with (
   item.text, length);
stuff.high := call GetString with (
   item.text, length);
stuff.low := call GetString with (
   item.text, length);
stuff.last := call GetString with (
   item.text, length);
stuff.change := call GetString with (
   item.text, length);
stuff.volume := call GetString with (
   item.text, length);

// name of stock follows after more "---"
position := StrPos(item.text,
   kCharactersBeforeQuote, position);
if position then begin
   stuff.name := Substr(item.text,
      position + StrLen(kCharactersBeforeQuote),
```

```
            kNameLength);
         TrimString(stuff.name);
      end;

      return stuff;
   end
```

Once we've got a quote frame, it's a simple matter to set the various static texts in a floater. Here is the `SetResponse` method of the floater which does that:

```
func(quote)
begin
   SetValue(headline, 'text, headline.text &&
      quote.name);
   SetValue(symbol,   'text, quote.symbol);
   SetValue(high,     'text, quote.high);
   SetValue(low,      'text, quote.low);
   SetValue(last,     'text, quote.last);
   SetValue(change,   'text, quote.change);
   SetValue(volume,   'text, quote.volume);
end
```

Handling the Unregistration

The application must be unregistered if there are no more outstanding requests. The `RemoveRequestFromQueue` method does that:

```
func()
begin
   numberOutstandingRequests :=
      numberOutstandingRequests - 1;
   if numberOutstandingRequests = 0 then begin
      // Time to unregister our AppSymbol from UMS
      motorola:UnregisterMailPackage(kAppSymbol);
   end;
end
```

In addition to unregistering when there are no more requests, we must unregister if our application is removed. Here is the application's `RemoveScript`:

```
RemoveScript := func(partFrame)
begin
   // in case we are currently registered
   motorola:UnregisterMailPackage(kAppSymbol);
end;
```

Note that it is OK to call `UnregisterMailPackage` even if you aren't currently registered.

Communication Using Endpoints

Low-level communications on the Newton are accomplished using endpoints. An endpoint is an object that acts as an intermediary between an application and a communication service. Endpoints have a common API; thus, an application you've written that does communication using an endpoint for one service can easily use an endpoint for a different type of service.

Differences between the services include the configuration options which can be set as well as how and whether addressing is done.

Communication Services on the Newton

Several communication services are provided on the MessagePad Original/100/110/120.

Serial

This service supports serial communication. Configuration options include speed, start bits, stop bits, and parity. No addressing is necessary. Note that communication occurs with whatever is on the other end of the serial line.

Serial with Compression

This service is like the serial endpoint, but also provides MNP error correction and compression. Configuration options include those of the serial endpoint plus compression options. No addressing is necessary.

Modem

This service supports modem communications. It transparently handles PCMCIA modems and serial modems. Configuration options include:

- Speed

- Start bits
- Stop bits
- Parity
- Compression options
- A choice between tone or pulse dialing.

Addressing is in the form of a phone number to dial.

ADSP

This service supports ADSP sessions over AppleTalk. There are no configuration options. Addressing is in the form of an AppleTalk NBP address (which includes the name, entity type, and zone).

IR

This service supports half-duplex infrared communication. No addressing is necessary (communication occurs with whatever the IR port is pointed to).

Specifying Endpoints

All endpoints are frames that proto from `protoEndpoint`. To this endpoint frame is added a `configOptions` slot that specifies the following items within it:

- The type of service you want to use.
- Any service options you want to specify.
- The address can also be provided, if the service requires one.

The `configOptions` slot is an array of frames. Each element in the array specifies either a service, service options, or an address. Each frame in this array will contain either three or four of the following slots:

`label`	A string specifying the option identifier.
`type`	A symbol specifying the kind of option. The possibilities are `'service`, `'option`, and `'address`.

`opCode` This slot specifies how the option request should be handled. The possibilities are: `opSetRequired`—if the request cannot be satisfied, the request should fail; `opSetNegotiate`—if the request cannot be satisfied, a reasonable value can be substituted.

`data` This slot specifies details about the option. Not needed if the type is `'service`.

Note that service frames don't contain a `data` slot; only option and address frames do. Given the above slots in a typical endpoint, here is a sample to look at:

```
mySerialEndpoint := {
   _proto:protoEndpoint,
   configOptions:[
      {
         label:   kCMSAsyncSerial,
         type:    'service,
         opCode:  opSetRequired
      },
      {
         label:   kCMOSerialIOParms,
         type:    'option,
         opCode:  opSetRequired,
         data: {
            bps:     k2400bps,
            parity:  kEvenParity,
            dataBits:k7DataBits,
            stopBits:k1StopBits,
         }
      },
   ]
};
```

This endpoint will do serial communications at a speed of 2400 bps with 7 data bits, 1 stop bit, and even parity.

Lifecycle of an Endpoint

Once you've created your endpoint frame, you need to `Instantiate` the endpoint. After you've done this, you will actually establish communication. For an active connection, you'll use `Connect`. For a passive connection (receiving from IR, answering a modem, or serial with compression), you'll just call `Listen`. Either of these methods returns when the connection has been established.

When your communication is finished, you'll call `Disconnect` to close the connection (or `Release`, which closes the connection after all pending input and output is finished). Then, you'll call `Dispose` to uninstantiate your endpoint. FIGURE 2.5 shows this graphically.

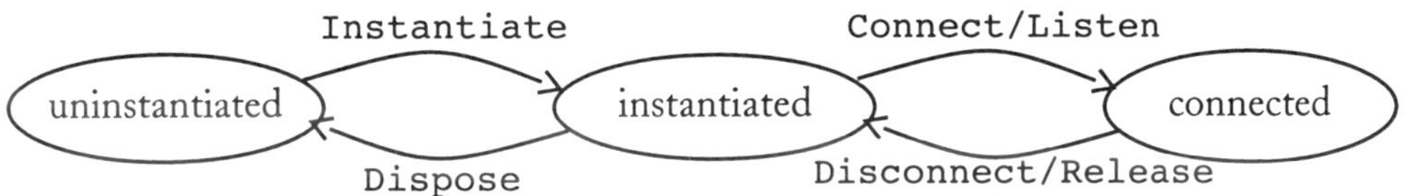

FIGURE 2.5 Transitions among the three states of an endpoint.

There are a number of endpoint methods. We will detail each in this section, but here is a brief listing in the order in which they are covered:

```
endPoint:Instantiate(endPoint, options);
endPoint:Connect(address, options);
endPoint:Listen(options);
endPoint:Disconnect();
endPoint:Release();
endPoint:Output(data, flags);
endPoint:OutputFrame(frame, flags);
endPoint:FlushOutput();
endPoint:SetInputSpec(inputSpec);
endPoint:FlushPartial();
endPoint:Partial();
endPoint:Input();
endPoint:ExceptionHandler(error);
endPoint:Abort();
```

Lifecycle Endpoint Methods

Now that you have an understanding of the life of a communication endpoint, let us go over the endpoint methods that deal with creating, destroying and connecting.

`endPoint:Instantiate(endPoint, options)`

This method creates data structures and does initialization for the specified service. If *options* is `nil`, options are used from *endPoint*`.configOptions`. Otherwise, the options in *options* are used, while the options found in the *endPoint*`.configOptions` are ignored. Thus, this parameter is a useful way of overriding default settings.

With the current Newton OS, only one endpoint can be instantiated at a time. If one is already instantiated when this call is made, then an error is returned.

`Instantiate` will throw an exception on any error.

Note: Instantation will normally power-on any necessary hardware. For instance, instantiating a serial service will turn on the serial chip. Since this takes power, keeping an endpoint instantiated may reduce the battery life of a Newton. Consider instantiating only as necessary, rather than as your application opens.

`endPoint:Connect(address, options)`

This method opens a connection. If *address* is `nil`, the address from *endPoint*`.configOptions` is used (if this service requires an address); otherwise *address* is an address frame which specifies the connecting address. The *options* parameter is not used and should be `nil`.

This method may be time consuming (for instance, it may require dialing the modem). For this reason, it should normally be executed from within a deferred action so that your user interface continues to operate and thus allows the user to abort, if desired.

`endPoint:Listen(options)`

This method waits for a connection to be made to this endpoint. The *options* parameter is unused and should be `nil`.

This method may be time consuming (for instance, it may require waiting for a phone call, or for an IR transmission to arrive). For this reason, it should normally be executed from within a deferred action. This way your user interface continues to operate and the user can abort, if necessary.

`endPoint:Disconnect()`

This method closes the connection (the opposite of `Connect` or `Listen`).

`endPoint:Release()`

This method closes the connection (the opposite of `Connect` or `Listen`) after waiting for all pending input and output to complete.

Data Sending Endpoint Methods

There are two endpoint methods you can use for sending data: `Output` and `OutputFrame`. The first can output arrays, characters, integers, or strings. The second sends a flattened frame.

`endPoint:Output(data, flags)`

This method sends *`data`*; the type of *`data`* will determine how it is sent. The *`data`* parameter is one of the following four possibilities:

`integer`	Sends the low 8 bits of the integer as a byte.
`character`	Sends that character. Each character is converted from Unicode to ASCII using the Macintosh character encodings (for example '…' is converted from Unicode `\u2026` to ASCII `0x7e`). Characters that don't have an equivalent in the Macintosh character encoding are converted to `0xff`.

`string`	Sends the characters of the string (each character is converted to ASCII).
`array`	Sends each element of the array (each element must be an integer). The low 8 bits of each integer is sent.

The *`flags`* parameter is normally `nil` and is ignored by most endpoints. The IR endpoint uses this parameter, but the details of this parameter aren't covered in this book.

Note that data are buffered; the ADSP endpoint, in particular, allows at most three `Output` calls before you need to call `FlushOutput` to send the buffered data.

`endPoint``:OutputFrame(`*`frame, flags`*`)`

This method sends *`frame`* as a flattened sequence of bytes. The format of this sequence is undocumented.

`endPoint``:FlushOutput()`

This routine sends buffered data, returning when more data can be sent.

Data Receiving Endpoint Methods

Before detailing the data receiving methods that you can use, let us first look at the frame you need to create to handle the incoming data.

The Input Spec

In order to receive data, you must create an input spec frame which specifies the kind of input expected. Slots in the input spec are:

`byteCount`	The number of bytes to read. If you provide this slot you do not provide an `endCharacter` slot.
`discardAfter`	The input buffer size. If more than `discardAfter` characters are received, only the most recent `discardAfter` ones are saved. This slot is optional.
`endCharacter`	The character that terminates input. If you provide this slot you do not provide a `byteCount` slot.
`inputForm`	The format of the incoming data. Possible values are: `'array` each incoming byte is converted to an integer and stored as successive elements of an array. *This is the default value.* `'frame` the incoming sequence of bytes is a flattened frame. The `byteCount`, `discardAfter`, and `endCharacter` slots are ignored; `'raw` no conversion to Unicode takes place and a binary objects is returned; `'string` ASCII input is converted to a Unicode string.
`inputScript`	A function that is called when input has been received (as specified by the `byteCount` or `endCharacter` values, or, if the `inputForm` is `'frame`, when the entire frame has arrived).
`partialFrequency`	The frequency (integer number of milliseconds) at which the `partialScript` is called. Provide this slot only if a `partialScript` slot is provided.

`partialScript`	A method that is called with the data received so far. The frequency at which the method executes is specified in `partialFrequency`. This slot is optional.

To make this input spec the active input spec, use `SetInputSpec`.

Note: Another way to specify the active input spec is to add a `nextInputSpec` slot to the endpoint. However, for future compatibility, you are encouraged to use the explicit call to `SetInputSpec` instead.

endPoint:`SetInputSpec(`*inputSpec*`)`

This method makes *inputSpec* the active input spec. If *inputSpec* is `nil`, this aborts the current input spec.

Input Scripts

An `inputScript` is called whenever the termination condition of the input spec occurs. The `inputScript` expects the following parameters:

`endPoint`	The endpoint that had input.
`data`	The data that was input. The format of the data depends on the value of `inputForm` (the slot in the input spec).

Commonly, an `inputScript` will specify a new input spec using `SetInputSpec`. This provides a way to chain input specs. Note, however, that if an `inputScript` doesn't specify a new input spec, the current input spec remains in use.

Partial Scripts

A `partialScript` provides a way to look at input that has been received *before* the termination condition has occurred. The `partialScript` is executed based on the specified frequency (no more often than specified). It is called if new input has arrived since the last time it was called.

The `partialScript` expects the same parameters as the `inputScript`:

`endPoint`	The endpoint that had input.
`data`	The data that was input. This is all the data in the buffer.

After the `partialScript` is called, the data remains in the buffer (as opposed to calling the `inputScript`). To remove data, you can use `FlushPartial`. To see what data has arrived since the last call to `partialScript`, use `Partial`.

`*endPoint*:FlushPartial()`

This method removes all the buffered input through the end of the last call to the `partialScript`.

`*endPoint*:Partial()`

This method returns the data since the last call to the `partialScript`. `Partial` doesn't remove data from the input buffer.

Other Input, Error, and Abort Methods

`*endPoint*:Input()`

This method returns the data in the input buffer, formatted according to the current input spec. It terminates the current input spec.

Note that the `inputScript` method of the current input spec is not called.

endPoint:ExceptionHandler(*error*)

This method gets called whenever a communications error occurs. The *error* parameter is a frame with a name slot that contains the exception type, along with a data slot containing an integer error code.

Your endPoint can have an ExceptionHandler method. If this method is present, it will be called whenever a communications exception or error occurs.

endPoint:Abort()

If a Connect or Listen hasn't yet completed, this method will abort it. If they have completed, Abort will abort any input or output in progress. To abort input, you should use the following lines of code:

```
endPoint.nextInputSpec := nil;// no next input spec
ep:SetInputSpec(nil);// replace present input spec
ep:Abort();           // abort present input spec
```

To abort input, output, or an in-progress Connect or Listen, use the Abort method. Abort will cause your exception handler to execute. This exception can usually be ignored.

Service Types, Options, Addressing

As we previously stated there are several service types supported on the Newton. In review, these types are:

- Serial
- Serial with Error-Correction and Compression
- Modem
- ADSP
- IR.

Now, let us walk through each service type. We will detail the various configuration options that you can set and how you create each type as a frame in the `configOptions` slot of an endpoint frame. You might also be using these frames when you instantiate or connect that particular type of endpoint.

Serial

The following is the frame you would use for the serial service:

```
{
   label:    kCMSAsyncSerial,
   type:     'service,
   opCode:   opSetRequired }
```

There are three possible options that you can set for serial service. The first is `kCMOSerialIOParms`. If you choose this option, then the `data` slot in the `configOptions` frame should be a frame with the following slots:

`bps`	This contains the speed of the connection as an integer.
`dataBits`	This should be one of `k5DataBits`, `k6DataBits`, `k7DataBits`, or `k8DataBits`.
`stopBits`	This should be one of `k1StopBits`, `k1pt5StopBits`, `k2StopBits`.
`parity`	This should be one of `kNoParity`, `kEvenParity`, `kOddParity`.

For example, here is the element that you might see in the `configOptions` array for this type of option:

```
configOptions:[ {…}, {…},
     {
        label:    kCMOSerialIOParms,
        type:     'option,
        opCode:   opSetRequired,
        data: {
           bps:      k9600bps,
           dataBits:k8DataBits,
           stopBits:k1StopBits,
           parity:   kNoParity,
        }
     },
   ]
```

The second and third options are `kCMOInputFlowControlParms` and `kCMOOutputFlowControlParms`. They specify the flow control to be used for input and output. In each of these cases, the `data` slot of each (in the `configOptions` array element) should be a frame with the following slots:

`xOnChar`	The character used to resume sending data.
`xOffChar`	The character to be used to stop sending data.
`useSoftFlowControl`	True specifies that software flow control should be used.
`useHardFlowControl`	True specifies that hardware flow control should be used.

Note: You should not use soft flow control if you will be sending or receiving binary data because the Xon/Xoff characters may be in the binary data itself and will confuse the endpoint.

An example of each type of option in the `configOptions` array might be:

```
configOptions:[ {…}, {…},
      {
         label:   kCMOInputFlowControlParms,
         type:    'option,
         opCode:  opSetRequired,
         data: {
            xOnChar: kDefaultxOnChar, //predefined
            xOffChar:kDefaultxOffChar, //predefined
            useSoftFlowControl:true,
            useHardFlowControl:nil,
         }
      },
      {
         label:   kCMOOutputFlowControlParms,
         type:    'option,
         opCode:  opSetRequired,
         data: {
            xOnChar:    kDefaultxOnChar,
            xOffChar:   kDefaultxOffChar,
            useSoftFlowControl:true,
            useHardFlowControl:nil,
         }
      },
   ]
```

Serial with Compression

The following is the service option for the serial with compression service:

```
{
   label:   kCMSMNPID,
   type:    'service,
   opCode:  opSetRequired
}
```

The options are the same as those for the serial service plus the following option labels:

`kCMOMNPAllocate` — This option specifies whether a buffer should be allocated for compression. If you are requesting compression, you must allocate a buffer. The `data` slot should contain either `kMNPDoAllocate` or `kMNPDontAllocate`.

`kCMOMNPCompression` — This option specifies the type of compression to use. The `data` slot should contain one of `kMNPCompressionNone`, `kMNPCompressionMNP5`, or `kMNPCompressionV42Bbis`. If you request V42.bis and it cannot be supported (due to negotiation with the party on the other end of the line), fallback will occur to MNP5 compression. If that can't be supported, fallback will occur to no compression.

`kCMOMNPDataRate` — The `data` slot is an integer specifying the data rate in bps. This slot is optional.

Modem

The following is the frame you would use for the modem service:

```
{
   label:   kCMSModemID,
   type:    'service,
   opCode:  opSetRequired
}
```

The configuration options are the same as those for the serial with compression service (except that the `kCMOMNPDataRate` option is not supported). There are also two additional options.

The first additional option is `kCMOModemECType`. For this option, the `data` slot should contain one of the following three choices:

`kModemECProtocolNone` Specifies no error correction.

`kModemECProtocolMNP` Specifies software MNP error correction.

`kModemECProtocolExternal` Specifies error correction done by the modem (some modems may support error correction; others may not).

The second option is `kCMOModemDialing`. In this case, the `data` slot should contain a frame with the following slots:

`speakerOn` True specifies the speaker should be on while dialing. The default is `true`.

`detectDialTone` True specifies the modem should listen for a dial tone before dialing. The default is specified by the user.

`detectBusy` True specifies the modem should detect a busy signal after dialing. The default is `true`.

`dtmfToneDialing` True specifies the modem should dial using tones rather than pulses. The default is specified by the user.

`manualDial` True specifies the user should dial the number manually rather than having the modem dial automatically. The default is `nil`.

`speakerVolume` This contains one of these possible constant values: `kSpeakerVolumeLow`, `kSpeakerVolumeMedium`, and `kSpeakerVolumeHigh`. The default is specified by the user.

`waitForCarrier` This integer specifies the number of seconds to wait for a carrier signal. The default is 55.

`waitBeforeBlindDial`	If `detectDialTone` is `nil`, this specifies the number of seconds to wait before dialing. The default is 3.
`commaDelay`	This integer specifies the number of seconds to wait when encountering a comma in the phone number (this is usually used to wait for a second dial tone when accessing an outside line; a phone number might be "9,5551212"). The default is 1.
`ringToAnswerAfter`	This integer specifies how many rings should occur before the modem answers the phone. This option is only used while a `Listen` has been issued to the endpoint. The default is 2.

Caution: There is a bug in the code which reads the modem dialing options. Rather than looking for these slots in the `data` frame, they are looked for directly in the option frame. In order to work around this bug, duplicate each of the slots in the `data` frame in the option frame. Here is an example that works around the bug:

```
{
   label:kCMOModemDialing,
   type:'option,
   opCode:opSetRequired,
   data: {
      speakerOn: nil,
      manualDial: true,
   },
   // slots from data frame are duplicated to
   // work around a bug
   speakerOn: nil,
   manualDial: true,
}
```

The Modem item in the Prefs application allows the user to select three different preferences (see FIGURE 2.6) for the modem. In turn, the `userConfiguration` contains three slots, each of which holds one of these preferences: `blindDialing`, `modemSoundVolume`, and `dialing`. Further, the modem service automatically takes into account these values in the `userConfiguration` frame.

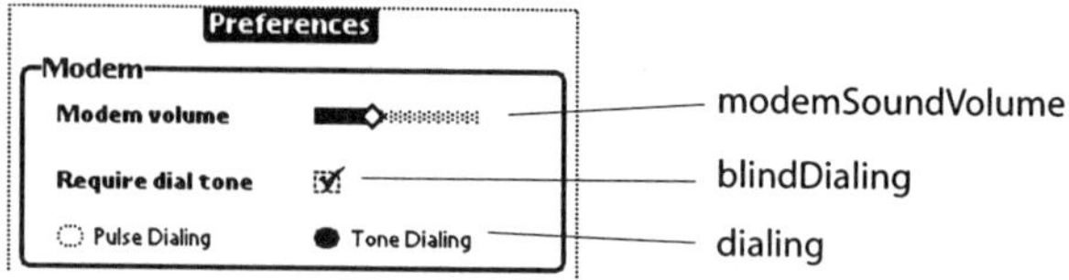

FIGURE 2.6 The Modem preferences.

IR

This is the frame you would use for the IR service:

```
{
    label:    kCMSSlowIR,
    type:     'service,
    opCode:   opSetRequired
}
```

There are no configuration options or addresses for this service.

Note: An IR service must have a `recvFlags` slot in the endpoint set to the value `kFrame`.

ADSP

The following is the service frame used for the ADSP service:

```
{
    label:    kCMSAppleTalkID,
    type:     'service,
    opCode:   opSetRequired
}
```

The configuration options for an ADSP service are required as is; don't change them. They are:

```
{
    label:kCMSAppleTalkID,
    type: 'option,
    opCode:opSetRequired,
    data: kCMOAppleTalkADSP
},
```

```
{
   label:kCMOEndpointName,
   type: 'option,
   opCode:opSetRequired,
   data: kADSPEndpoint
}
```

Addressing is in the form of an NBP entity stored in the `addressData` slot of the `data` frame:

```
{
   type:    'address,
   label:   kCMARouteLabel,
   opCode:  opSetRequired,
   data : {
      addressType:kNamedAppleTalkAddress,
      addressData: "entityName:entityType@zoneName"
   }
}
```

AppleTalk Functions

When you are using an ADSP endpoint, you may end up needing to use some of the following functions (they are AppleTalk-specific). If you are using any other type of endpoint, you can cheerfully skip this section.

Note: For more information on AppleTalk Data Session Protocol (ADSP), Name Binding Protocol (NBP), or other AppleTalk information, see the book *Inside Macintosh: Networking.*

```
OpenAppletalk()
```

This opens AppleTalk. You would call this routine before any other AppleTalk global functions.

`CloseAppletalk()`

This closes AppleTalk. You would call this routine after you are finished calling all other AppleTalk global functions.

`HaveZones()`

This function returns `true` if there are zones on the network. It returns `nil` if there are none to be found.

`GetMyZone()`

This function returns the name of the current AppleTalk zone (or '*' if the zone has no name).

`GetZoneList()`

This function returns an array containing the names of all the available zones. If there are no named zones, then it returns an array containing the string '*'. If this zone list exceeds the memory available, this function may run out of memory.

`GetNames(`*`entityOrEntities`*`)`

This function extracts the entity name from the NBP address *`entityOrEntities`*. If *`entityOrEntities`* is an array of NBP address entities, this function returns an array of extracted entity names.

For example, **`GetNames("Ralph:Pager@*")`** returns **`"Ralph"`**. **`GetNames(["Ralph:Pager@*", "Bob:Pager@Fastzone"])`** returns the array **`["Ralph", "Bob"]`**.

`GetZoneFromName(`*`entity`*`)`

This function extracts the zone name from the NBP address entity.

For example, `GetZoneFromName("Ralph:Pager@*")` returns '*'. `Get-ZoneFromName("Bob:Pager@Fastzone")` returns `"Fastzone"`.

`NBPStartLookup(`*`entity`*`)`

This function begins a lookup for entities as specified by *`entity`*. The entity parameter can be a complete entity (specifying a name, type, and zone), or a wildcard '=' as the name. Such a wildcard would specify all entities of the given type found in the given zone. Since you cannot also specify a wildcard for the zones, this call only finds entities within a given zone.

For example, `NBPStartLookup("Ralph:Pager@*")` looks for the entity named "Ralph" of the type "Pager" in the current zone. On the other hand, `NBP-StartLookup("=:Pager@*")` looks for all entities of the type "Pager" in the current zone.

`NBPLookupCount()`

This function returns the number of entities that have been found so far by the current NBP lookup. Since the NBP lookup is dynamic (some entities take longer than others to appear, and entities may appear and disappear from the network), this function may return different values on successive calls.

`NBPGetLookupNames()`

This function returns an array of the entity names that have been found so far by the current NBP lookup. Since the NBP lookup is dynamic (some entities take longer than others to appear, and entities may appear and disappear from the network), this function may return different values on successive calls.

```
NBPStopLookup()
```

This function stops a lookup started by `NBPStartLookup`.

Sample Program

Now that you have learned how to construct an endpoint and how to specify what type of service you wish to use, it is time to look at some code that actually does this. In this section, we'll present a sample application which will metamorphize three times to support different services:

- The initial application will use the serial service.
- Next, we'll modify the application to use a modem instead.
- Last of all, our application will use ADSP.

What the Program Does

This application sends an alphanumeric page to a pager service. It allows the user to send alphanumeric pages from a Newton to anyone with an alphanumeric pager.

Most pager services have a phone number which you can dial with your modem. The typical type of conversation that one has with a pager service is as follows (output from pager service is in bold):

```
<CR>
ID=
M<CR>
Pager ID?
a pager number<CR>
Message:
The message to send<CR>
Page accepted.
Pager ID?
<CR>
Thank you.
```

This is, in fact, the protocol used for the Newton PCMCIA Messaging Card pager service. Note that other pager services may vary the protocol slightly. We, however, will use the Apple protocol in our sample.

The user interface for the application is fairly simple. FIGURE 2.7 shows the application. In this view you can see an input line for the pager ID, and another multiline one in which you write a message. Tapping Send sends the page.

FIGURE 2.7 Pager application.

While the page is being sent, the user also gets some status. FIGURE 2.8 shows the status slip that appears.

FIGURE 2.8 The status slip while a page is being sent.

To cancel the send, the user just taps the close box of the application. FIGURE 2.9 shows the application in a closed state and the status slip showing the progress of the disconnect. Note that this application's status slip will also show particular information at different times, thereby letting the user know what is happening as the page is sent.

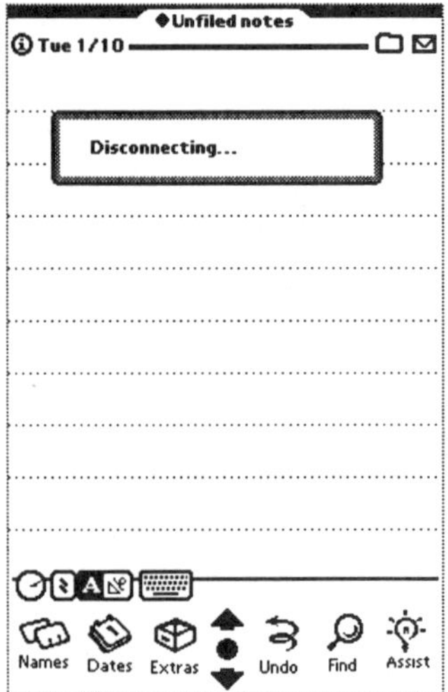

FIGURE 2.9 The status slip if the application is closed while a page is being sent.

Creating the Endpoint

In the `viewSetupFormScript` of our application base template, we'll create the endpoint. We will save it in a slot named `fEndPoint` in the view (you can't create it in the template because some of the slots are set at runtime). Our endpoint will also have an `_parent` slot pointing to the application base view. In this way, the endpoint can access methods and slots from the application using inheritance. Further, we get to avoid the awkward syntax associated with sending messages to the application base view—the endpoint would have had to explicitly reference the application base view using `GetRoot().(kAppSymbol)`.

Here's our `viewSetupFormScript`:

```
func()
begin
   self.fEndPoint := {
      _proto:             protoEndPoint,
      _parent:            self,
      exceptionHandler: ExceptionHandler,
      configOptions:      kEndPointConfig,
      fConnectAddress: DeepClone(kEndPointAddress),
      floater:            nil,
   };
end
```

There are two slots (`fConnectAddress`, `floater`) that aren't used by `protoEndPoint`, but are here as a convenient place to store information relevant to the endpoint.

We `DeepClone` the address because we will need to change it to work with ADSP. At that point, we'll be setting a slot within that frame at runtime. Notice that if we didn't `DeepClone` it, we couldn't change a slot because `kEndPointAddress` is a constant defined in Project Data. The `ExceptionHandler` will be discussed in "Exception Handling" on page 95. The `configOptions` and `fConnectAddress` will differ depending on the service used; `kEndPointConfig` and `kEndPointAddress` will be defined in the Project Data file.

Defining the Endpoint States

The endpoint can be in a variety of states that we need to know about. We'll keep track of these by creating a set of constants; and the `status` slot in the application base view will then specify the current state. Here are the constants we will use:

`kState_Disconnected`	Not connected.
`kState_Connect`	Preparing for an asynchronous connect.
`kState_Connecting`	In the process of an asynchronous connect.
`kState_Connected`	Connected.
`kState_Disconnecting`	In the process of an asynchronous disconnect.

The `status` slot will be initialized to hold `kState_Disconnected`.

Starting the Connection

When the user taps the Send button, the `SendPage` method of the application is called:

```
func()
begin
   :DoConnect();
end
```

The `DoConnect` method is located in the application base template:

```
func()
begin
   if status <> kState_Disconnected then
      return;

   // we do the connection in a deferred action so
   // that the user can abort (by closing the app).
   :SetEndPointState(kState_Connect);

   AddDeferredAction(ConnectAction, [fEndPoint]);
end;
```

The method starts a deferred action which actually does the instantiation and connection. In this way, the user interface is still active, and the user can cancel the connection by closing the application.

Note: Deferred actions do not normally allow user interface and other events to be processed. The endpoint methods have been specifically architected to provide this feature.

We must check the state to make sure we are disconnected. Otherwise, we could end up in the messy situation of trying to connect when we are already connected (the user might tap the Send button twice in succession).

Now, we set the endpoint state to hold the new `kState_Connect` value. Note that the `SetEndPointState` method of the application base template just sets the status slot of the application base view:

```
func(newState)
begin
   GetRoot().(kAppSymbol).status := newState;
end
```

Providing Status

Before going too much farther, it is important to provide a status window. This is necessary not only for the user of the application, but for us as well. It provides an indispensable tool for debugging communications code. Remember, only one service can be active at a time; this means that the Inspector can't be connected while any other service is active. Thus, our debugging is reduced to printing somewhere other than to the Inspector. Possibilities include writing to a soup (which has the advantage that information is saved even on a restart), writing to the screen, and beeping or making other sorts of sounds.

To debug this application, we chose to write various strings to the status window. Thus, getting this window working is high on our list of important things to do early. To create our window, we will use a user proto called `status` which provides a floating window. It has one method, `SetMessage` which displays the string passed to it. Now, we'll need to modify `DoConnect` to open the status window and display the initial status:

```
func()
begin
   if status <> kState_Disconnected then
      return;
   fEndPoint.floater := BuildContext(pt_status);
   fEndPoint.floater:Open();
   fEndPoint.floater:SetMessage("Connecting...");

   :SetEndPointState(kState_Connect);
   AddDeferredAction(ConnectAction, [fEndPoint]);
end;
```

Our new debugging floater

We do the connection in a deferred action so that the user can abort (by closing the application).

We'll store the floater view in the endpoint itself for easy access.

Connecting

Now, let us look at the `ConnectAction` function that we call as a deferred function from within the last line of `DoConnect`:

```
func(theEndPoint)
begin
   local err := kError_EndpointInUse;

   try
      err := theEndPoint:Instantiate(
         theEndPoint, NIL);
   onexception |evt.ex| do begin
      theEndpoint:NotifyError(err);
      :SetEndPointState(kState_Disconnected);
      return;
   end;

   :SetEndPointState(kState_Connecting);
```

```
      err := theEndPoint:Connect(
         theEndPoint.fConnectAddress, NIL);
      theEndPoint:ConnectCompletionProc(err);
   end
```

This code uses an exception handler around the call to `Instantiate` (for information on exception handlers, see "Exceptions" on page 245). This way, if `Instantiate` fails, we can notify the user of the error with the application method `NotifyError`, and set the state back to `kState_Disconnected`.

If we sail through with no errors, then we set the state to `kState_Connecting`, and do a connect to the address specified by `fConnectAddress`. Connecting may take some time to complete, but remember that this function is executing as a deferred function. Therefore the task executing our user interface continues to run (allowing a cancel). Once the `Connect` completes, we call our `ConnectCompletionProc` routine.

Completing the Connection

The `ConnectCompletionProc` routine is in the application base template:

```
func(err)// called from a deferred action
begin       // self is the endpoint frame
   if not err then begin
      :SetEndPointState(kState_Connected);
      floater:SetMessage("Connected");
      :SetInputSpec(inputSpecInitial);
      AddDelayedAction(func(s) begin
         s:DoOutput(""); s:DoOutput("");
      end, [self], 1000);
   end
   else if err <> -10039 then begin
      // -10039 means :Abort() was called to kill
      // an in-progress (asynchronous) :Connect()
      // so do NOT call :DoDisconnect() when
      // this error occurs(!)  Just ignore it...

      :DoDisconnect();
      :NotifyError(err);
   end;
end
```

If no error occurred, then we are connected; we set the state to reflect that, along with the status. Next, we set our input spec to `inputSpecInitial`, which is on the lookout for "ID=". We send two carriage returns using `DoOutput`, but wait 1 second before sending them (the paging service seems to ignore input sent immediately after connecting).

If an error occurs, we must disconnect. If an `Abort` aborts an in-progress connection, however, we should just ignore the error (since the `Abort` will precede a disconnect).

Outputting

We've written a routine `DoOutput` which outputs the string passed to it along with a carriage return:

```
func(data)
begin
   if status <> kState_Connected then begin
      :DoNotify("Not connected.");
      return;
   end;
   fEndPoint.floater:SetMessage("Sending" && data);
   fEndPoint:Output(data & unicodeCR, NIL);
   fEndPoint:FlushOutput();
end
```

We check to make sure we're connected first. If we aren't, we issue an error. Otherwise, we output the string along with a carriage return. Then we call `FlushOutput` to make sure we don't exceed our buffers.

For debugging purposes, it is also useful to show the data we are sending in the status window.

Handling ID=

Our `inputSpecInitial` slot holds the first input spec we'll be using:

```
{
   inputForm:'string,
   endCharacter:$=,
   discardAfter:256,
   inputScript:func(ep, data)
      begin
         if StrPos(data, "ID=", 0) then
            ep:DoInputInitial();
      end,
}
```

We're waiting for a string that ends in the '=' character. Notice also, that we only want the last 256 characters that come in (if we call the wrong service which starts spitting out *War and Peace*, we don't want to overflow our input buffer).

If a string terminated by '=' comes in, our `inputScript` will be called. We'll make sure the string contains "ID=". If it does, we'll send a `DoInputInitial` message to our endpoint and that will find the corresponding method in the application base view due to parent inheritance. If the string doesn't contain "ID=", we'll just wait patiently until we get another string ending in '='.

The `DoInitialInput` method in our application base template looks like this:

```
func()
begin
   :DoOutput("M");
   :SetInputSpec(inputSpecPagerID);
end
```

We simply output an 'M' character and then change to a different input spec.

Handling Pager ID

The `inputSpecPagerID` slot we reference in `DoInitialInput` contains the following input spec:

```
{
   inputForm:'string,
   endCharacter:$?,
   discardAfter:256,
   inputScript:func(ep, data)
      begin
         if StrPos(data, "Pager ID?", 0) then
            ep:DoInputPagerID();
      end,
}
```

The `inputScript` will be called with a string when a '?' character is seen. The script verifies that the string contains "Pager ID?" and then sends the `DoInputPagerID` message to the endpoint.

The `DoInputPagerID` method is found in the application base template:

```
func()
begin
   :DoOutput(pin.text);
   :SetInputSpec(inputSpecMessage);
end
```

The code sends the text from the `pin` view (this is a numeric pager ID number specifying the particular pager to which the message is being sent). It then switches to use the `inputSpecMessage` input spec.

Handling Message

The `inputSpecMessage` input spec is found in the application base template:

```
{
   inputForm:'string,
   endCharacter:$:,
   discardAfter:256,
   inputScript:func(ep, data)
      begin
         if StrPos(data, "Message:", 0) then
            ep:DoInputMessage();
      end,
}
```

When a string comes in which ends in a ':', the `inputScript` is called. It verifies that the string contains "Message:" and then sends the `DoInputMessage` to the endpoint. The `DoInputMessage` is found in the application base template:

```
func()
begin
   :DoOutput(message.text);
   :SetInputSpec(inputSpecDone);
end
```

The method sends the text in the message view (the actual view which contains the desired message). It then switches to the final input spec: `inputSpecDone`.

Handling Page Accepted

The `inputSpecDone` input spec is also located in the application base template:

```
{
   inputForm:'string,
   endCharacter:$?,
   discardAfter:256,
   inputScript:func(ep, data)
      begin
         if StrPos(data, "Page accepted", 0) then
            ep:DoInputDone();
      end,
}
```

If a string comes in which ends in a '?', the `inputScript` is called. It verifies that the string contains "Page accepted". If it does, our spec sends the `DoInputDone` message to the endpoint. If it doesn't, it waits for another string ending in '?'.

Note that it is quite possible that a string ending in '?' comes in that doesn't contain "Page accepted". Remember that the pager server echoes any characters which are typed to it. Therefore, if a message is sent which contains an embedded '?', the `inputScript` will fire. Of course, if the message itself contains "Page accepted" as well as a '?', we'll will most certainly get royally confused.

Here's the `DoInputDone` method of the application base template:

```
func()
begin
   :DoOutput("");
   fEndPoint:DoDisconnect();
end
```

We send a final carriage return, and then disconnect and uninstantiate the endpoint by sending it the `DoDisconnect` message. We don't need to wait to receive a "Thank you" since we aren't obliged to be courteous by responding "You're welcome".

Starting the Disconnect

The `DoDisconnect` method is located in the application base template:

```
func()
begin

   if status <> kState_Connected
      and status <> kState_Connecting then
      return;

   local currentState := status;
   :SetEndPointState(kState_Disconnecting);
   fEndPoint.floater:SetMessage("Disconnecting…");

   fEndPoint.nextInputSpec := NIL;
   fEndPoint:SetInputSpec(NIL);
   fEndPoint:Abort();

   AddDelayedAction(DisconnectAction,
      [fEndPoint, currentState], 2500);
end
```

First, we verify that we are either connected or in the middle of connecting (otherwise, there's no reason to disconnect). Then, we set our status message and our state to show that we are disconnecting.

Next, we must abort current input and output. Thus, we must handle the setting of `nextInputSpec`, the call to `SetInputSpec`, and the call to `Abort`. Treat these as a magical incantation that should always be followed in exactly this order.

Finally, we issue a delayed call to actually disconnect. We delay for two reasons:

1. Since disconnecting takes time, we want to do it asynchronously so that the user interface continues to operate.

2. You are required to wait a little while after calling `Abort` before calling `Disconnect`.

Disconnecting

The `DisconnectAction` method is in the application base template:

```
func(ep, state)
begin
   if state = kState_Connected then
      ep:Disconnect();
   ep:Dispose();
   ep:DisconnectCompletionProc();
end
```

The method calls `Disconnect` if we were connected (the `DoDisconnect` method might have been called while we were in the process of disconnecting). Then, it sends the `Dispose` method to the endpoint to uninstantiate it. Finally, we send the `DisconnectCompletionProc` to the endpoint to do whatever other cleanup is needed.

Completing the Disconnect

The `DisconnectCompletionProc` is a method in the application base template:

```
func()
begin
   if floater then begin
```

```
            floater:Close();
            floater := nil;
        end;

        :SetEndPointState(kState_Disconnected);
    end
```

When this routine is called, `self` is the endpoint. We close the floater and `nil` out the `floater` slot in the endpoint. Then, we set the state to `kState_Disconnected`.

Handling User Abort

The user interface we've chosen for aborting is for the user to close the application. Therefore, we've got a `viewQuitScript` in the application base template that handles this feature:

```
func()
begin
    :DoDisconnect();
end
```

If we're in the middle of a connection when the user quits, this will disconnect. If we're already disconnected, `DoDisconnect` will do nothing.

Time-out

We have also got a protocol problem to worry about. What if the pager server never sends us a '='? We'll twiddle our thumbs forever waiting for the termination condition of our input spec to occur. The same thing will happen if at any point we don't terminate. Our chosen solution to both of these problems is to set a watchdog timer. If the timer expires and we're still connected, something's gone wrong and we'll go ahead and disconnect. If the timer expires and we are no longer connected, we'll do nothing.

We'll modify the `DoConnect` method of the application base template to add our little Lassie:

```
func()
begin
    if status <> kState_Disconnected then
        return;
```

```
      fEndPoint.floater := BuildContext(pt_status);
      fEndPoint.floater:Open();
      fEndPoint.floater:SetMessage("Connecting...");
      // we do the connection in a deferred action so
      // that the user can abort (by closing the app).
      :SetEndPointState(kState_Connect);

      // give it 2 minutes
      AddDelayedAction(TimerExpired, [fEndPoint],
         120000);
      AddDeferredAction(ConnectAction, [fEndPoint]);
   end;
```

The `TimerExpired` function will execute after 2 minutes and is created as a slot in the application base template:

```
func(theEndPoint)
begin
   if theEndPoint.floater then begin
      theEndPoint:DoDisconnect();
      theEndPoint:DoNotify("Timed out");
   end;
end
```

Notice that if the floater is up, then we must be connected. Therefore, we disconnect and put up a message to the user.

In other applications, you might want to consider setting up watchdog timers while waiting for a particular piece of input (after you `SetInputSpec` to a particular input spec) rather than for the complete session. In our case, we can estimate the time of the entire session fairly accurately; if that time is exceeded, something must have gone wrong. Thus, we get our watchdog to herd our errant endpoint back into a proper state.

Exception Handling

In case of an error, we should report an error and cleanup. The `ExceptionHandler` function in the application base template handles this:

```
func(exceptionFrame)
begin
// -16005 and -16013 is an exception generated by
// ep:Abort() while connected -- really not an error

   if exceptionFrame.data exists
      and exceptionFrame.data <> -16005 and
```

```
        exceptionFrame.data <> -16013 then begin

        :NotifyError(exceptionFrame.data);
        :DoDisconnect();
    end;

    return true;
end
```

Errors -16005 or -16013 are caused by an `Abort` call while connected. As this is not really an error (this is part of our cleanup), we will just ignore it. In the case of any other error, we notify the user of the error and then disconnect.

The `NotifyError` method of the application base template is:

```
func(error)
begin
    if error = kError_EndpointInUse then
        :DoNotify(kMessage_EndpointInUse);

    else if error = -18003 then
        :DoNotify(kMessage_BufferOverrun);

    else
        :DoNotify("An error has occurred.  Error:" &&
            NumberStr(error));
end
```

This method relies on some constants that we define in the Project Data file. These constants are defined there as:

```
constant kError_EndpointInUse := -666;
constant kMessage_BufferOverrun := "A parity or bit-frame
error has occured, or the communication data buffer was
overrun and has been reset.";
constant kMessage_EndpointInUse := "Another application seems
to be using the communications port.";
```

`NotifyError` calls the `DoNotify` method in the application base template:

```
func(errString)
begin
    :Notify(kNotifyAlert,
        EnsureInternal(kAppName),
        EnsureInternal(errString)
    );
end
```

Preventing Power Off

There is another situation that we should take into account: *What happens if the power goes off while we're communicating?* The user could press the sleep switch, or the automatic sleep could go off. In such a case, we would then lose our connection. Here are the two options we for dealing with the situation:

1. Prevent the Newton from sleeping while we're connected.
2. Disconnect our connection on sleep and `Connect` again when the Newton wakes up (using a deferred action).

We'll use the first option since our communication doesn't take too long (we won't be around long enough to drain the batteries), and because we would have to start from scratch if we reconnected.

Thus, to handle power-offs we'll use `AddPowerOffHandler` and `RemovePowerOffHandler` (see "AddPowerOffHandler(view)" on page 340 as well as "RemovePowerOffHandler(view)" on page 369 for more details).

We're connected only if the status floater is up; therefore we'll add the power-off handler in the `viewSetupFormScript` of the floater:

```
func()
begin
   AddPowerOffHandler(self);
end
```

We'll remove the power-off handler in the `viewQuitScript` of the floater:

```
func()
begin
   RemovePowerOffHandler(self);
end
```

We add a `powerOffScript` method in the floater that will be called if the Newton tries to go to sleep:

```
func(what)
begin
   if what = 'okToPowerOff then
      return nil
   else
      return true;
end
```

For more information on the `powerOffScript` see "powerOffScript(what)" on page 332. If the Newton decides to sleep (on its own or due to the user sliding the sleep switch) it'll call our `powerOffScript` passing `'okToPowerOff`. We'll return `nil` saying that it is not OK to go nighty-night.

Note: If you are using a MessagePad 110, make sure you are using a release of the 1.3 system greater than or equal to 345025. Earlier versions of the system had a bug with the powerOffScript (the symptom was no longer being able to power off). To determine the release, see the bottom of the status bar in the Prefs application. To obtain a new release, contact Apple at (800) sos-appl, or see the eWorld or Applelink online services.

The Serial Version of the Application

Before attempting to use the modem to connect to the actual paging service, it is simpler to use a serial connection to connect to a desktop computer. Then, being quite tricky from a terminal emulation program, we can pretend we are the computer. We'll type by hand the strings "ID=", "Pager ID?" and so on.

We need to define the following configuration options constant and address constant in our Project Data file:

```
DefConst('kEndPointConfig, [
   {
      label:kCMSAsyncSerial,
      type: 'service,
      opCode:opSetRequired
   },

   {
      label:kCMOSerialIOParms,
      type: 'option,
      opCode:opSetRequired,
      data: {
         bps:     k9600bps,
         parity:kNoParity,
         dataBits:k8DataBits,
         stopBits:k1StopBits,
      }
   },
```

```
        {
            label:kCMOInputFlowControlParms,
            type: 'option,
            opCode:opSetNegotiate,
            data: {
                useHardFlowControl:nil,
                useSoftFlowControl:true,
                xonChar:    unicodeDC1,
                xoffChar:   unicodeDC3,
            }
        },

        {
            label:kCMOOutputFlowControlParms,
            type: 'option,
            opCode:opSetNegotiate,
            data: {
                useHardFlowControl:nil,
                useSoftFlowControl:true,
                xonChar:    unicodeDC1,
                xoffChar:   unicodeDC3,
            }
        },
    ]
    );

    constant kEndPointAddress := nil;
```

We're specifying a 9600-bps connection with control-S and control-Q software flow control in both directions.

These two constants are used in creating the endpoint in the `viewSetupFormScript` of the application base template (see "Creating the Endpoint" on page 84).

At this point, we can run the application. Here are the steps to go through:

1. Connect a serial cable between the Newton and a desktop machine.

2. Set the desktop machine to the appropriate bps, stop bits, start bits, and parity.

3. Tap Send on the Newton.

4. On the desktop computer, type "ID=". You should see an "M".

5. Type “Pager ID?”. You should see a pager number.

6. Type “Message:”. You should see the message.

7. Finally, type “Page accepted.<CR>Pager ID?”. The Newton application should now disconnect and the status floater should disappear.

Test user cancelling by doing a Send and then tapping the close box of the application. The application view should disappear, while the status window should change to say: “Disconnecting…” (see FIGURE 2.9).

Test time-out by doing a Send and then doing nothing for 2 minutes.

Modem Version

To use a modem instead of a serial line, all that we must change are the configuration options and address in the Project Data file:

```
DefConst('kEndPointConfig, [
   {
      label:kCMSModemID,// use a modem,
                        // PCMCIA or external
      type: 'service,
      opCode:opSetRequired
   },

   {
      label:kCMOSerialIOParms,
      type:    'option,
      opCode:  opSetRequired,
      data:    {
         bps:     k1200bps,
         parity:kEvenParity,
         dataBits:k7DataBits,
         stopBits:k1StopBits,
      }
   },
   {
      label:   kCMOModemECType,
      type:    'option,
      opCode:  opSetNegotiate,
      data:    kModemECProtocolNone
   },
```

```
    {
       label:   kCMOMNPAllocate,
       type:    'option,
       opCode:  opSetRequired,
       data:    kMNPDontAllocate
    },

    {
       label:   kCMOMNPCompression,
       type:    'option,
       opCode:  opSetNegotiate,
       data:    kMNPCompressionNone
    },

    {
       label:   kCMOInputFlowControlParms,
       type:    'option,
       opCode:  opSetNegotiate,
       data: {
          useHardFlowControl:NIL, // Don't specify
                          // hardware flow control
                          // with modem endpoints
          useSoftFlowControl:TRUE,
          xOnChar:    unicodeDC1,
          xOffChar:   unicodeDC3,
       }
    },

    {
       label:kCMOOutputFlowControlParms,
       type: 'option,
       opCode:opSetNegotiate,
       data: {
          useHardFlowControl:NIL,// Don't specify
                          // hardware flow control
                          // with modem endpoints
          useSoftFlowControl:TRUE,
          xOnChar:    unicodeDC1,
          xOffChar:   unicodeDC3,
       }
    },
]
);
```

```
DefConst('kEndPointAddress,
    {
        label:kCMARouteLabel,
        type: 'address,
        Code: opSetRequired,
        data: {
            addressType:kPhoneNumber,
            addressData:"1 800 946 4644"
        },
    }
);
```

The options are 1200 bps, even parity, 7 data bits, 1 stop bit, no error control, no compression buffer allocated, no compression, Control-S and Control-Q software flow control.

The address is a phone number; this particular phone number is the toll-free U.S. number for the paging service. Note that you may need to prepend this phone number with whatever sequence is necessary to obtain an outside line.

Once we've done this, we can test by attaching a supported external modem or a supported PCMCIA modem. Then, tapping Send will connect by dialing the phone number specified in the address. Once we're done, the user whose pager ID matches the one that you've provided will receive the given page.

Caution: Pager owners have a limited number of free pages they receive per month. Beyond that, they must pay for each received page. Don't send to any pager id without permission (express or implied) because it is very bad manners.

ADSP version

For ADSP connections, slightly more work is needed. Due to a quirk in the ADSP implementation, you must call `CloseAppletalk` after disposing of an endpoint. Therefore, the `DisconnectAction` method of the application base template changes to:

```
func(ep, state)
begin
    if state = kState_Connected then
        ep:Disconnect();
    ep:Dispose();
```

```
        if kUseAppletalk then
            CloseAppletalk();
        ep:DisconnectCompletionProc();
    end
```

We've defined a new constant, `kUseAppletalk`, in the Project Data file:

```
    constant kUseAppletalk := true;
```

Here are the configuration options and addressing in the Project Data file:

```
    DefConst('kEndPointConfig,
        [
            {
                label:kCMSAppleTalkID,
                type: 'service,
                opCode:opSetRequired
            },

            {
                label:kCMSAppleTalkID,
                type: 'option,
                opCode:opSetRequired,
                data: kCMOAppleTalkADSP
            },

            {
                label:kCMOEndpointName,
                type: 'option,
                opCode:opSetRequired,
                data: kADSPEndpoint
            },
        ]
    );
    DefConst('kEndPointAddress,
        {
            type: 'address,
            label:kCMARouteLabel,
            opCode:opSetRequired,
            data : {
                addressType:kNamedAppleTalkAddress,
                addressData: "Ralph:Pager@fastzone"
            }
        }
    );
```

The configuration specifies an ADSP connection. The address specifies that we'll be connecting to an entity named "Ralph" whose type is "Pager" in the "fastzone" zone. Obviously, hardwiring the address makes it very difficult to choose an AppleTalk entity, especially if the entity is on another zone; later on we'll provide a way to have the user choose the entity.

Now we can test. From the Newton side, everything is as it was before. On the desktop side, however, things are a little more challenging. Here is what we do:

1. On a Macintosh, we can use a comm toolbox compliant terminal emulator with the AppleTalk ADSP tool.

2. We'll specify that tool for our connection.

3. We'll configure it with the name "Ralph" and the type "Pager" as shown in FIGURE 2.10. Then, we must listen for a connection.

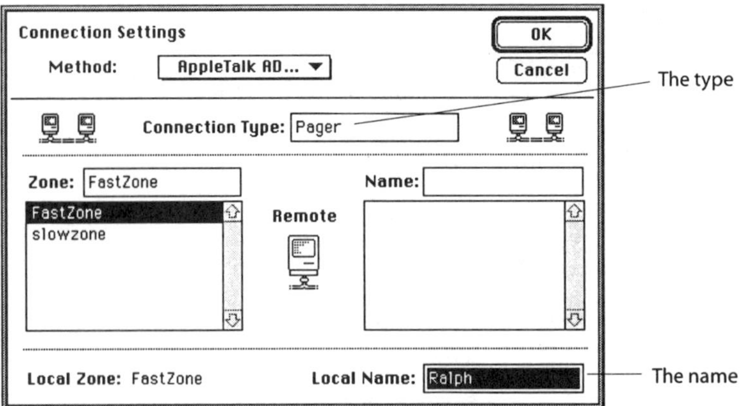

FIGURE 2.10 Configuring the AppleTalk ADSP connection tool.

4. Using MacTerminal, we choose the Wait for connection menu item (see FIGURE 2.11).

   Other terminal programs should have a similar item. Once you've waited for a connection, you should be able to tap Send from the Newton application just as you did with all the other services.

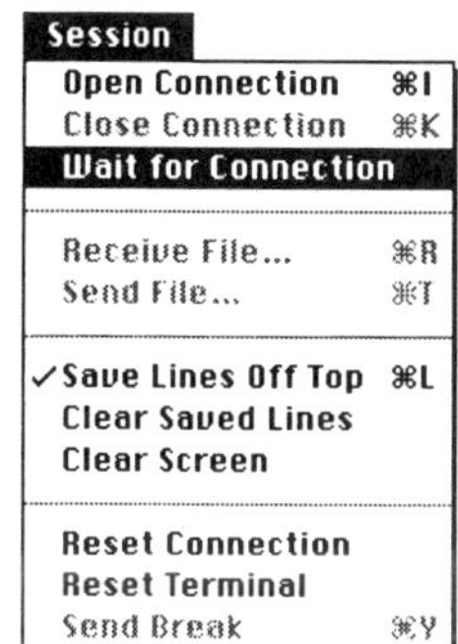

FIGURE 2.11 The Wait for Connection menu item in MacTerminal.

Now, it is time to remove our hardwired entity. We will put up the Chooser so that the user can select a connection entity. To begin with, we'll change `SendPage` in the application base template:

```
func()
begin
   if kUseAppletalk then
      :GetAddressThenConnect()
   else
      :DoConnect();
end
```

If we're using the ADSP service (remember we defined a constant `kUseAppletalk` to true in the Project Data file), then we'll call `GetAddressThenConnect` to put up a Chooser, and, once the user selects an entity, do the connection.

Here's the `GetAddressThenConnect` method of the application template:

```
func()
begin
   GetRoot().NetChooser:OpenNetChooser(
      lastZone,// current zone
      "=:" & kEntityType &
         "@",// look for kEntityType entities
      lastEntity,// item to select initially
      GetRoot().(kAppSymbol), // send message here when done
      "Send Page",// text in button
      "Pager",// Select a Pager
      "pagers"// Looking for pagers
   );
end
```

The method uses two slots from the base view, `lastZone` and `lastEntity`, which are defined in the base template and initialized to `nil`. It also uses a constant defined in the Project Data file, `kEntityType`:

```
constant kEntityType := "Pager";
```

You should also read the additional information we have on "rootView:NetChooser:OpenNetChooser(zone,entityType,entity, notifyView, buttonText, singularType, pluralType)" on page 313 which describes the parameters to `OpenNetChooser`. We could pass `nil` for the first and third parameters. If we did that, however, the Chooser would always open to the local zone, and with no entity selected. By keeping track of the last entity selected and the last zone, we can preselect for them. This is the same behavior you see in the Chooser when selecting a network printer (it's likely you'll want to select the same printer you used last time). It would certainly be nice if the Toolkit App did the same thing when doing a Download Package over AppleTalk.

`OpenNetChooser` will return as soon as it is open. When the user closes the Chooser (either by tapping "Send Page" or by tapping the close box), the frame specified in the fourth parameter will receive the message `NetworkChooserDone`.

Here is the `NetworkChooserDone` method of the application base view:

```
func(currentSelection, currentZone)
begin
   if currentSelection and
      not StrEqual(currentSelection, "") then begin

      if currentZone = nil or
         StrEqual(currentZone, "") then
         currentZone := "*";// local zone

      fEndPoint.fConnectAddress.data.addressData :=
         currentSelection & ":" &
         kEntityType & "@" & currentZone;
      self.lastEntity :=
         fEndPoint.fConnectAddress.data.addressData;
      self.lastZone := currentZone;

      :DoConnect();
   end;
end
```

If the user cancelled, then `currentSelection` will be `nil`. We also need to make sure that they actually selected an entity (the user could have tapped "Send

Page" while having nothing selected). If, in fact, there is a current selection, then we make sure a zone has been provided (an empty or `nil` zone should be treated as the local zone, and represented as "*").

We construct the address as *name:entityType@zone* and save it in the address frame. We've got the name and zone as parameters; the `entityType` we've got as a constant. Then, we save the entity and zone in slots in the application base view; that way, the next time the user uses the Chooser in this application, we'll preselect that zone and entity.

Finally, we'll send ourselves the `DoConnect` message, which will connect using the address we just constructed. That is all there is to the ADSP implementation of this program.

Summary

First, we talked about the various communications options available on the Newton now or possibly in the future. Next, we showed you how to post items programmatically to the Out Box. Then, relying on this information, we created a mail-based Marco application. Finally, we discussed the ins and outs of programming with endpoints.

Chapter 3

Filing

Have a place for everything and keep the thing somewhere else. This is not advice, it is merely custom.

—Mark Twain

Overview

In this chapter, you will learn about filing on the Newton. First, we will take a look at the filing user interface elements. Second, we will implement filing in a small application, walking you through this process step by step. In this section, we will provide you with all of the necessary code. Following the implementation, we address the most commonly asked programming questions concerning filing. Last we provide a quick checklist of the tasks involved in filing. Once you are familiar with filing, you can use this checklist as an easy method for ensuring that you address each implementation point in your own programs.

User Interface of Filing

Let's look at the user interface of filing in a Newton application. The basic Newton concept of filing is that there are folders (at most 12) in which the user can file items. The folders are shared across all applications and are created and named by the user. The items in an application are organized by folder, although the user can also see the items in all folders simultaneously.

User Interface Elements

There are a number of user interface elements associated with filing. The folder tab and filing button are the most visible of these elements (see FIGURE 3.1).

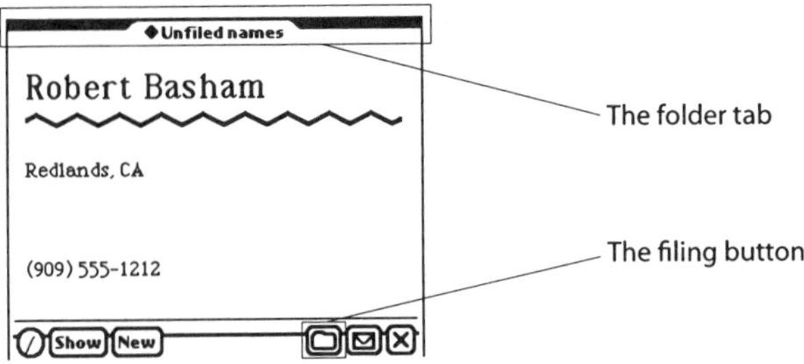

FIGURE 3.1 The user interface elements of filing.

The Folder Tab

The folder tab serves two purposes. The first is to display the name of the current folder. The second is to determine which folder will be displayed (see FIGURE 3.2).

The "Unfiled *items*" folder is a special folder used for items that have not yet been assigned a folder. The "All *items*" folder is a special folder which displays all items regardless of their individual folders. The "*items*" portion of those two folder names is specific to the application. The Names application uses "Unfiled names" and "All names", a compact-disc catalog application might use "Unfiled CDs" and "All CDs".

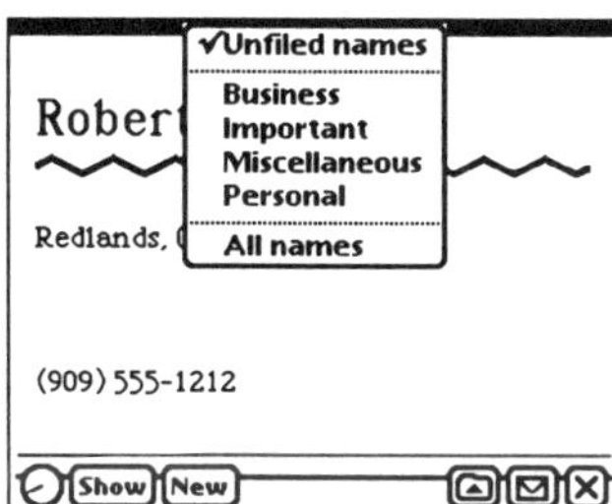

FIGURE 3.2 Choosing a folder from the folder tab.

Notice that the folder tab is displayed in both a detail view (to affect what items will be displayed as the user scrolls) and also in an overview (see FIGURE 3.3).

◆Unfiled names	
Information	408 974-1010
Reservations	(800) 527-0700
Information	(408) 255-2525
Information	(309) 862-1704
Information	(408) 554-8649
Information	
A Airborne Express	(800) 562-3133
Alamo	(800) 354-2322
American Airlines	
American Airlines	(800) 433-7300
B Robert Basham	(909) 555-1212
Bookstar	(619) 721-0706
C Canine Companions fo	(619) 756-1012
ABCDEFGHIJKLMNOPQRSTUVWXYZ	

FIGURE 3.3 The folder tab as it displays in the overview.

The Filing Button

The filing button serves three purposes. The first is to display the store (i.e., internal or card) location of the current item. The second is to allow the user to file the current item in a different folder. Notice that since the filing button usually operates on the current item, it is usually not shown in the overview (see FIGURE 3.3). Last of all, the filing button is used by the user to get to the Edit Folders slip where the names of the folders can be changed.

The triangle is the user interface element that shows the user where an item is stored. When the triangle is missing, the item is stored on an internal store. If the button shows a triangle within it, the current item is on an external store (see FIGURE 3.4).

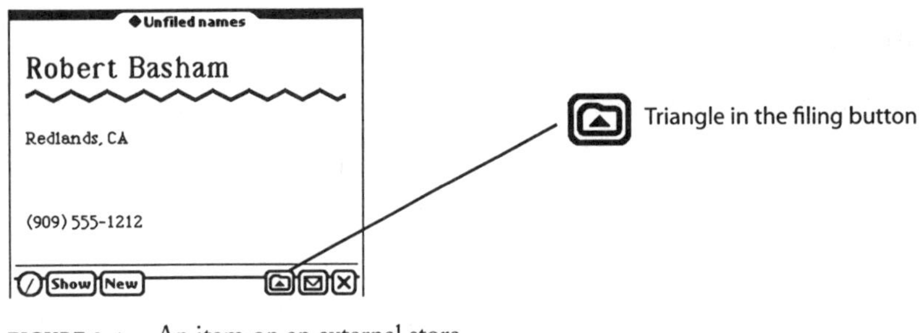

FIGURE 3.4 An item on an external store.

Tapping the filing button brings up the Filing slip (see FIGURE 3.5). Selection of a button in this list then determines in which folder the item is filed.

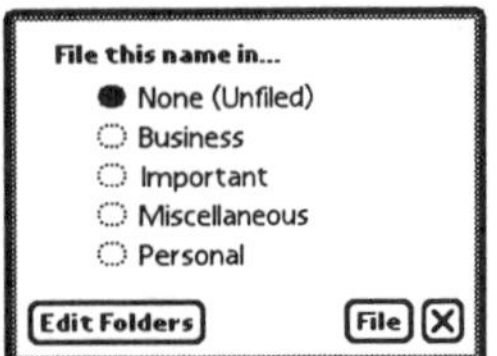

FIGURE 3.5 The Filing slip.

If the user taps on the Edit Folders button, the Newton brings up the Edit Folders slip. Here the user can change the name of the folders, add new folders, or remove ones that are no longer useful. The user can have up to 12 folders in which to file items (see FIGURE 3.6).

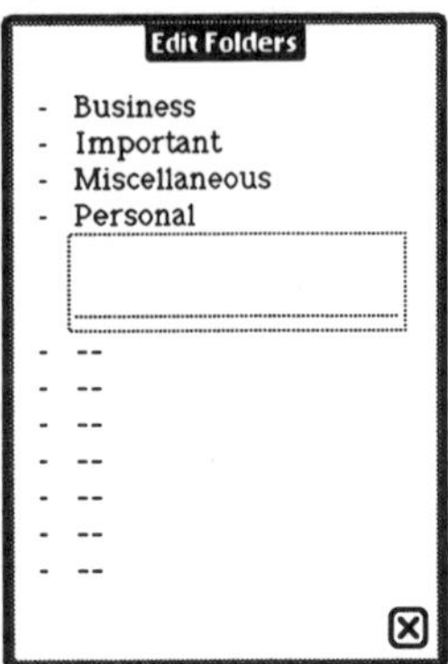

FIGURE 3.6 The Edit Folders slip.

Handling Particular Events

Creating a New Item

When a new item is created, it is automatically placed in the currently displayed folder (or if "All *items*" is displayed, it is placed in the "Unfiled *items*" folder).

Filing an Item

When the user taps on the filing button, it brings up the filing slip, and the folder in which the item is currently filed is already selected. When the user selects a different folder for an item and then taps the File button, it is placed in this new folder.

The exception to this is when the user taps on the filing button and the current item is on a read-only store. In this case, the filing slip is not even displayed. Instead, the user sees a notification as shown in FIGURE 3.7.

FIGURE 3.7 The notification slip when trying to change the folder of a write-protected item.

Changing the Folder

When the user changes the folder (using the folder tab), only items from the newly chosen folder are displayed. If there are no items in that folder, the application displays a notice to that effect (see FIGURE 3.8).

FIGURE 3.8 The display of an empty folder.

In the detail view, when a new folder is displayed, the contents will need to change. The old item being displayed is not usually from the new directory. Thus, a new item from the new folder will need to be shown instead. There is one exception to this:

- When switching to or from the "All *items*" folder, the current item is still in the new folder, so it can still be displayed.

Editing Folders

In the Edit Folders slip, the user can add, delete, or modify folders. The process is simple when the user adds a new folder. Renaming and deleting folders is not quite so straightforward. If the user modifies a folder name, then a confirmation slip is presented (see FIGURE 3.9). If the user OKs this confirmation, the Newton updates all the soups in the current stores (obviously stores which aren't available aren't updated). The Edit Folders slip displays the status at the bottom during the updating (see FIGURE 3.10).

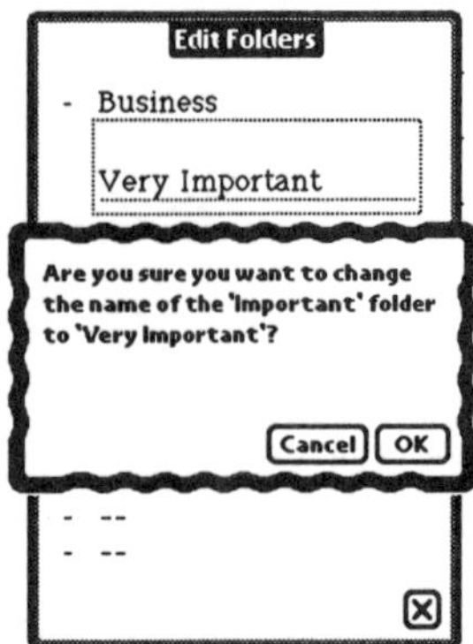

FIGURE 3.9 Confirmation slip when modifying a folder name.

Deleting a folder is similar to modifying a folder, except that any items in the deleted folder are now put in the "Unfiled *items*" folder instead. Likewise, the confirmation slip is slightly different (see FIGURE 3.11). If the user OKs this slip, again all soups in all available stores are updated, with the status of the changes still displayed at the bottom of the Edit Folders slip.

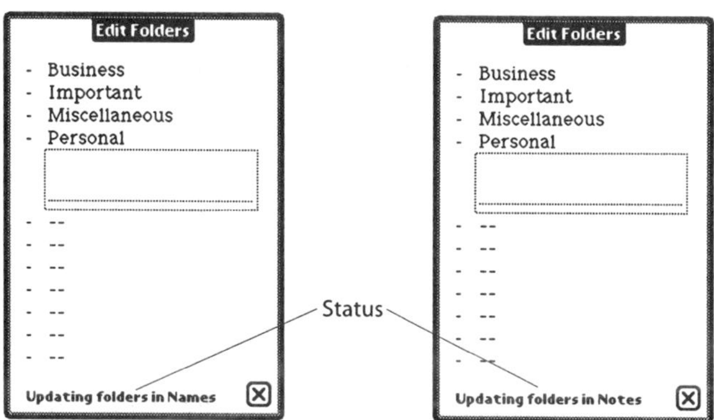

FIGURE 3.10 Status when updating a folder name.

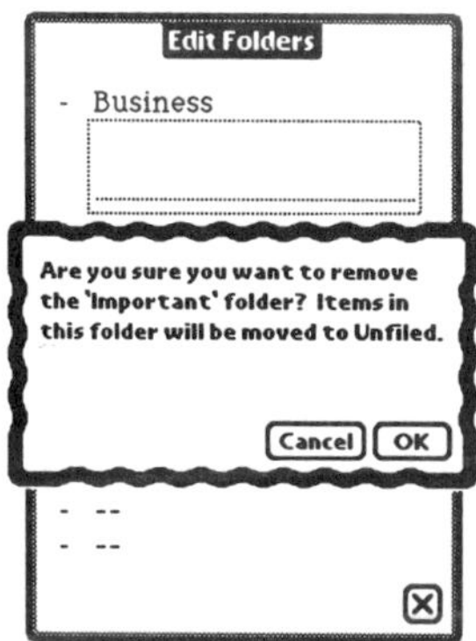

FIGURE 3.11 Confirmation slip when removing a folder.

Inserting a Card with Items in Nonexistent Folders

It is possible to insert a card into a Newton which contains folders that aren't on the Newton. This can happen if a user modifies or deletes a folder when the card wasn't inserted. It can also happen if a user inserts a card from another Newton.

What is the right thing to do in such cases? Nonexistent folders might be dealt with in a number of ways:

- The nonexistent folders are added to the folder tab.
- The items in the card that are in nonexistent folders are placed in the "Unfiled *items*" folder.

- The items in the card are kept in their folders and are only shown when viewing the "All *items*" folder.

The first option isn't useful because there are a maximum number of folders allowed. This is based on a picker size limitation–like the folder tab, it can't exceed the screen height—and on a folders limitation–only 12 of them are possible.

The items in the card are not modified to be in the "Unfiled *items*" folder either. Here is why: Imagine your mother lends you the card from her Newton. Do you think she wants you to move all of her stuff to unfiled? Her custom folders and their contents should be left alone so that her card returns to her in the same condition in which she lent it.

What Newton applications do, then, is the third possibility. They only display items from nonexistent folders when looking at the "All *items*" folder. Likewise, for these types of items, the filing slip shows that no folder has been selected (see FIGURE 3.12). Unfortunately, this behavior may not be obvious to users; see "How Do I Make Items in Nonexistent Folders Work Better?" on page 132 for ideas on ways to modify this behavior.

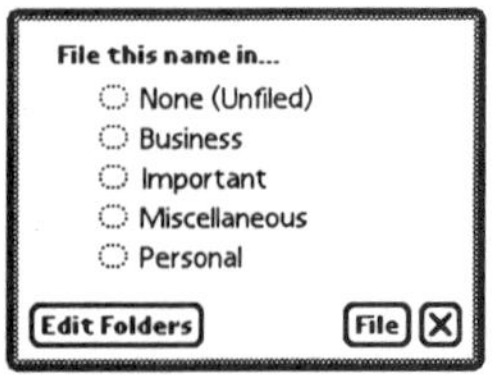

File this name in...
None (Unfiled)
Business
Important
Miscellaneous
Personal
Edit Folders
File

FIGURE 3.12 The filing slip for an item in a nonexistent folder.

Putting Away an Item from a Nonexistent Folder

What happens when an item from the In Box is put away? It is usually placed in the same named folder from which it originated (on the sending Newton). This only works, of course, if that folder exists on the receiving Newton. If the folder does not exist, the item is placed in the "Unfiled *items*" folder. On the surface this may seem at odds with behavior that occurs when inserting a card, but that is because the ownership of the item is clear. An item that is put away is a copy sent from another Newton. This copy now belongs to the receiving Newton. Items on a card, however, may actually belong to a different Newton (for example, the card may be merely borrowed from another Newton).

How Filing Is Implemented

The implementation of filing is done by marking each item (normally entries in a soup) with a symbol specifying its filing label. For instance, if an entry is in the "Personal" folder, it would have a `labels` slot with the value `'personal`. Unfiled items are represented with the value `nil`. Entries are not actually stored in different folder directories–that is an illusion of the operating system. In reality, all entries for an application are stored together in a common area (the soup) and are tagged separately.

The user can choose to display a different folder from the folder tab; this act specifies a different specific value to use as a filter for items displayed in the view. This filter can be `nil` (to specify the unfiled items); it can be the symbol of a folder (like `'personal` or `'travel`); or it can be `'_all` (indicating that all items will be shown).

Adding Filing to Your Application

We'll start with an application that supports soups, scrolling, and routing. The application is called Words, since the content of each item is one word.

Adding the Filing UI Components

A `protoFolderTab` and a `protoFilingButton` are the only required user interface (UI) elements for filing. In review, the `protoFolderTab` allows us to display different folders, and the `protoFilingButton` allows us to specify the folder in which an item is filed and modify the names of the folders.

Adding the New Protos

Do the following to add the `protoFilingButton`:

- Add a `protoFilingButton` as a child of the `protoStatus`.
- Ensure that the `protoActionButton` is the first child of the protoStatus.
- Make the `protoFilingButton` the second child. (The `proto-FilingButton` uses sibling relative justification; it assumes the `protoActionButton` is its sibling.)

Do the following to add the `protoFolderTab`:

- Add a `protoFolderTab` as a child of the base view. If you currently have a `protoTitle` at the top of the base view, delete it.
- Add a `viewFlags` slot to the template.
- Set the `viewFlags` to `vClipping` and `vVisible` (otherwise, when you switch from a long folder name to a short one, there will be some bits left on the screen from the long folder name as shown in FIGURE 3.13).

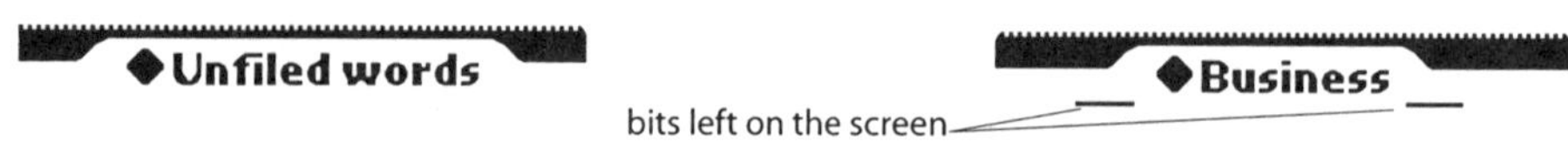

FIGURE 3.13 Switching from a long folder to a short folder without `vClipping` in the protoFolderTab.

If you look at FIGURE 3.14, you will see the Words application as it should appear in NTK after deleting the protoTitle and adding the necessary filing UI elements.

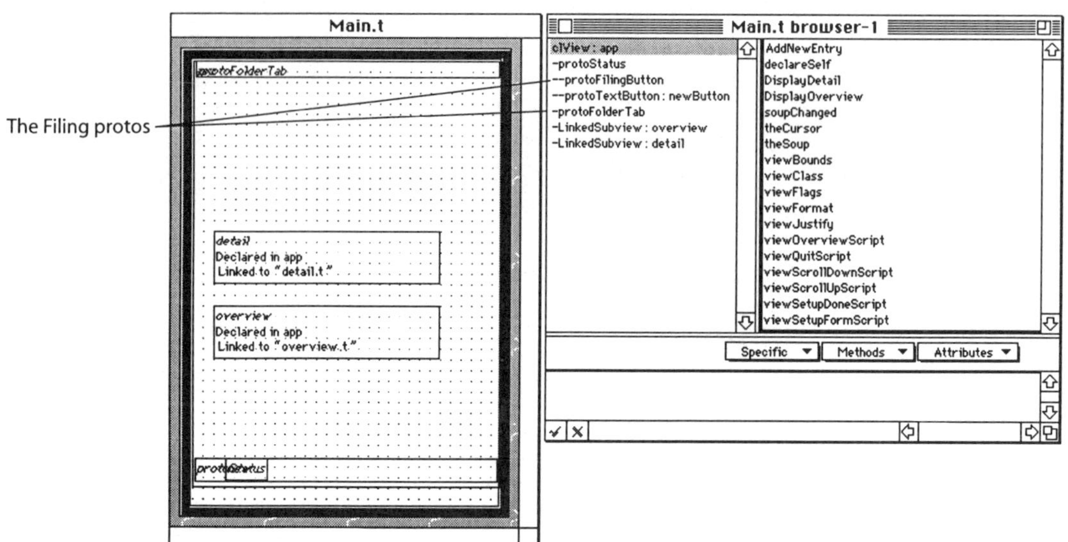

FIGURE 3.14 The Words application with the filing protos added.

Adding Required Slots

Now, you need to add the following slots to implement Filing:

`appAll`	The "All *items*" string which is displayed at the bottom of the `protoShowBar`.
`appObject`	An array of two strings representing the singular and plural form of the items to be filed.
`labelsFilter`	The symbol representing the currently selected folder (for example, `nil`, `'_all`, `'personal`). This slot is maintained by the `protoFolderTab`.
`target`	The frame to be filed. This slot should be `nil` if no object has been selected to file.
`targetView`	The view that receives the filing messages. This view must contain the `appAll` and `labelsFilter` slots. (The `targetView` behaves differently for filing than it does for routing; in routing, the `targetView` is passed as a parameter, whereas in filing, the `targetView` receives messages.)

The `target` and `targetView` slots are the same as those used for routing (see "Required Application Slots" on page 5). As a result, you need to make sure that these slots are located in a common ancestor of the `protoFolderTab` and `protoFilingButton`. The `appAll` and `labelsFilter` slots must be located in the `targetView`. Normally, all five of the slots are in the application base view, and the `targetView` points to the application base view.

Supporting the Diamond

Before actually supporting filing, let's look at the mechanism used for displaying the diamond in the folder icon. When you send a `protoFilingButton` the `Update()` message, it modifies its appearance (either displaying or not displaying the diamond) based on `target`. As a result, every time `target` is changed, the `Update` message should be sent to the `protoFilingButton`.

Supporting the Diamond in Words

You must do a number of things to support the diamond display. First, let's look at a one-line checklist of each of these items and then at each item in detail.

1. Name the filingButton and declare it to the base view.
2. Add a `ChangeTarget` method to the base view.
3. Call the `ChangeTarget` method from your display method.
4. Call `ChangeTarget` from the `viewShowScript` of the overview.

Now, let us look at the implementation details for these items. First, name the `protoFilingButton` that you added to the layout "folderButton". Make sure to declare it to the application base view so that you can send messages to it.

Second, add a `ChangeTarget` method to the base view:

```
func(newTarget)
begin
   target := newTarget;
   folderButton:Update();
end;
```

Now, handle the display of the correct items in a detail view by calling `ChangeTarget` from the `DisplayWord` method of detail:

```
func(entry)
begin
   :SaveCurrentEntry();

   local theWord := Clone(currentEntry.word);
   SetValue(wordEdit, 'text, theWord);

   GetRoot().(kAppSymbol):ChangeTarget(entry);

   self.entryHasChanged := nil;
end
```

Last of all, handle the overview display by calling `ChangeTarget` from the `viewShowScript` of overview:

```
func()
begin
   :DisplayWordsFromSoup();

   GetRoot().(kAppSymbol):ChangeTarget(nil);
end
```

Hiding the Filing Button in the Overview

In many applications, it does not make sense to let the user file an item from the overview. For those applications, the filing button should not appear.

In the Words application, when the overview is shown it hides the filing button, and when the detail view is shown it shows the filing button.

Here is the overview's `viewShowScript`:

```
func()
begin
   filingButton:Hide();
   actionButton:Hide();
   :DisplayWordsFromSoup();

   GetRoot().(kAppSymbol):ChangeTarget(nil);
end
```

Here is the detail's `viewShowScript`:

```
func()
begin
   filingButton:Show();
   actionButton:Show();
   :DisplayWord(theCursor:Entry());
end
```

Displaying Entries from the Correct Folder

To display entries from the folder the user has selected, you must modify your `Query` so that your cursor skips soup entries which are not part of the current folder. All your entries will have a `labels` slot (if an entry doesn't, reading its `labels` slot will yield `nil`, and it will be treated as unfiled). You will be checking for two situations:

- If the value of the `labels` slot matches that in the `labelsFilter` slot of the target view.

- If the `labelsFilter` slot is equal to `_all`.

In either of these cases, the entry should be part of the cursor. Otherwise, the entry should not be part of the cursor.

Note: In the original design of filing, soup entries could belong to multiple folders. This is why the `labels` slot is plural rather than singular (`label`). The design was changed so that soup entries could belong to only one folder; unfortunately, the old name stuck.

Commonly, this testing is done using a `validTest` in conjunction with the `Query`. Here is a sample `Query` made as part of the `viewSetupFormScript` of the application base view:

```
app.viewSetupFormScript := func()
begin
   ...
   self.targetView := self;
   self.theCursor := Query(theSoup,
      {indexPath: 'theSortSlot,
         validTest: func(e)
         begin
            local theFilter := labelsFilter;
            return theFilter = '_all or
               theFilter = e.labels;
         end
      });
end
```

The `validTest` checks the value of the `labelsFilter` slot located in the application base view. If it is `_all`, then all entries in the soup are part of the cursor. Otherwise, an entry is part of the cursor only when it matches the `labelsFilter`.

Now, let us modify the script in which we will call this query. In the case of the Words application it is the `viewSetupFormScript` of the application base view:

```
func()
begin
   self.targetView := self;

   theSoup := call kRegisterCardSoupFunc with
      (kSoupName, kSoupIndexes,
      kAppSymbol, kAppObject);
   theCursor := Query(theSoup, {type: 'index,
      indexPath: 'word,
```

```
      validTest: func(e)
      begin
      local theFilter := labelsFilter;
      return theFilter = '_all or
         theFilter = e.labels;
      end
   });
   if not Query(theSoup,{type:'index}):Entry()
then
   begin
      for i := 1 to 7 do
         :AddNewEntry();
            // move cursor to the first soup entry
      theCursor:Reset();
   end;


   // to be notified of changes to the soup
   call kRegisterForSoupNotify with ();
end
```

Adding New Entries in the Current Folder

New entries should be created in the current folder, unless the current folder is "All *items*". In that case, the new entry should be created in the unfiled folder.

Now modify the `AddNewEntry` method of the application base view so that new entries are labeled with the right folder label:

```
func()
begin
   local folder;

   if labelsFilter <> '_all then
      folder := labelsFilter;

   local randomWordEntry := {
      word: GetRandomWord(3, 10),
      changeDate: Time(),
      labels: folder,
   };
   theSoup:AddToDefaultStore(randomWordEntry);
   return randomWordEntry;
end
```

What Happens When the User Changes Folders

When the user uses the folder tab to change the displayed folder, the `protoFolderTab` changes the value of the `labelsFilter` slot in the target view and then sends itself the `FilterChanged` message. You usually implement the `FilterChanged` method in your application base view and the folder tab inherits this method via parent inheritance.This method must do whatever is necessary so that the view displays items from the newly chosen folder.

It is rarely necessary to issue another query; the `validTest` looks at the current value of `labelsFilter` each time it is called. Therefore, the cursor will automatically iterate over those items in the new folder. Usually only a redisplay is required.

In most cases, if a folder changes, the current entry will no longer be displayed. There are two exceptions to this:

- The folder is switched to "All *items*".
- The switch is from "All *items*" to the folder containing the current entry.

In all other cases, the display will need to change. The question, then, is what should be displayed. In most cases, it is probably best to display the next entry that follows the current one (the one obtained by moving the cursor forward).

This is similar to what should happen when the user is looking at an entry on a card when the card gets removed.

For the Words application, the `FilterChanged` method of the target view calls the `Redisplay` method, just as the `soupChanged` method does. Here is our `FilterChanged` method:

```
func()
begin
   :Redisplay();
end
```

Now, looking at the `Redisplay` method we can see how the correct display is determined:

```
func()
begin
   // change the cursor so it doesn't
   // point at 'deleted
   local e := theCursor:Entry();
   if e = 'deleted then begin
      e := theCursor:Next();
```

```
            if e = nil then
                e := theCursor:Prev();
        end else if e and not theCursor:Goto(e) then
        begin
            // current entry wasn't in the cursor
            e := theCursor:Entry();
        end;
        if e = nil then begin
            theCursor:Reset();
            e := theCursor:Entry();
        end;

        if Visible(overview) then
            overview:DisplayWordsFromSoup()
        else begin
            if e = nil then
                :DisplayOverview()
            else
                :DisplayDetail(e);
        end;
    end
```

Once you have implemented this code, you should be able to test the Words application by switching folders and creating new entries. You should see that as you switch from folder to folder, only the items from that folder should display.

Displaying an Empty Folder

If a folder has no items in it, your application should display some message that informs the user of such a situation. In the Words application, we'll do that by:

- creating a static text that has an appropriate message.
- never displaying the detail view for an empty folder, otherwise, what item would be displayed?
- opening and closing the static text as necessary.

Start by creating a new layout named "noWords.t" which contains a static text reading "There are no words in this folder". Create a linked layout in the main layout that is linked to this static text. Name it "noWords" and declare it to the application base template.

The Words application is already careful to display only the overview when a folder is empty. Notice that the last lines in the **`Redisplay`** method switch from the detail view to the overview if necessary:

```
if Visible(overview) then
   overview:DisplayWordsFromSoup()
else begin
   if e = nil then
      :DisplayOverview()
   else
      :DisplayDetail(e);
end;
```

The best place to open and close the `noWords` static text is in the overview's `DisplayWordsFromSoup` method:

```
func()
begin
   local cursorCopy := theCursor:Clone();
   if cursorCopy:Entry() then
      noWords:Close()
   else
      noWords:Open();
   foreach row in childrenViews do begin
      row:DisplayWordInRow(cursorCopy:Entry());
      cursorCopy:Next();
   end;
end
```

What Happens When the User Files an Item

Now, let us examine the case of the user filing an item. When the user taps the filing button, the system automatically displays the Folder slip with the radio button set to the currently displayed folder. This is automatic, you can just sit back and watch. Here is how this works: The filing button looks at the current `target`, and at the `labels` slot of that `target`. If the user changes the folder and then taps "File" the `labels` slot of the `target` is changed to the new folder, and then your `targetView` is sent the `FilingChanged` message. The `FilingChanged` method must then save the entry to the soup and then redisplay.

At redisplay time, if the current folder is "All *items*", the newly filed entry will still be displayed. Otherwise, the next qualified entry should be displayed.

For the Words application, the `Redisplay` method will do the appropriate thing. The `FilingChanged` method calls `EntryChange` to save the modified entry and then calls `Redisplay`:

```
func()
begin
   EntryChange(target);
   :Redisplay();
end
```

If the target is on a write-protected card, your application should notify the user that the entry can't be changed. Here is a `buttonClickScript` for the filing button view that checks whether the target is on a write-protected store.

```
func()
begin
   constant kProtectedCard :=
"This Word card is on a write protected card and cannot be
changed";
   if target and
      EntryStore(target):IsReadOnly() then
      :Notify(kNotifyAlert,
         EnsureInternal(kAppName),
         EnsureInternal(kProtectedCard))
   else
      inherited:buttonClickScript();
end
```

If the target cannot be changed, then the preceding script gives a notification slip to the user. If the target can be changed, then the script just passes the job off to its `protoFilingButton`'s `buttonClickScript` to handle:

What Happens When the User Edits a Folder Name

If the user deletes a folder (from the Edit Folders slip) or renames a folder, all applications registered in the `soupNotify` array receive a `FolderChanged` message. Each application should then update its soup entries to reflect the change. In the case where the user adds a folder, no such notification is made (no changes need to be made to soup entries).

Registering in the `soupNotify` Array

Your application may not be open when the user edits folders, but it must still update its entries. Therefore, you must make sure that your application is registered in the `soupNotify` array when it gets installed, rather than when it is opened.

In the Words application, the registration in the `soupNotify` array was made in the `viewSetupFormScript` (using a compile-time function, `kRegisterForSoupNotify`, in the Project Data file). This must now be done in the `InstallScript`. Thus, you will modify `viewSetupFormScript` of the application base view by removing a couple of lines:

```
func()
begin
   self.targetView := self;

   theSoup := call kRegisterCardSoupFunc with
      (kSoupName, kSoupIndexes,
      kAppSymbol, kAppObject);
   theCursor := Query(theSoup, {type: 'index,
      indexPath: 'word,
      validTest: func(e)
      begin
      local theFilter := labelsFilter;
      return theFilter = '_all or
         theFilter = e.labels;
      end
   });
   if theCursor:Entry() = nil then begin
      for i := 1 to 7 do begin
         AddNewEntry();
      end;
            // move cursor to the first soup entry
      theCursor:Reset();
   end;

   // to be notified of changes to the soup
   call kRegisterForSoupNotify with ();
end
```

You now add those lines to the `InstallScript` of Words:

```
InstallScript := func(partFrame)
begin
   // to be notified of changes to the soup
   // (including a changed folder)
   call kRegisterForSoupNotify with ();

   local appBaseTemplate :=
      partFrame.theForm;
```

```
   routing.(kAppSymbol) :=
      appBaseTemplate.routingFrame;
   local formatFrame :=
      appBaseTemplate.formatFrame;
   GetRoot().(kFormatSymbol) :=
      BuildContext(formatFrame); end;
end;
```

In tandem, you need to move some code that handles the unregistration. Originally, this had been done in the `viewQuitScript` of the base view (using a compile-time function, `kUnRegisterForSoupNotify`, in the Project Data file):

```
func()
begin
   theSoup := nil;
   theCursor := nil;

   // for notifications of changes to the soup
   local soupNotifyPos :=
   call kUnRegisterForSoupNotify with ();
end
```

You will move these lines instead to the `RemoveScript` of Words where the unregistration will now be handled:

```
RemoveScript := func(partFrame)
begin
   // for notifications of changes to the soup
   local soupNotifyPos :=
   call kUnRegisterForSoupNotify with ();

   RemoveSlot(routing, kAppSymbol);
   RemoveSlot(GetRoot(), kFormatSymbol);
end;
```

Now that the Words application is registered even when the application is closed, the `SoupChanged` method may be called when the application is closed. Thus, you must modify your `SoupChanged` method to handle both an open and closed application state. It is only when the application is open that it needs to redisplay:

```
func(theSoupName)
begin
   if call kViewIsOpenFunc with (self) then
      :Redisplay();
end
```

Handling the **FolderChanged** Message

The FolderChanged method will look very similar from one application to another in most cases. The method is passed the old and the new labels. Each entry whose labels slot matches the old label must be updated to the new label. The exception, of course, is entries on a read-only store. In addition, this method must send a message to the folder tab so that it updates the displayed folder name.

Here is the FolderChanged method for the Words application:

```
func(soupName, oldFolder, newFolder)
begin
   // could signify a new folder was added
   if oldFolder = nil then
      return;

   local s := GetUnionSoup(kSoupName);

   // no soup yet, so nothing to do
   if not s then
      return;

   local cursor := Query(s,
      {type: 'index,
         validTest: func(e)
            return e.labels = oldFolder
      });

   local e := cursor:Entry();

   while e do begin
      if not EntryStore(e):IsReadOnly() then begin
         e.labels := newFolder;
         EntryChange(e);
      end;
      e := cursor:Next();
   end;

   if oldFolder = labelsFilter and folderTab then
       folderTab:UpdateFilter(oldFolder,
         newFolder);
end
```

Putting Away an Item

When an item is received in the In Box (beamed or mailed), it can be put away. An item should be put in the same folder as it came from, if that folder exists on this new Newton. If there is no corresponding folder, the item should be put in the unfiled folder.

There is a global function, `CheckThatFolderExists`, which will check the value of the `labels` slot of its parameter. If that folder exists, it does nothing; otherwise it sets the `labels` slot to `nil`.

You should call `CheckThatFolderExists` in your `PutAway` method:

```
func(item)
begin
   local newEntry := item.body;
   local theSoup := nil;
   local appIsOpen := call
      kViewIsOpenFunc with (self);

   CheckThatFolderExists(newEntry);
   if appIsOpen then
      theSoup := self.appSoup
   else begin
      theSoup := call kRegisterCardSoupFunc with
         (kSoupName, kSoupIndexes,
         kAppSymbol, kAppObject);
   end;

   theSoup:AddToDefaultStore(newEntry);
   if not appIsOpen then
      call kUnRegisterCardSoupFunc with
         (kSoupName);
   BroadcastSoupChange(kSoupName);
end
```

Common Questions about Filing

Why Is It So Slow?

Whenever you have a large soup with many entries in a particular folder, displaying that folder is slow. The reason for this is `validTest`; it is called for each item in a soup. We discuss ways to speed up the display of items in folders in "Indexing on Two Slots" on page 290.

How Do I Make Items in Nonexistent Folders Work Better?

Some developers have requested a way to make items in nonexistent folders display when the user views the "Unfiled *items*" folder rather than only in the "All *items*" folder (see "Inserting a Card with Items in Nonexistent Folders" on page 115).

The `validTest` in the `Query` must be modified to:

```
func(e)
   begin
   local theFilter := labelsFilter;
   local folders;

   if theFilter = '_all or theFilter = e.labels then
      return true
   else if theFilter = nil then begin
      folders := userConfiguration.userFolders;
      if folders.(e.labels) then
         return nil
      else
         return true;// non-existent folder
   end else
      return nil;
end
```

The `validTest` continues to return `true` if the use is displaying "All *items*" or if the chosen folder matches the `labels` slot in the item. If the user is displaying "Unfiled *items*", the `labels` slot of the item is checked to see if it exists in the global frame that defines the folders. If it does exist, then the user can display this item by showing the folder. If the folder doesn't exist, the `validTest` returns `true` so that it will display among other unfiled items.

Filing Checklist

1. Add a `protoFolderTab` (with `vClipping` on) and a `protoFilingButton` (as second child of `protoStatus`). Name them and declare them to the base template.

2. Add `appAll`, `appObjects`, `labelsFilter`, `target`, and `targetView` to the application base template.

3. Write code to set `target` and `targetView` appropriately.

4. Send the `Update` message to the filing button when the target changes.

5. Modify your `Query` to add a `validTest` which checks `labelsFilter`.

6. Use the value of `labelsFilter` to set the initial value of the `labels` slot for new items.

7. Write a `FilterChanged` method in the application base view.

8. Write code to display a message if there are no items in a folder.

9. Write a `FilingChanged` method.

10. Write a `buttonClickScript` for the filing button to handle items on a read-only card.

11. Register and unregister in the `soupNotify` array in your `InstallScript` and `RemoveScript`.

12. Write a `FolderChanged` method.

13. If you have a `PutAway` method, add a call to `CheckThatFolderExists`.

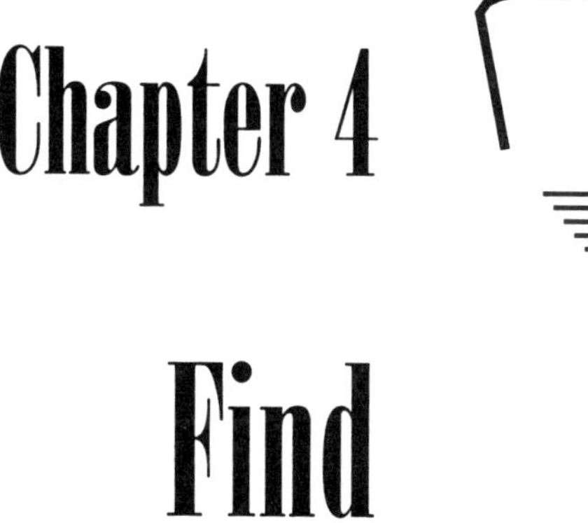

Chapter 4

Find

Seek, and ye shall find.

—Matthew 7:7

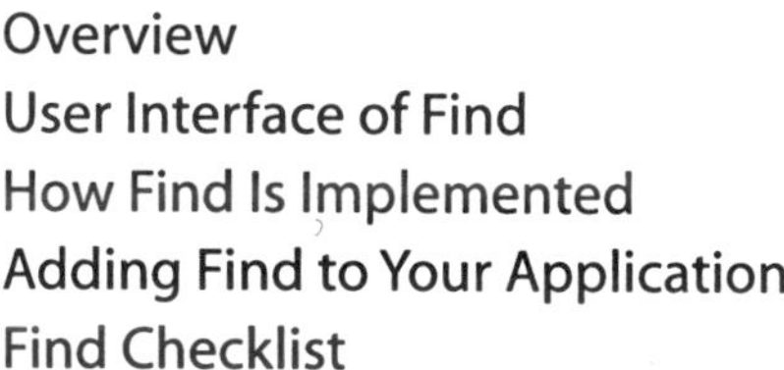

Overview

In this chapter, we will show you how to implement Find in your application. There are two flavors of Find to support, text Find and date Find. The first looks for word matches and the second looks for entries before or after a certain date. Before we look at the Find methods that you will need to add to your application, however, we will describe the standard user interface of Find.

User Interface of Find

In a Newton application, a user can search through either one application or all applications for something. In either case, the same mechanism is used: The user taps the Find button on the Newton, and this brings up a Find slip (see FIGURE 4.1). In this Find slip, the user writes a word or phrase to find. Tapping the "Find" button searches the frontmost application for the entry. Tapping the "All" button searches all applications (whether open or not).

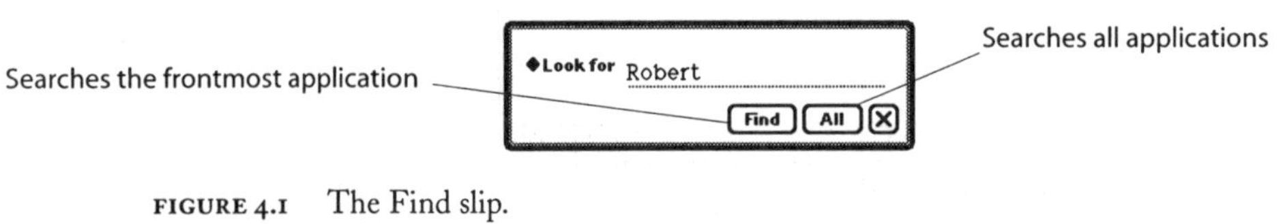

FIGURE 4.1 The Find slip.

While the Newton is searching for the entry, status information is displayed at the top of the Find slip (see FIGURE 4.2). Once the search is done, the matching entries are displayed in the Find Results slip (see FIGURE 4.3).

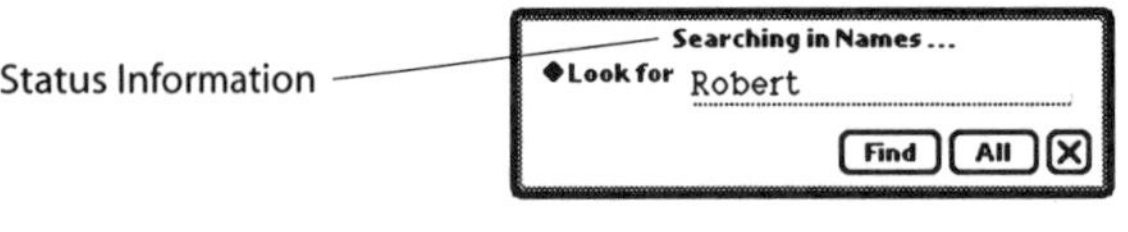

FIGURE 4.2 Status in the Find slip while searching.

Navigating the Find Results Slip

Within the Find Results slip, if you tap on an application name the Find Results slip toggles the hiding and showing of found items for that application (see FIGURE 4.4). When an application's items are showing, each item is displayed with a brief, but useful, summary beside it.

Tapping on a *particular item* in an application's list takes you to that item (opening the application if necessary). The Find slip is still displayed in front of the tapped-on item, however (see FIGURE 4.5). You can scroll up and down in the Find slip and this takes you to the previous or next items that were found in that application. Tapping the overview button takes you back from the Find slip to the Find Results slip.

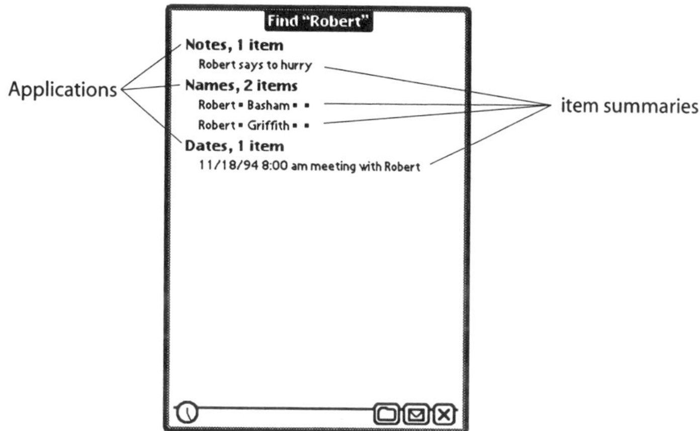

FIGURE 4.3 The Find Results slip.

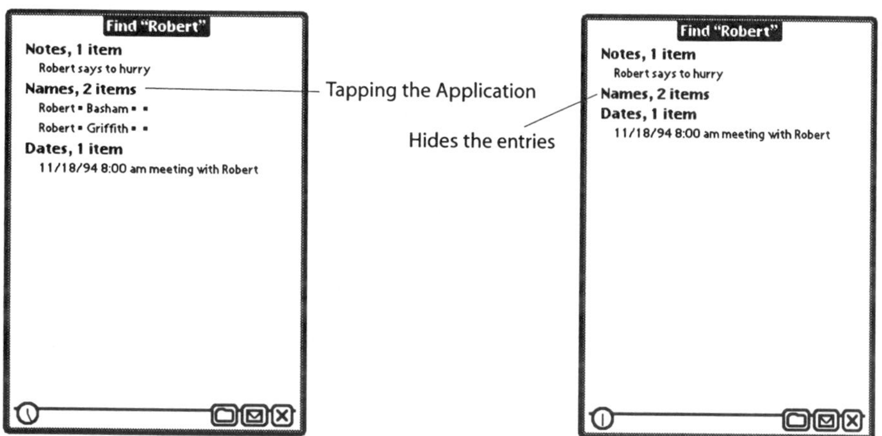

FIGURE 4.4 Toggling the results for a particular application.

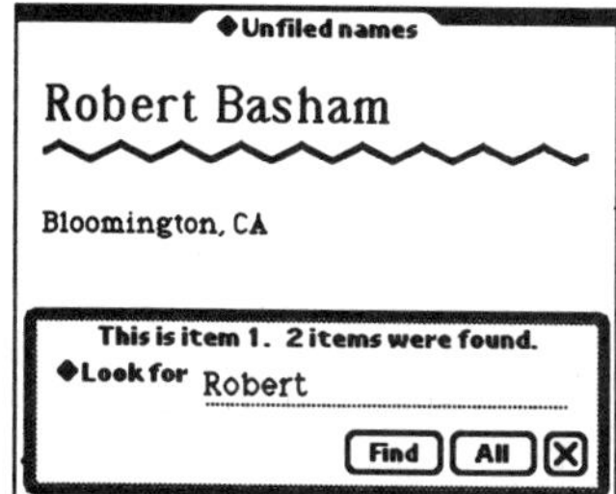

FIGURE 4.5 After tapping on an item in the Find Results slip.

Displaying the Right Folder

An application may have to switch folders to display the found item. The correct user interface for this switch is to change from the current folder to the "All *items*" folder and from there display the found item.

When Only One Item Is Found

There is one case in which the user never sees the Find Results slip. When a find yields only one item, the display of the Find Results slip is circumvented; the found item is immediately shown (the application is opened if necessary).

Local versus Global Find

If the user does a local Find (searching only the frontmost application) and no matching items are found, then the Find slip displays a message that the find was unsuccessful (see FIGURE 4.6). On the other hand, if a global Find ("All") is unsuccessful, then Find slip displays a slightly different message (see FIGURE 4.6).

FIGURE 4.6 An unsuccessful Find.

Searching by Date

Users can search not only for text, but also by date (see FIGURE 4.7). By tapping on the picker in the Find slip, the user can switch from a text Find to searching before or after a particular date. Notice that no provision is made for searching on a particular date.

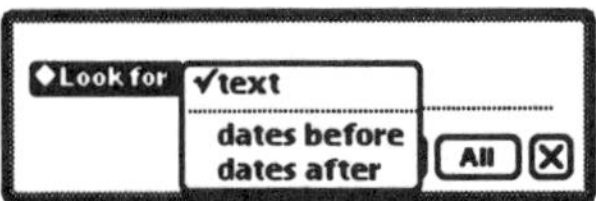

FIGURE 4.7 Searching by date.

Some applications don't support finding text; others don't support finding by date. Some applications don't support Find at all. FIGURE 4.8 shows the notifications a user receives when trying to find text or find by date in applications that don't support it.

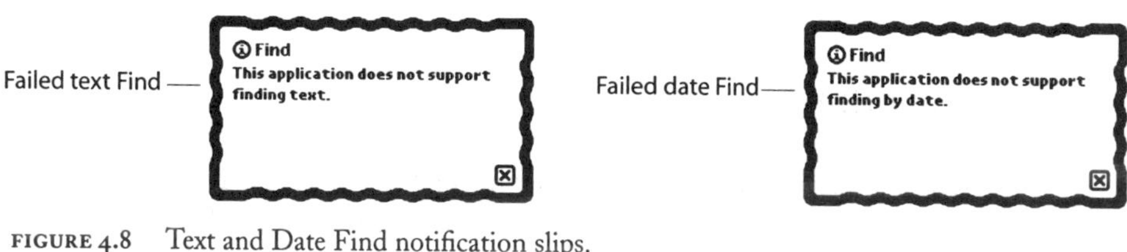

FIGURE 4.8 Text and Date Find notification slips.

How Find Is Implemented

Now that you have seen the elements of Find from the user's perspective, it is time to learn how Find works within the Newton operating system. Executing Find within a Newton application involves the interaction of four methods. These application base view methods are the complete building blocks of Find and are named `Find`, `DateFind`, `FindSoupExcerpt`, and `ShowFoundItem`. Here is a brief description of these functions.

`Find`	The method responsible for returning found text items.
`DateFind`	The method responsible for returning found date items.
`FindSoupExcerpt`	This method gets called to create a one-line description of the item in the Find Results slip.
`ShowFoundItem`	The method that handles the display of a found item.

 *Note:* We'll be looking at how to implement finding soup entries. Although you can also support finding items which aren't soup entries, we won't show how to do that.

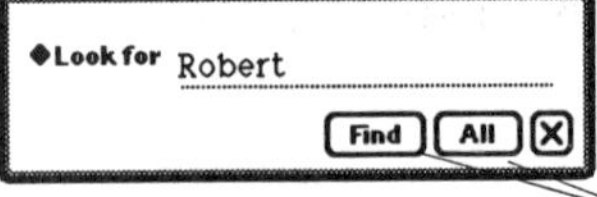

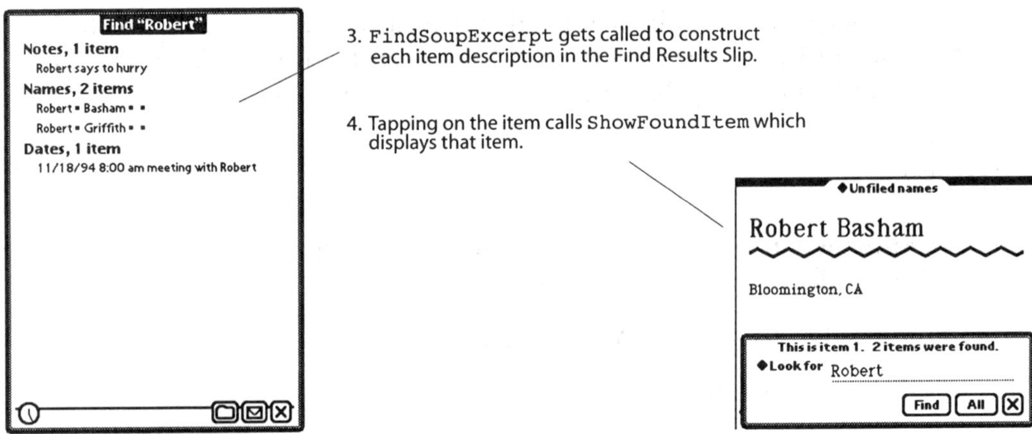

FIGURE 4.9 Relationship of `Find/DateFind`, `FindSoupExcerpt`, and `ShowFoundItem`.

FIGURE 4.9 shows the relationship between each of these methods as they interact in an application.

How Find Returns Multiple Items

Of course, one application may need to return many different items. Since those items are probably entries in a soup, however, commonly you'll return the items with a cursor that iterates over the found items. The Find Results slip can then use the cursor to retrieve each found item. An important benefit of using a cursor is that all of the found items need not be in memory at once.

What Happens When `Find` Is Not Implemented

When a user taps Find from the Find slip, the system looks for the `Find` method in the frontmost application. If the method doesn't exist, the system displays the alert shown in FIGURE 4.8. Similarly, if the user is doing a date Find for the frontmost application, and it doesn't have a `DateFind` method, the system displays the alert shown in FIGURE 4.8.

Adding Find Support to Your Application

From the programmer's point of view, Find is implemented once all four methods (`Find`, `DateFind`, `FindSoupExcerpt`, `ShowFoundItem`) are added to the base view template and the application is registered with the system.

The Checklist

The easiest way to implement Find, however, is to implement the methods in a particular order, testing as you go. Here is the recommended order:

1. Add a `Find` method to the application base template.
2. Make sure the application has a `title` slot.
3. Test with local text Find.
4. Add a `ShowFoundItem` Method to display one item.
5. Test by tapping on an item in the Find Results slip.
6. Add a `DateFind` method to the application base template.
7. Add a `FindSoupExcerpt` method to display the items in the Find Results slip.
8. Test by doing a local date Find.
9. Register and unregister the application for Find in the `InstallScript` and `RemoveScript`.
10. Test by doing a global Find.

1. Adding a `Find` *Method*

Remember that the `Find` method is responsible for returning all found text items. It does this based on the information passed to it in its four parameters.

The `Find` Method and Its Parameters

Here is the structure of a typical `Find` method:

```
func(what, results, scope, statusForm)
begin
...
end
```

Here are the details of `Find`'s four parameters:

`what`	The word, or words, to look for. This parameter is a string or an array of strings.
`results`	An array to which a return result is added. Your `Find` method will add a frame to this array if it successfully finds something.
`scope`	This has a value of either `'localFind` or `'globalFind` depending on whether the user tapped "Find" or "All". You can use this to instigate a broader search if the user chooses a local Find.
`statusForm`	This is a view to which the method will report the status information concerning a search.

The `Find` method will create a cursor based on the contents of the `what` parameter. This means, in most cases, that the application will do a words `Query`. This query creates a cursor that iterates over all entries that contain any one of the strings in the `what` parameter.

Here is a sample `Find` method:

```
func(what, results, scope, statusForm)
begin
   if statusForm then
   statusForm:SetStatus("Searching in" &&        // Set status to "Searching in MyApp…"
      kAppName & $\u2026);

      local theSoup := GetUnionSoup(kSoupName);
      if theSoup then                             // Necessary for global Find, because the application may not have been run
      begin
         if classof(what) <> 'array then          // Query needs an array of words
            what := SplitString(what);
```

```
                local theCursor := Query(theSoup,
                   {indexPath: 'sortOrderSlot,
                   type: 'words, words: what});

                if theCursor:Entry() then
                begin
                   theCursor contains the result of
                   the successful search
                end
             end
     end
```

What the **Find** Method Returns

If the `Find` method successfully finds something, it adds a frame to the `results` array. Here is a typical `results` array frame:

```
{
   _proto: soupFinder,
   owner: GetRoot().(kAppSymbol),
   title: kAppName,
   cursor: theCursor,
   findType: 'text,
   findWords: what,
}
```

Here are the descriptions of the slots in the frame:

`_proto`	Applications returning a cursor will use `GetRoot().soupFinder`. On the other hand, applications that return an array of items use `ROM_CompatibleFinder`.
`owner`	The view that is sent the `FindSoupExcerpt` and `ShowFoundItem` messages. This view *must* be the application base view.
`title`	This holds the name of the application (its title) that is used in the Find Results slip. If no `title` slot is provided, the Find Results slip uses the `title` slot from the owner instead.
`cursor`	This contains the cursor that will iterate through the found items.

`findType`	This contains the symbol `'text`.
`findWords`	This holds the array of the words that were searched for in the application.

The portion of the `Find` method that constructs this frame will commonly look like this:

```
func(what, results, scope, statusForm)
begin
...
      if theCursor:Entry() then begin
         AddArraySlot(results,
            {_proto: soupFinder,
            owner:   GetRoot().(kAppSymbol),
            title:   kAppName,
            cursor:  theCursor,
            findType:'text,
            findWords:what,
            }
         );
      end
end
```

Note: The owner slot needs to contain the value `GetRoot().(kAppSymbol)` rather than the value `self`. This is because there are certain applications that set the `vApplication` bit for views other than the base view (usually in order to directly receive `viewScrollUpScript`, `viewScrollDownScript`, and `viewOverviewScript` messages). When the user does a local Find, however, the `Find` message is sent to `GetView('viewFrontMost)`—the frontmost view with the `vApplication` bit set. The `Find` method in the base view will be found (due to parent inheritance), but `self` will not be the base view. Because it is crucial that the `owner` slot be an application base view, the owner is specifically set to `GetRoot().(kAppSymbol)`. This is necessary since `self`, as you can see, may not necessarily be the correct view.

The Complete `Find` Method

Here is the `Find` method in completed form for your reference:

```
func(what, results, scope, statusForm)
begin
   if statusForm then
      statusForm:SetStatus("Searching in" &&
         kAppName & $\u2026);

   local theSoup := GetUnionSoup(kSoupName);
      if theSoup then
      begin
         if classof(what) <> 'array then
            what := SplitString(what);
         local theCursor := Query(theSoup,
            {indexPath: 'sortOrderSlot,
            type: 'words, words: what});

         if theCursor:Entry() then begin
            AddArraySlot(results,
               {_proto: soupFinder,
               owner:   GetRoot().(kAppSymbol),
               title:   kAppName,
               cursor:  theCursor,
               findType: 'text,
               findWords:what,
               }
            );
         end
      end
end
```

The `Find` Method and Searching for Nonstring Values

Notice also that the code presented for `Find` only works if the searched for text is stored as strings in the soup entry slots. Some applications might need to support searching for data stored differently. For instance, a checkbook program might need to support Find for check numbers. This is complicated by the fact that particular check numbers are probably stored as integers in the soup entry slots. In such a case, you would need to modify the `Query` to use a **`validTest`** that converts check numbers to strings and does string comparison. Even faster, you could convert the string to an integer, and then do an indexed Query on the check number slot.

Using the Newton's `StandardFind` Method

From the preceding descriptions, it should be clear that most applications support `Find` in pretty much the same way. Since each application should not have to maintain large amounts of duplicate code in a memory-tight environment, Newton provides a standard method that captures the commonality. This method is called `StandardFind` and is a root view method. `StandardFind` takes these five parameters:

`what`	Identical to the `what` parameter of `Find`.
`soupName`	The soup name in which to conduct Find searches.
`results`	Identical to the `results` parameter of `Find`.
`statusForm`	Identical to the `statusForm` parameter of `Find`.
`indexPath`	The path to use in the `Query`. Items will be found sorted on this path (an index must exist for this path). If you don't wish to specify a sort order, simply pass `nil`.

Here is a sample `Find` method that uses `StandardFind`:

```
func(what, results, scope, statusForm)
begin
   GetRoot().(kAppSymbol):StandardFind(what,
      kSoupName, results, statusForm, 'word);
end
```

It is important to know that the `StandardFind` method sets the `owner` slot in the results frame to point to `self`. As we discussed earlier, it is crucial that `self` actually be the application base view in this case, and this is why the `StandardFind` message is sent to the application base view. As you can see, using `StandardFind` can make implementing the `Find` method much simpler.

2. Make Sure the Application Has a `Title` Slot

As we said earlier, if the result frame contains a `title` slot, it is used as the title in the Find Results slip. It is important to know, however, that unsuccessful local finds cause the system to display a notice that includes the application name (see

FIGURE 4.6). This leads to a potential problem: Since the Find was unsuccessful, there is no result frame from which to obtain a `title`. So, for this notice, the system reads the `title` slot from the frontmost view (the same view that was sent the `Find` message). This will work as long as it can find the slot using both proto and parent inheritance. It can do so if your application base template has a `title` slot containing the name of your application. In that case, you are home free.

If you are using `protoApp`, you already have a `title` slot. If you are using a `clView` instead, you need to make sure to add a `title` slot to your application base template.

If a result frame doesn't have a `title` slot, the system will look for a `title` slot in the owner view. Since the owner view should be the application base view, and should contain a `title` slot, it is optional for the result frame to have a `title` slot. In fact, `StandardFind` doesn't even provide a `title` slot in the result frame.

3. *Test with Local Text Find*

Once you added the `Find` method to your application, you should be able to test it with a local text Find. Here are the points to cover in your tests:

- Make sure that your test finds two or more items, because you want the Find Results slip to be displayed.
- Test an unsuccessful Find. When finding no items, the Find slip should correctly specify that no items could be found in your application.

Don't do a Find that finds exactly one item (since you have not implemented `ShowFoundItem`). In addition, make sure that you don't tap on an item in the Find Results slip.

4. *Add a* `ShowFoundItem` *Method*

Now it is time to fix your application so that it can display a found item. As we said earlier, the `ShowFoundItem` method is responsible for displaying an item, so let us discuss that code. The system will send the application the `ShowFound-Item` message when a user taps on a item to display it or when only one item is found. Thus, this method will need to handle the display of one item appropriately.

First, let us look at the two parameters that `ShowFoundItem` takes:

`entry`	The soup entry to display.
`finder`	The frame from the results array. Normally, you'll have no use for this parameter.

When writing this base view method there are several things that you should take into account. First, remember that when `ShowFoundItem` is called, the application will already be open (if the user does a global Find, the system will open the application automatically). Next, remember that `ShowFoundItem` will normally need to open to the logical place in your application to display the found item. In most applications, this is some type of detail view.

Last of all, remember that you need to deal with folders. Notice that if your application supports folders, your code will need to make sure that the current folder contains the item. If it doesn't, your method should switch the current folder to "All *items*".

This sample `ShowFoundItem` method takes into account all of these issues:

Checking the item's folder

Calling the app's detail view display method

```
func(entry, finder)
begin
if labelsFilter <> '_all and
        labelsFilter <> entry.labels then
        folderTab:UpdateFilter(labelsFilter, '_all);
    :DisplayDetail(entry);
end
```

5. Test by Tapping on an Item in the Find Results Slip

After you have added the `ShowFoundItem` method to your application base view, you can test it to see if it is working correctly. Here is what to check for:

- Do a Find and then tap on an item in the Find Results slip.
- Do a Find that finds exactly one item; you should see that the Find Results slip is not displayed and that the found item is displayed directly.
- Display a folder and find exactly one item from that folder. The folder should not change.

- Display a folder and find exactly one item not in that folder. The folder should change to "All *items*".

6. *Add a* `DateFind` *Method*

If your application will not be supporting date Find, don't bother adding a `DateFind` method. In this case, users will see a slip stating that the application doesn't support date Find when they try to do a local date Find (see FIGURE 4.8).

As with your `Find` method, `DateFind` must be in your application base view. Unlike `Find`, however, `DateFind` takes five parameters, the first two of which are unique to it:

`findTime`	The time to compare against in number of minutes since midnight January 1, 1904.
`findType`	Either `'dateBefore` or `'dateAfter` depending on what was specified in the Find slip.
`results`	An array to which a return result is added. Your `DateFind` method will add a frame to this array if it successfully finds something.
`scope`	This has a value of either `'localFind` or `'globalFind` depending on whether the user tapped "Find" or "All". You can use this to instigate a broader search if the user chooses a local Find.
`statusForm`	This is a view to which the method will report the status information concerning a search.

Note that the last three parameters are the same as the corresponding parameters to `Find`.

What the DateFind Method Returns

Like `Find`, if the `DateFind` method successfully finds something, it adds a frame to the `results` array. The frame it adds should have the following slots:

`_proto`	Applications returning a cursor will use `GetRoot().soupFinder`. On the other hand, applications that return an array of items use `ROM_CompatibleFinder`.
`owner`	The view that is sent the `FindSoupExcerpt` and `ShowFoundItem` messages. This view *must* be the application base view.
`title`	This holds the name of the application (its title) that is used in the Find Results slip. If no `title` slot is provided, the Find Results slip uses the `title` slot from the owner instead.
`cursor`	This contains the cursor that will iterate through the found items.
`findType`	This contains either the symbol `'dateBefore` or `'dateAfter` depending on the type of search.
`findTime`	This contains the time that is used in the find search. This should be set to the `findTime` parameter.

The first four slots are the same as those in the `Find` method.

Supporting `DateFind` If the Date Slot Is Not Indexed

Most applications support date Find by storing a date in each entry in the soup. The date is stored as a standard Newton date (number of minutes since midnight January 1, 1904). If the slot containing the date isn't indexed, then the `Query` you write in `DateFind` will need a `validTest`.

Here is a sample `DateFind` that does this:

```
func(findTime, findType, results, scope, statusForm)
begin
   if statusForm then
      statusForm:SetStatus("Searching in" &&
         kAppName & $\u2026);

   local theSoup := GetUnionSoup(kSoupName);
   if theSoup then begin
```

```
            local theCursor := Query(theSoup,
               {type: 'index,
               indexPath: 'sortOrderSlot,
               validTest:
                  if findType = 'dateBefore then
                     func(e)
                        return e.changeDate < findTime
                  else
                     func(e)
                        return e.changeDate > findTime
            });

            if theCursor:Entry() then begin
               AddArraySlot(results,
                  {_proto:  soupFinder,
                  owner:    GetRoot().(kAppSymbol),
                  title:    kAppName,
                  cursor:   theCursor,
                  findType: findType,
                  findTime: findTime,
               });
            end
         end
      end
```

This code will search through each entry in the soup, looking for ones that match the date criterion. When a successful match is made, it will construct a proper frame and add it to the `results` array.

Supporting `DateFind` If the Date Slot Is Indexed

If the slot containing the date is indexed, the search will be much quicker. By using that slot as the `indexPath`, the entries will be visited in ascending date order. Here is what you do to support finding dates before and after certain points:

Dates Before	In this case, you will use a `Query` that has an `endTest` that checks the date slot. Once the end test finds a date that exceeds the required date, no more entries need to be visited.
Dates After	To find dates after, the `Query` can have a `startKey` which starts the search at the

desired date. An end test also needs to be provided to prevent the cursor from ever moving before the start date.

Here is the code that implements `DateFind` given an indexed date:

```
func(findTime, findType, results, scope, statusForm)
begin
    if statusForm then
        statusForm:SetStatus("Searching in" &&
            kAppName & $\u2026);

    local theSoup := GetUnionSoup(kSoupName);
    if theSoup then begin
        local queryFrame := {
            type:     'index,
            indexPath:'changeDate,
        };
        if findType = 'dateBefore then begin
            queryFrame.endTest := func(e)
                return e.changeDate >= findTime
        end else begin
            queryFrame.startKey := findTime;
            queryFrame.endTest := func(e)
                return e.changeDate <= findTime;
        end;
        local theCursor;
        theCursor := Query(theSoup, queryFrame);

        if theCursor:Entry() then begin
            AddArraySlot(results,
                {_proto:  soupFinder,
                owner:    GetRoot().(kAppSymbol),
                title:    kAppName,
                cursor:   theCursor,
                findType: findType,
                findTime: findTime,
            });
        end
    end
end
```

You will have much quicker finds with this `DateFind` method, especially if the application has many entries that don't fall into the desired date range.

If you have been waiting for us to show you the equivalent for `DateFind` of `StandardFind`, you better not be holding your breath. Every application must have its own `DateFind`, even though each method is very similar.

Note: You would not normally want to use the system slot `_modtime`. It only becomes a part of soup entries after an entry has been modified (after calling `EntryChange`); it isn't in new entries freshly added to a soup.

---

7. *Add a* `FindSoupExcerpt` *Method*

If you were to test your `DateFind` at this point, you would find the Find Results Slip display very unsatisfying. This display contains a useless summary of each entry (see FIGURE 4.10).

This summary is obtained for each entry by sending the application the `FindSoupExcerpt` message. If an application doesn't provide a `FindSoupExcerpt` method, inheritance causes the root view's `FindSoupExcerpt` to be used instead. Although the system's method may have handled text Find summaries quite well, it does a lackluster job with dates.

FIGURE 4.10 Find Results slip with default soup entry summaries.

For entries from a text Find, the root view's `FindSoupExcerpt` works well because it uses the global function `FindStringInFrame`, along with the `findWords` that are part of the results frame. It uses these to find the entry's text slot that contains the right word. That way, the Find Results slip will also show that part of each entry that is relevant to the Find.

For date Find, however, a "---" is returned by the `FindSoupExcerpt` method of the root view. Thus, your application needs a `FindSoupExcerpt` that will call the root view's method for a text Find, and create its own summary for a date Find. This summary should probably include not only the date in the entry, but some other useful snippet of information, so that the entry will be identifiable to the user. Here is a sample `FindSoupExcerpt` that does just that:

```
func(entry, finder)
begin
   if finder.findType = 'text then
      return GetRoot():FindSoupExcerpt(entry,
         finder)
   else begin
      return DateNTime(entry.changeDate) &
         ":" && entry.word;
   end;
end
```

Now, with this elegant little piece of code, you will get a Find Results slip similar to the one shown in FIGURE 4.11.

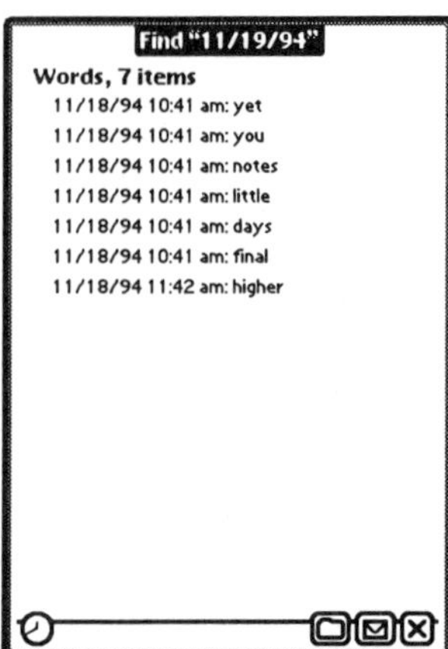

FIGURE 4.11 Find Results slip with customized soup entry summaries.

8. Test Local Date Find

At this point, you can test date Find. Make sure you test it with both dates before and dates after. Make sure the Find Results slip displays the correct summary information.

9. Register and Unregister for Find

To support global Find, you must register your application with the system. The global variable `findApps` is an array of application symbols for those applications that support global Find. Since your application needs to support global Find even when closed, it must add its application symbol to the `findApps` array in its `InstallScript`. Similarly, it must remove its application symbol in its `RemoveScript`.

Because the exact mechanics of registration could change in the future, we recommend that you encapsulate the code that handles the registering and unregistering in compile-time functions. Thus, your Project Data file should contain:

```
DefConst('kRegisterForFind,
   func()
   begin
      AddArraySlot(findApps, kAppSymbol);
   end);

DefConst('kUnregisterForFind,
   func()
   begin
      SetRemove(findApps, kAppSymbol);
   end
);

InstallScript := func(partFrame)
begin
   ...

   call kRegisterForFind with ();
end;

RemoveScript := func(partFrame)
begin
   ...

   call kUnregisterForFind with ();
end;
```

10. Test Global Find

Now, you have Find fully implemented and you should be able to test it anyway you wish. Make sure, however, that you test the new additions by exercising the global Find code. Here are some tests to try:

- Close your application and do a Find All that will find at least one item in the application. The Find Results slip should show results for all applications in which an item was found, including your own.
- Tap on an item in your application's list. Verify that your application correctly opens and displays that item.
- Remove your application from the Newton and try another Find All. This tests to make sure that the application symbol has been successfully removed from the `findApps` array. If it hasn't, you'll get an error when the system tries to reference the nonexistent application.

Find Checklist

Here is a quick checklist that you can use as a guide when you implement Find in your applications.

1. Add a `Find` method to the application base template.
2. Make sure the application has a `title` slot.
3. Test with local text Find.
4. Add a `ShowFoundItem` method to the application base template.
5. Test by tapping on an item in the Find Results slip.
6. Add a `DateFind` method to the application base template.
7. Add a `FindSoupExcerpt` method to the application base template. This method handles the display of items in the Find Results slip.
8. Test by doing a local date Find.
9. Register and unregister the application for Find in the `InstallScript` and `RemoveScript`.
10. Test by doing a global Find.

Chapter 5

Newton Connection

There's no limit to how complicated things can get, on account of one thing always leading to another.

—*E. B. White*

Newton Connection is the vehicle provided to users for exchanging data between the Newton and a desktop computer. After users synchronize their Newton to their desktop machine using Newton Connection (NCK), they can both view and edit data from their Newton applications on the desktop (Mac or Windows). They can also import and export data from within NCK to the desktop. For example, names and phone numbers can be updated within Newton Connection and then exported to a desktop application for use there as well.

An Overview of Newton Connection

In this chapter, we explain Newton Connection and provide step-by-step directions for adding this feature to your application. This support requires giving Newton Connection information about the structure of your soup entries and supplying some functions which do validation and conversion. You place this material in your application's meta data frame. Further, these functions are written in NewtonScript and execute on the desktop machine within the NewtonScript interpreter that is running within NCK.

We approach Newton Connection from a sequential viewpoint. You will be adding some functionality and then refining and testing it. Then you will repeat this process with additional functionality. Before looking at the code in detail, however, we will first discuss the user interface elements and give you a general understanding of how Newton Connection works.

Newton Connection's User Interface

When the user launches Newton Connection from the desktop she or he will already have a Newton connected or tethered to the desktop computer via a serial cable or a LocalTalk connection. Once launched, Newton Connection presents the user with the dialog shown in FIGURE 5.1. As you can see the user has a choice between restoring previously saved information, installing a package, and Synchronizing (the latter is important for our purposes).

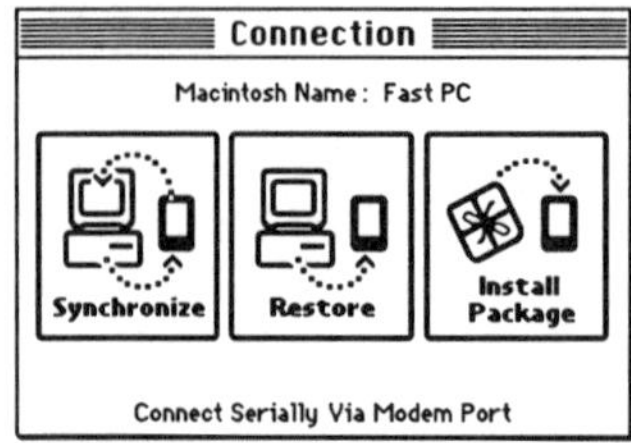

FIGURE 5.1 Newton Connection initial dialog.

When the user selects "Synchronize", Newton Connection will request a connection to the Newton via the Connection application in the Extras drawer. Once the connection is established, Newton Connection will proceed sequentially through the data of all of the applications found on the Newton (both on the internal store and PCMCIA card, if present). Once NCK has synchronized the data for all of these applications, it will present the user with the window shown in FIGURE 5.2. This window contains a list of all applications which have data that can be viewed within Newton Connection.

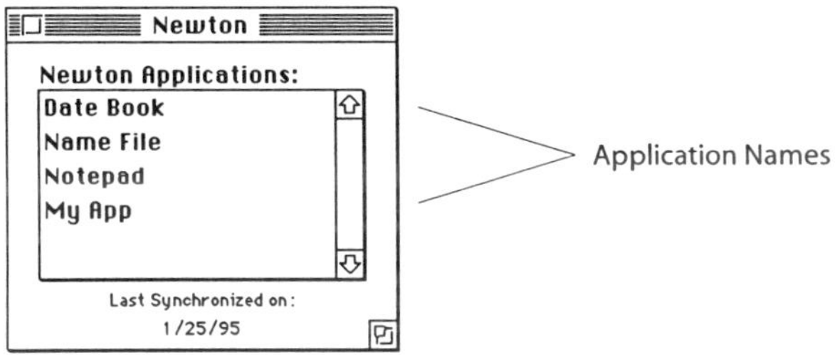

FIGURE 5.2 Newton Connection's list of applications.

When the user double-clicks on a name within the application list, Newton Connection brings up another window which displays the application's data (see FIGURE 5.3). Within this window the user can change the display viewing criteria (for example, sorting by last name instead of company), change which columns of data are displayed, or access an individual entry to modify it.

Newton:Name File

Click to create a brand new entry

13 Items New Entry

Last Name	First Name	Company	Phone 1	Key 1	Phone 2	Key 2
		Conglomerated Credit	800 555-1000	Work	415 555-8123	Fax
		Sears	1(800)-566-2333	Phone		
		Becky's Cafe	456-7890	Phone		
		MacConnection	1-800-800-2222	Phone	603-446-7791	Fax
Anderson	Bob		315 555-4476	Home		
Claus	Santa		1-800-SantaOK	Phone		
Ferry	Brian	Warner Brothers	(408) 844-5678	Phone		
Lucas	George	Industial Light and Magic	415-799-0440	Phone		
McKeehan	Julie	Calliope Enterprises	(909)793-5995	Phone		
Rhodes	Neil	Calliope Enterprises	(909)793-5995	Phone		
Thomas	Linda	Newton Cafe	617 555-1000	Phone	617 555-8123	Fax
Walthrop	Royce		419 555-3543	Home		

FIGURE 5.3 Data Browser window in Newton Connection.

If the user wants to change which columns of information are displayed, the dialog shown in FIGURE 5.4 is used. Simply checking or unchecking items will change the display.

To edit an individual entry in the Data Browser window, the user double-clicks on it (see FIGURE 5.5). The resulting window is called the Detail window and within it the user can modify any part of the entry. The user can also bring up this window for a brand new entry by selecting "New Entry" in the Data Browser window.

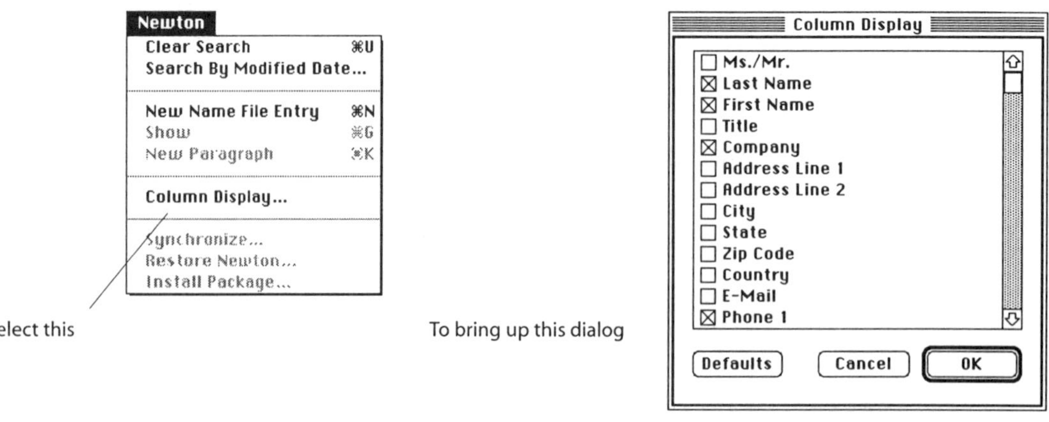

FIGURE 5.4 The Column Display dialog.

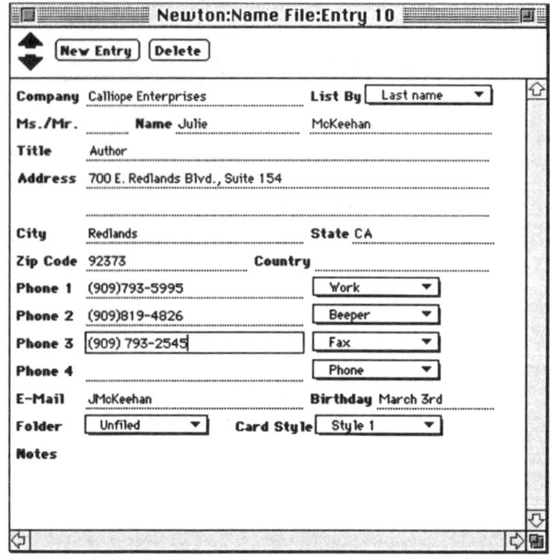

FIGURE 5.5 The Detail window in Newton Connection.

If the user wants to import text from a desktop application, then the Import menu shown in FIGURE 5.6 is used. Newton Connection uses translators to import and export from another application's native file format.

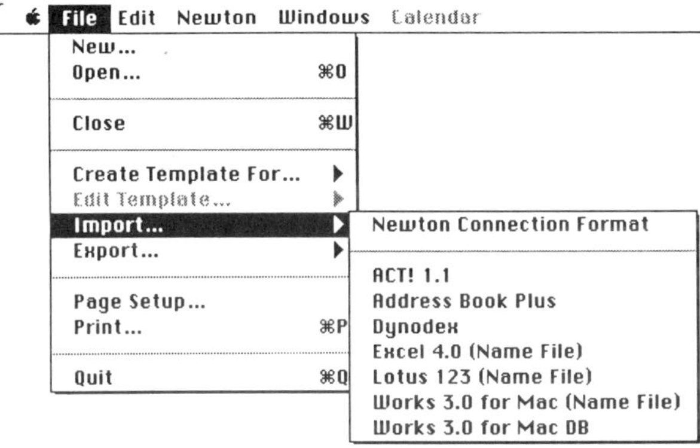

FIGURE 5.6 The Import menu in Newton Connection.

If the user wants to export from NCK to a desktop application he or she uses the Export menu instead (see FIGURE 5.7). Once a file type is selected then the data is saved into that file format and is available for use from within that application.

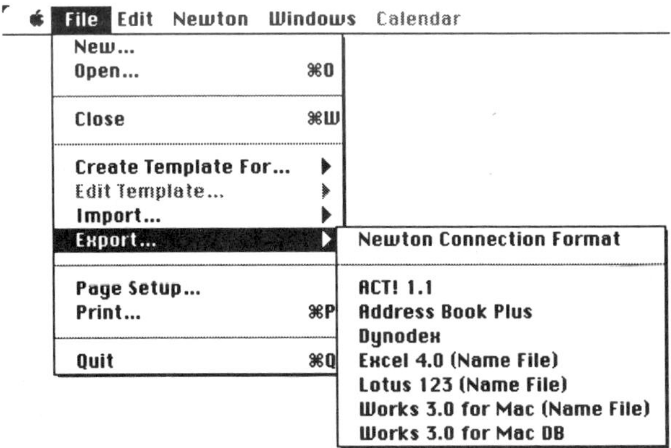

FIGURE 5.7 The Export menu in Newton Connection.

From this overview of the user interface elements in NCK, it should be clear that you have a variety of displays that you might want to handle. Now that you have a better idea of what occurs within NCK, let us turn to a discussion of the code involved in this support. The logical first stop is the meta data frame, the location of all your application's Newton Connection code.

Introduction to the Meta Data Frame

The first essential element in supporting Newton Connection is understanding the meta data frame that you must provide in your application. The meta data frame is a frame that gives information about your soup to the system so that it can be edited within Newton Connection. This frame describes each of the slots in your soup entries and how to edit those entries. This frame also contains all the additional information necessary to handle the importing and exporting of soup entries to the desktop machine.

Where the Meta Data Frame Is Stored

The meta data frame is stored in a soup called the Meta Data soup. You add your meta data frame to this soup when you create your application soup and remove it when you destroy your application soup (although most applications should never remove their soups). The frame has a `soup` slot and the Meta Data soup is indexed on that slot. In addition, the meta data frame has a `version` slot which can be used for version control.

Here is sample code which adds the meta data frame to the Meta Data soup. The meta data frame is defined as a slot (or constant) named `nckMetaData`:

```
local soup := GetStores()[0]:GetSoup(ROM_MetaSoupName);
local q := Query(soup,
     {type: 'index,
     indexPath: 'soup,
     startKey: kSoupName,
     endTest: func(item) NOT StrEqual(item.soup, kSoupName)
     }
);
// upgrading an existing meta data frame
local e := q:Entry();
if e and e.version < nckMetaData.version then begin
     EntryRemoveFromSoup(e);
     e := nil;
end;
```

```
if not e then begin
    soup:Add(Clone(nckMetaData));
end;
```

Special code is needed to deal with new versions of an application. These new versions may have added slots to the soup and therefore may need to update the meta data frame. Thus, we have code that removes an existing meta data frame if it has a version number less than that of the new frame.

Installing the Meta Data Frame in Your Application

Your meta data frame resides as a slot in your application base template. We usually name it `nckMetaData`. Alternatively, you can create your meta data frame as a constant in your Project Data file.

Slots in the Meta Data Frame

For the purposes of our current discussion, you do not need to understand each slot in the meta data frame. Thus, we have simply given you the basic structure of the meta data frame, including the slot names and which features they effect.

```
nckMetaData := {
  soup: kSoupName,
  name: "Application Name",
  DisplayInfo: [{column1}, {column2}, … ]
  BrowserExport: func(entry, soupStore)
  editInfo: [{item1}, {item2} … ]
  EditImport: func(textFrame, soupStore)
  EditExport: func(entry, soupStore)
  exportInfo:[{item1}, {item2} … ]
  TabImport: func(textFrame, soupStore)
  EntryValidation: func(entryFrame, soupStore)
  TabExport: func(entry, soupStore)
  NativeImport: func(nativeFrame, soupStore)
  NativeExport: func(entry, state, soupStore)
}
```


You will be working your way down the preceding meta data frame. Initially, we will talk about the first five slots in this frame, these being the only required slots. The third, fourth, and fifth slots (`DisplayInfo`, `BrowserExport`, `editInfo`) are responsible for the display of items in Newton Connection. The next three slots, `EditImport`, `EditExport`, and `exportInfo`, are then discussed. They are the slots that allow the user to edit soup entries. Finally, we will study the last six slots and show how they control the exporting and importing of files on the desktop machine.

Displaying Soup Entries

Five slots are required in the meta data frame:

```
nckMetaData := {
   soup: "Soup Name",
   name: "App Name",
   version: 5,
   DisplayInfo: [{column1}, {column2}, … ],
   BrowserExport: func(entry, soupStore),
}
```

Let us look at each slot in turn.

soup: string

This slot holds your application's soup name. The user does not normally see this name. You'll usually have already defined your soup name as a constant in your Project Data file (see FIGURE 5.8):

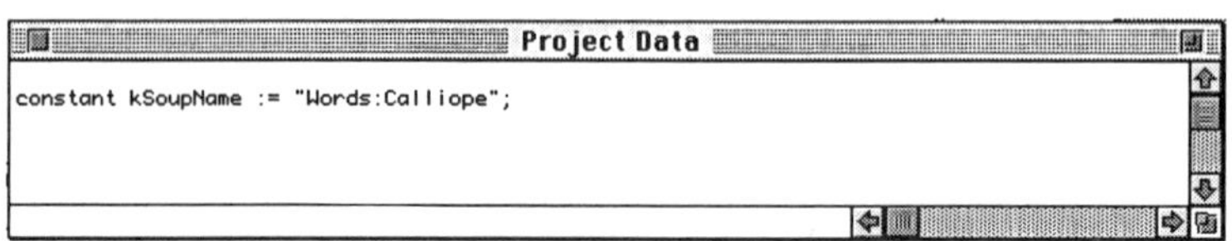

FIGURE 5.8 Defining a soup constant in the Project Data file.

name: string

This is the name of your application that shows up in Newton Connection's Application list (see FIGURE 5.9). Commonly, your application name is defined as a constant in your Project Data file.

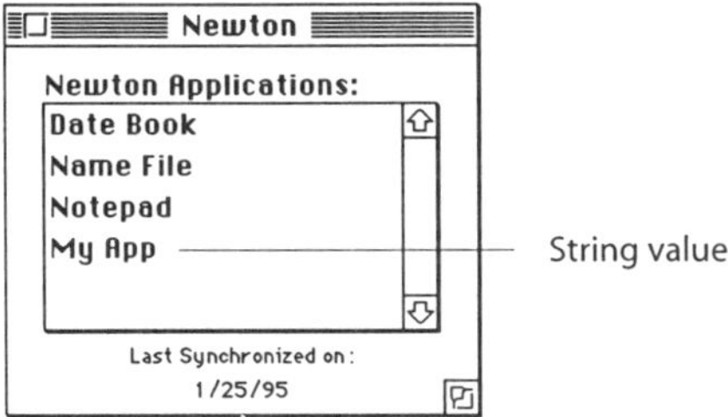

FIGURE 5.9 Showing the name of an application in the NCK application list.

version: integer

The version number is used to allow modification of the structure of the soup. When you update your application you may wish to modify the structure of soup entries, and therefore provide a new meta data frame. The code you use to add the meta data frame to the meta data soup should thus check the version number of any existing meta data frame. If the version number is lower, your code should replace the existing meta data frame; if it is higher, it should leave it alone.

displayInfo: [{column1}, {column2}, …]

This slot contains an array of frames. Each frame in the array describes the layout characteristics of a column full of data in NCK's Application Browser window.

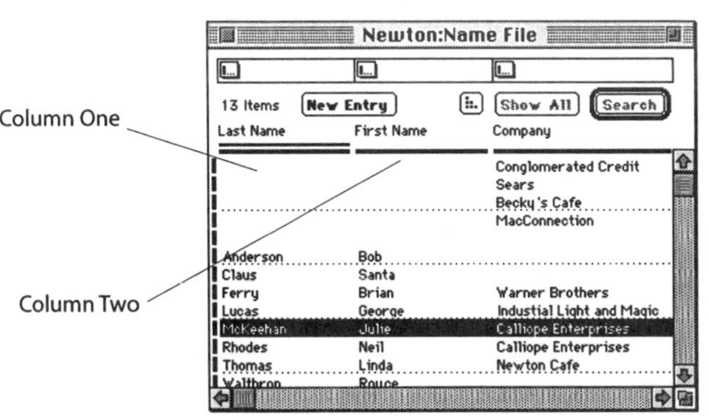

Each column of information has its own ordered frame in this array. In the example at the left, column one (containing last names) would be the first frame in this array. Likewise, column two (containing first names) would be handled by the second frame.

The `displayInfo` array contains one frame for each displayable column of information even if it is not initially shown.

Definitions of the Slots in Each `displayInfo` frame

Each frame in the `displayInfo` array contains a number of slots that together describe the characteristics of that particular column of information. Here are those slots with a brief description of each.

`slot`	This contains the slot symbol used in the text frame (see "The Text Frame" on page 167 for a description of the text frame).
`label`	This contains the string used for the column heading.
`size`	This is the integer value specifying the width (each unit is the width of a 0 character) of the column.
`type`	This designates the type of value used for searching and sorting on the column. The possible values are `'string`, `'char`, `'text`, `'integer`, `'real`, `'picklist`, `'folder`, `'dateNTime`, `'date`, `'time`, `'day`, `'dateInMonth`, `'month`, `'year`, and `'duration`.
`path`	This is a path expression that specifies the location of the data in the soup entry. It is used for searching and sorting.
`hidden`	A boolean that designates whether the information is initially invisible.
`unsearchable`	A boolean that designates whether the column is searchable.
`CompareScript`	This contains the symbol of a script which can be used for custom sorting and searching.

```
BrowserExport: func(entry, soupStore)
```

The Text Frame

Your application is responsible for converting each soup entry to displayable text. The `BrowserExport` function does this by creating a *text frame* with one slot for each displayable string. It converts each slot in the soup entry to a string and then stores that string as a slot in the text frame.

Note: To distinguish slots in the soup entry from slots in the text frame, we usually name each slot in the text frame `text`*`OrigSlotName`*. That is, we preface the slot in the text frame with the word `text`.

The `BrowserExport` Function

In creating this function you should make sure that the text frame slot names are equal to the names specified by slot in the `displayInfo` frames. Thus, if you have a `displayInfo` frame similar to the one on the left in TABLE 5.1, you should make sure your `BrowserExport` function looks like the one on the right.

<table>
<tr><th>A displayInfo array:</th><th>A BrowserExport function:</th></tr>
<tr><td><pre>displayInfo: [
    {slot: 'textName,
    label: "Name",
    size: 15,
    type: 'string
    path: 'name,},

    {slot: 'textAge,
    label: "Age",
    size: 3,
    type: 'integer,
    path: 'age},
],</pre></td><td><pre>BrowserExport: func(entry, soupStore)
begin
  local textframe := {
    textName: entry.name,
    textAge: NumberStr(entry.age),
    originalEntry: entry,
  };

  return textframe;
end</pre></td></tr>
</table>

TABLE 5.1 Compatible `displayInfo` and `BrowserExport` slots.

Notice that you also have an `originalEntry` slot in the `BrowserExport` function. This is because it is common to store a copy of the original soup entry in a slot in the text frame. In this way, any extra slots that might not be converted to text are maintained after editing.

Example Meta Data Frame

Now that you have seen a brief description of the required slots in the meta data frame, here is a typical example of one. Notice that this meta data frame describes a soup whose entries contain two slots, `name` and `age`:

```
{
   soup: "myApp:Calliope",
   name: "My App",
   version: 1,
   displayInfo: [
      {
         slot: 'textName,
         label: "Name",
         size: 15,
         type: 'string,
         path: 'name,
      },
      {
         slot: 'textAge,
         label: "Age",
         size: 3,
         type: 'integer,
         path: 'age,
      },
   ],

   BrowserExport: func(entry, soupStore)
   begin
      local textframe := {
         textName: entry.name,
         textAge:  NumberStr(entry.age),
         originalEntry: entry,
      };
      return textframe;
   end,
}
```

To test the workability of your meta data frame at this point, you must make sure it is installed in the meta data soup. This is normally done by running your application (which will install your frame into the meta data soup). Then, you must synchronize using Newton Connection. If you have accomplished this much, you should see whatever value you used in the `name` slot in the Application list in Newton Connection (see FIGURE 5.10):

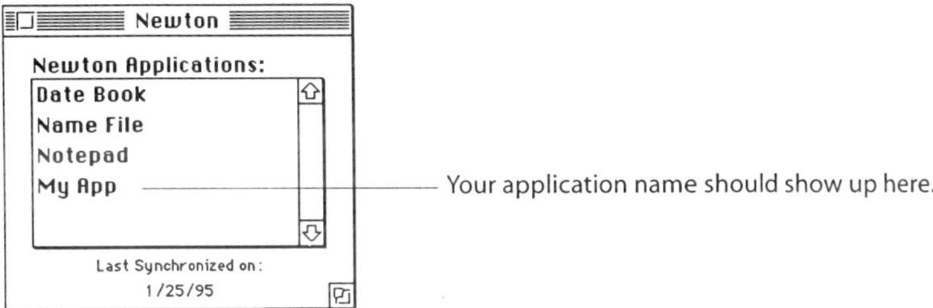

FIGURE 5.10 Your application listed in NCK's application list.

Now, if you double-click on the application name, it should open up a browser window that will display all of the items in the soup (see FIGURE 5.11):

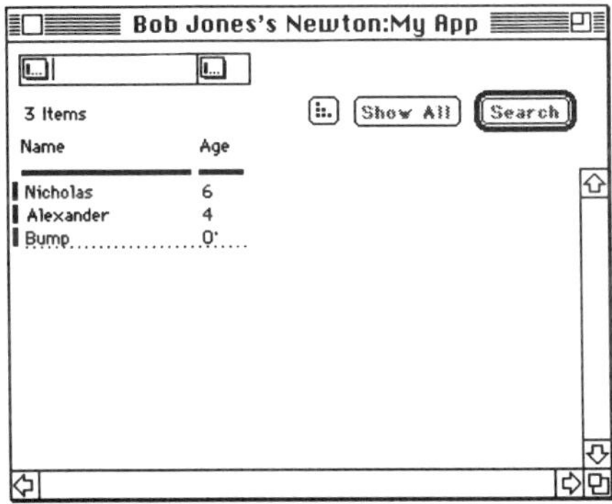

FIGURE 5.11 Your soup data displayed in the Browser window.

You can sort the entries on a particular slot (by clicking on column header), and you can search for values. Notice that there is no New Entry button, and no way to edit an entry as of yet.

User Editing of Soup Entries

Two slots are required in the meta data frame for the user to edit soup entries. A third slot is optional. These three slots are:

- `editInfo`
- `EditImport`
- `EditExport` *(optional)*

Let's look at each of these slots in turn.

`editInfo: [{item1}, {item2} ... ]`

This is an array of frames very similar to `displayInfo`. Each frame in the array corresponds to an editable item in the Detail window.

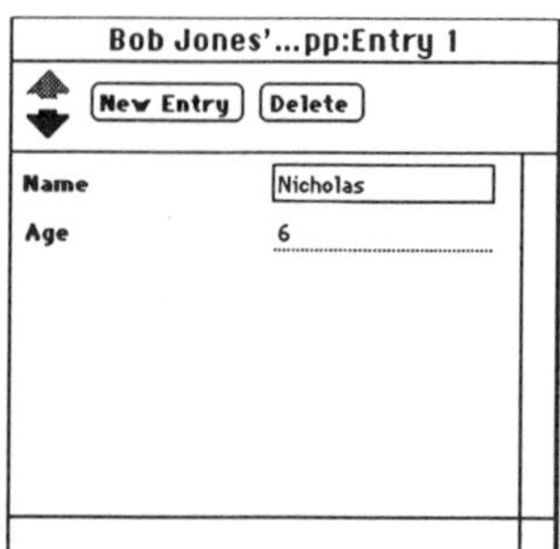

In the example at the left, the first frame in the array would correspond to the name item. Likewise, the second frame in the array would match the age.

There are various slots in each frame that among other things specify the width, input characteristics, and location of an item relative to other items.

Definitions of the `editInfo` Slots

Here is a brief description of the possible slots in each frame in the `editInfo` array. Notice that many of the slots are optional.

`slot`	This contains the slot symbol used in the text frame.
`label`	This contains the string used to label the slot.
`size`	This is the integer value specifying the width of the item.

`type` — This designates the type of value for the item (used for input validation). The possible values are the same as the `type` slot in `displayInfo`.

`sameLine` — This is a boolean value specifying if the field is on the same line as the previous field.

`choices` — This contains an array of strings if the value of the `type` slot is `'picklist`.

`uneditable` — This is a boolean that specifies if the field is read-only.

`numLines` — This is an integer value specifying the number of lines shown if the value of the `type` slot is `'text`.

An Example `editInfo` Array

Here is a sample `editInfo` array:

```
editInfo: [
   {
      slot: 'textName,
      label: "Name",
      size: 15,
      type: 'string,
   },

   {
      slot: 'textAge,
      label: "Age",
      size: 3,
      type: 'integer,
   },
]
```

EditImport: func(textFrame, soupStore)

Within Newton Connection, this function does the opposite of `BrowserExport` (which turns a soup frame into text); instead, `EditImport` takes the Newton

Connection text frame created by `BrowserExport` and edited by the user, and converts it back into a soup entry. The possible return values of this function follow:

`nil`	Newton Connection does not add the frame to the soup.
string	It does not add the frame and uses the string as an error message.
frame	This is added to the soup (if a new entry) or replaces a frame in the soup (if editing an existing entry).
array of frames	Each frame is added to the soup. However, if the first entry is nil, the frame being edited is deleted from the soup.

Here is a sample function:

```
EditImport: func(textFrame, soupStore)
begin
   local e := textFrame.originalEntry;
   if not e then
      e := {}; // brand-new entry
   if not textFrame.textName then
      return "the name was missing";
   if not textFrame.textAge then
      return "the age was missing";
   e.name := textFrame.textName;
   e.age := StringToNumber(textFrame.textAge);
   if not e.age then
      return "invalid age"
   else if e.age < 0 then
      return "the age was negative"
   else if e.age > 150 then
      return "the age was a wee bit high"
   else
      return e;
end
```

EditExport: func(entry, soupStore)

You will only be providing this function if you have different text frames for the Browser and Detail windows (see FIGURE 5.12). If you have the same text frame, then don't bother providing an `EditExport`; Newton Connection calls `BrowserExport` if `EditExport` isn't present. If you do use this function, note that the text frame slot names used here should equal those names used in `displayInfo` array (just as they did for `BrowserExport`). Here is a sample function:

```
EditExport: func(entry, soupStore)
begin
   local textframe := {
      textName: entry.name,
      textAge:  NumberStr(entry.age) && "years",
      originalEntry: entry,
   };
   return textframe;
end
```

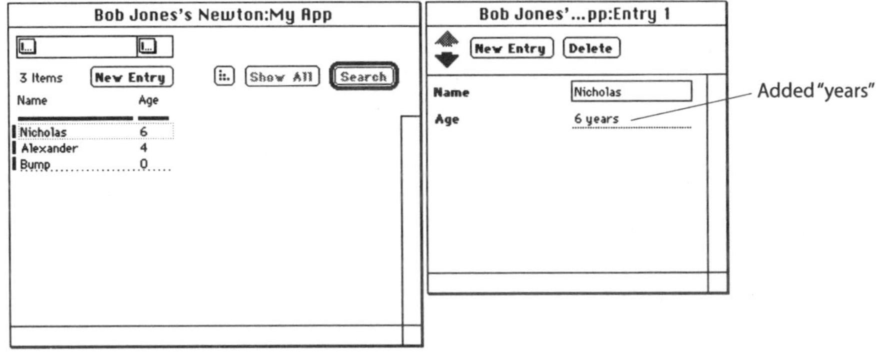

FIGURE 5.12 A Browser window and a Detail window with different text frames.

Testing the Meta Data Frame for Editing

Here are the steps you can take to test your newly created code:

1. Update the version number in your meta data frame.
2. Rerun your application so that the meta data soup is updated.
3. Synchronize so that the meta data is copied to Newton Connection.

4. From within Newton Connection, you should notice that the Browser Window now contains a "New Entry" button to create new soup entries.
5. Now, when you double-click on one of the entries to edit it, the Detail window should appear.

The Detail window does sanity checking on fields automatically. For instance, trying to enter the value "xyz" in an integer field (like `age`) yields an error. You see the type of error found in FIGURE 5.13 whenever you try to leave the field by tabbing or clicking on another field:

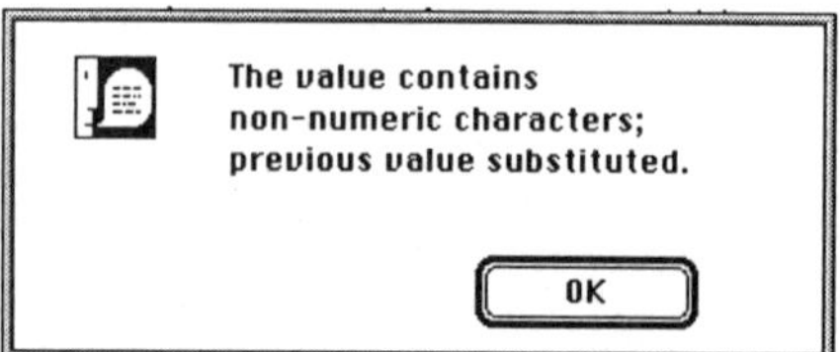

FIGURE 5.13 An error generated when you enter an incorrect data type.

6. You should check your `EditImport` code by trying both valid input and invalid input (like a negative age). You should also make sure that all the error strings your `EditImport` can possibly return have been tested.

Importing and Exporting of Data

The third and last task involved in your support of Newton Connection is being able to correctly export files to the desktop machine and being able to import them back again into Newton Connection. To handle simple file exporting you need to add two additional slots to your meta data frame. These two slots are:

- `exportInfo`
- `TabExport`

To handle importing, you need (in addition to the `exportInfo` slot):

- `TabImport`

Finally, to handle the importing and exporting of native format files to Newton Connection, you will be adding these slots to your meta data frame:

- `NativeExport`
- `EntryValidation`
- `NativeImport`

Now, let's look at each of these slot additions to the meta data frame.

Exporting Tab-Delimited Files

To export tab-delimited files, one slot, `exportInfo`, is required, and another, `TabExport`, is optional.

`exportInfo: [{item1}, {item2} … ]`

This slot contains an array of frames that controls what can be imported and exported to a text frame. The actual importing and exporting, however, are handled by `TabImport` and `TabExport`. Each `exportInfo` frame contains these three slots:

`slot`	This is a required slot that contains the symbol of the text frame.
`label`	This contains a string that is used as a descriptive label for the item.
`path`	This contains the path expression that specifies where the soup data are located.

Here is a sample `exportInfo` array:

```
exportInfo : [
   {slot: 'textName, label: "Name", path: 'name,},
   {slot: 'textAge, label: "Age", path: 'age,}
]
```

```
TabExport: func(entry, soupStore)
```

This function takes a soup entry and returns `nil` or a text frame. If it returns `nil` then the soup entry is not exported. If it returns a text frame, that frame's slots correspond to those found in `exportInfo`. Here is a sample function:

```
TabExport: func(entry, soupStore)
begin
   return {
      textName: entry.name,
      textAge: NumberStr(entry.age),
      textHonorific:
       if entry.honorific = 'mr then "Mr." else "Ms.",
   };
end
```

Note: If `TabExport` isn't present, `BrowserExport` will be called instead.

Testing File Export

After adding the `exportInfo` slot to your meta data, and synchronizing with Newton Connection, you can open the Template dialog (using the menu item "Create Template For...") to specify the organization of the tab-delimited file.

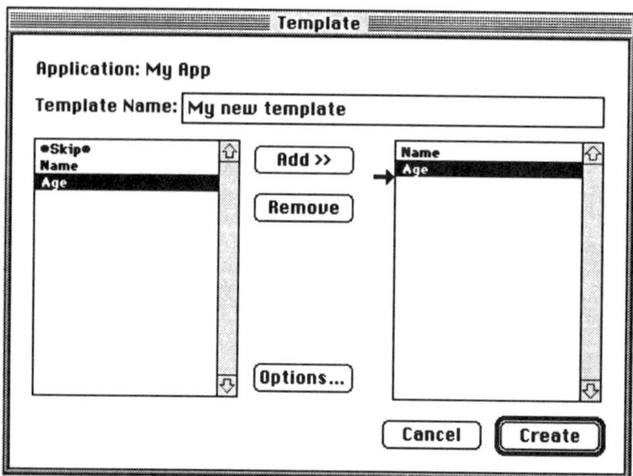

FIGURE 5.14 Creating a template for a new tab-delimited file in NCK.

The names on the left in FIGURE 5.14 correspond to the items in the `export-Info` array. Once you have created this new template, you can test exporting by using the Export menu as shown in FIGURE 5.15:

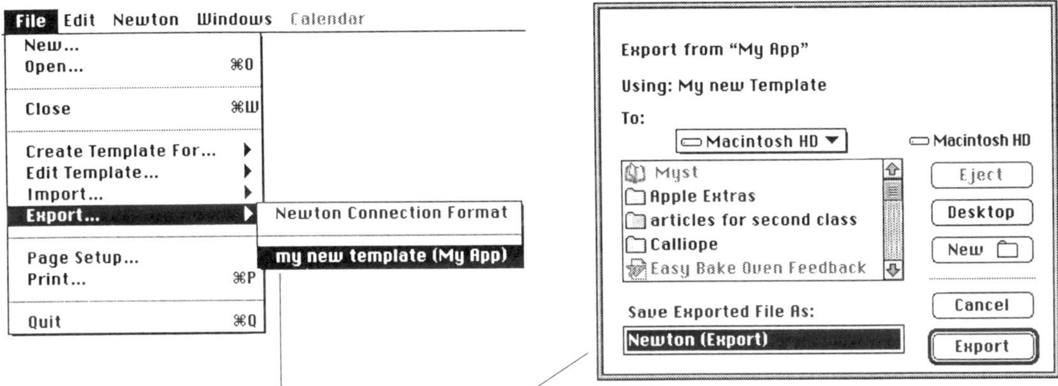

FIGURE 5.15 Exporting a file in NCK using a custom template.

Notice that once you export a new file using the custom template you created, NCK lets you know how many soup entries were exported as part of this file (see FIGURE 5.16).

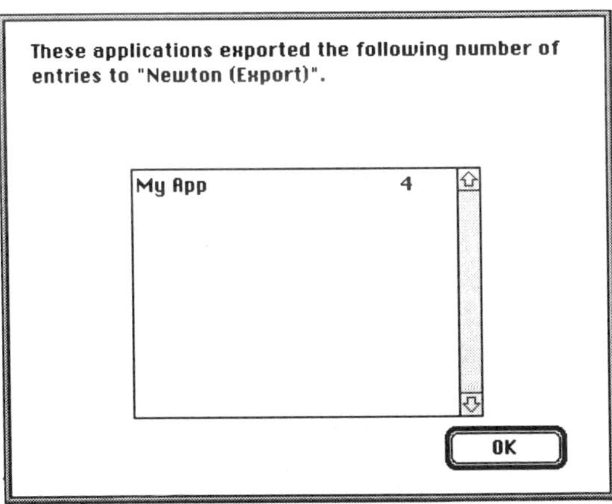

FIGURE 5.16 NCK status information about the export file.

The resulting file should have fields from each soup entry in the order specified in the Template dialog. The result should look something like this:

```
Nicholas     6
Alexander    4
Bump         0
```

Importing Tab-Delimited Files

To import a tab-delimited file, you must have an `exportInfo` slot (see "exportInfo: [{item1}, {item2} ...]" on page 175). In addition, you must provide a `TabImport` slot.

`TabImport: func(textFrame, soupStore)`

NCK creates a text frame for each row of the tab-delimited file. `TabImport` converts this to a soup entry. Here is a sample function that does just this:

```
TabImport: func(textFrame, soupStore)
begin
   if not textFrame.textName then
      return "the name was missing"
   else if not textFrame.textAge then
      return "the age was missing";

   local e := {
      name: textFrame.textName,
      age:  StringToNumber(textFrame.textAge),
   };
   if StrLen(e.name) = 0 then
      return "the name was blank"
   else if e.age = nil then
      return "the age was blank or not an integer"
   else if e.age < 0 then
      return "the age was negative"
   else if e.age > 150 then
      return "the age was a wee tad high"
   else
      return e;
end
```

Note: It is important for you to explicitly validate the incoming text frame yourself in the `TabImport` function. NCK doesn't do validation of the text frame.

Testing Importing

Here are the steps that you can take to test you new import capabilities.

1. Create a tab-delimited file to import as shown in FIGURE 5.17.

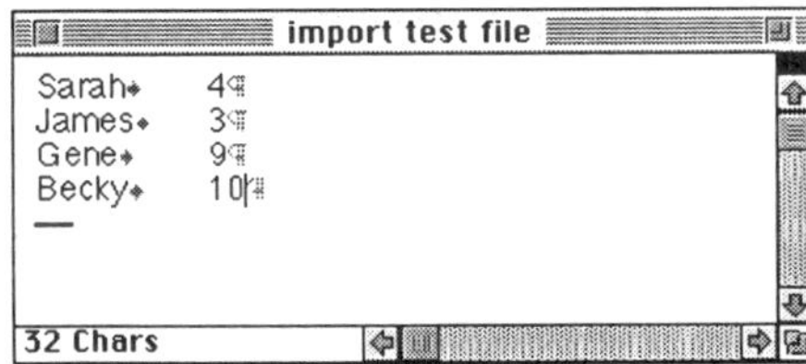

FIGURE 5.17 A sample tab delimited file.

2. Use the Import menu item to import that tab-delimited file as shown in FIGURE 5.18.
3. The new entries should appear in the Browser window (see FIGURE 5.18).

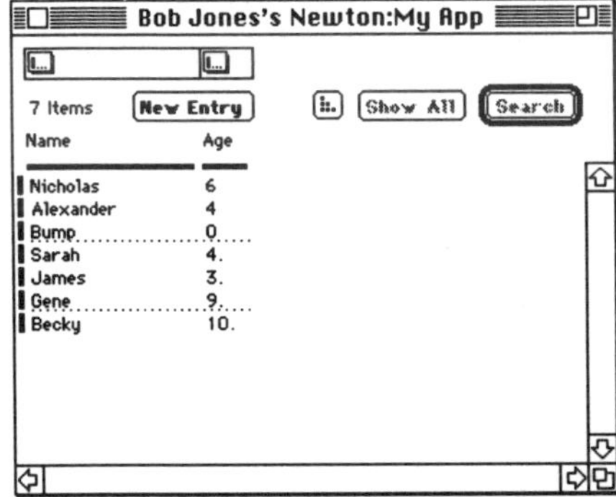

FIGURE 5.18 A newly imported file.

Make sure to test your error checking. Here are some suggestions:

- Change a numeric item in the tab-delimited file to something non-numeric.
- Remove a column or an item from one row.
- Try creating a template that specifies only one slot in your soup entries. What happens when you import it?

Exporting Native Files

Native files are text versions of the actual soup entries, which can be edited and then converted back (automatically by Newton Connection) to soup entries without losing any information. Some desktop applications may understand how to read these files and can synchronize using this information. There is one optional slot, `NativeExport`, used for the native export of files.

```
NativeExport: func(entry, state, soupStore)
```

Newton Connection calls this function for each entry that is exported from the soup. The `state` parameter is a frame used to save state information between calls. Here is a sample function:

```
NativeExport: func(entry, state, soupStore)
begin
   local e := Clone(entry);
   RemoveSlot(e, 'sortOn);
   return e;
end
```

The only reason to provide a `NativeExport` function is if you wish to modify the soup entries before actually exporting them. For example, you might have data that does not need to be exported that you want to remove.

Testing Exporting

To test native file exporting, you can use the Export menu (FIGURE 5.19).

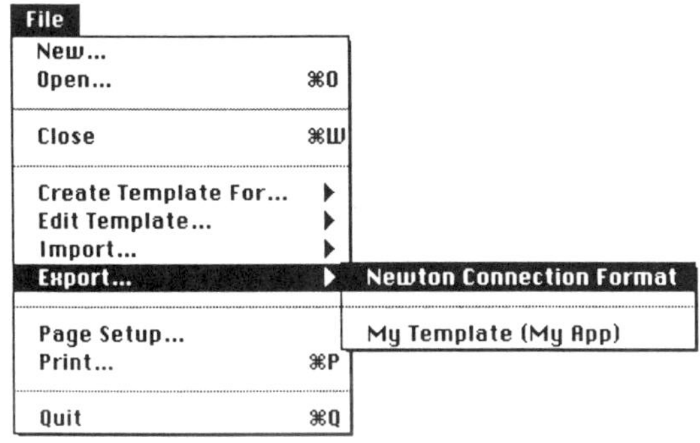

FIGURE 5.19 Export a file to native format.

An export file should look something like this:

```
FORMAT-VERSION 1
   {
   DATABASE "myApp:Calliope"
      {
      ADD /* ID=0, TIME=47890846 */
         { name: "Nicholas", age: 6, _exportid: 0 }
      ADD /* ID=1, TIME=47890824 */
         { name: "Alexander", age: 4, _exportid: 1 }
      ADD /* ID=2, TIME=47890824 */
         { name: "Bump", age: 0, _exportid: 2 }
      }
   }
```

Importing Native Files

There are two optional slots used for importing native files: `EntryValidation` and `NativeImport`. While you should always provide an `EntryValidation` routine in your meta data frame, it is uncommon to need a `NativeImport` slot.

`EntryValidation: func(entryFrame, soupStore)`

You validate entries within this function. It should return one of the following values:

- `nil`

- String
- Frame
- Array of frames.

`EntryValidation` is called automatically for `NativeImport`. It is important to note that this routine must check for the existence of required slots as well as the types of the values and the values themselves. The routine should assume nothing about the frame passed in (a user may have imported the wrong file) and it should completely verify it.

Here is a sample function:

```
EntryValidation: func(entryFrame, soupStore)
begin
   if ClassOf(entryFrame.name) <> 'string then
      return "the name was not a string"
   else if StrLen(entryFrame.name) = 0 then
      return "the name was blank"
   else if ClassOf(entryFrame.age) <> 'int and
      ClassOf(entryFrame.age) <> 'real then
      return "the age was not an integer"
   else if entryFrame.age < 0 then
      return "the age was negative"
   else if entryFrame.age > 150 then
      return "the age was a wee tad high"
   else
      return entryFrame;
end
```

`NativeImport: func(nativeFrame, soupStore)`

This function takes a frame that comes from a native format and saves it as a soup entry. You never do any entry validation here; you should be handling that in `EntryValidation` instead. Here is a sample function:

```
NativeImport: func(nativeFrame, soupStore)
begin
   nativeFrame.sortOn := nativeFrame.name &
      nativeFrame.age;
   return nativeFrame;
end
```

As is the case with `NativeExport`, it is uncommon to need a `NativeImport`. One use might be to add slots that were stripped by `NativeExport`.

Testing Native Importing and Entry Validation

In most cases, you'll only need to provide an `EntryValidation` routine to support both native import and export. When NCK creates a native export file, it automatically exports all slots in your entries. Similarly, on import all slots are automatically created from the data in the native file. The `EntryValidation` routine will make sure that all necessary slots are there, and contain reasonable values. Here are the steps you can take to test your new code:

1. Add an `EntryValidation` routine to your meta data frame.
2. Run your application so that it overwrites the old meta data entry and then synchronize with NCK.
3. Now, use the Export menu item to export your soup in native format.
4. Open the file that is created to inspect the text version of your soup entries.
5. Next, modify the native file to add some more entries.
6. Now, with anxious delight, use the Import menu item to import that file. The new entries should appear in the Browser window.

Remember to test your error-checking code by removing some slots from entries in the native file, or changing their types (for instance, instead of an integer value, create a string value). Make sure your routine correctly handles this situation.

Centralizing and Sharing Data

You may have realized by now that there are many slots in the meta data frame that hold either similar or identical code. Happily, there are some techniques you can use to reduce the amount of replicated code and data. Let's look at a few of these approaches.

Centralizing Validation Code

Rather than having validation code duplicated in your `EntryValidation`, `TabImport`, and `EditImport` routines, you can share code in this useful way:

- Have `TabImport` and `EditImport` call `EntryValidation`.

When using this approach, keep in mind that your meta data frame functions are called as methods within your meta data frame (`self` is the meta data frame). Therefore, you would write your `EditImport` as:

```
EditImport: func(textFrame, soupStore)
begin
   local e := textFrame.originalEntry;
   if not e then
      e := {};
   // make sure that StringToNumber will succeed
   if not textFrame.textAge then
      return "the age was missing";
   e.name := textFrame.textName;
   e.age := StringToNumber(textFrame.textAge);
   return :EntryValidation(e, soupStore);
end
```

You would likewise write your `TabImport` as:

```
TabImport: func(textFrame, soupStore)
begin
   // check for age to make sure
   // that StringToNumber will succeed
   if not textFrame.textAge then
      return "the age was missing";

   local e := {
      name: textFrame.textName,
      age:  StringToNumber(textFrame.textAge),
   };
   return :EntryValidation(e, soupStore);
end
```

Sharing Information in Info Arrays

You may have noticed that the information in the `displayInfo`, `editInfo`, and `exportInfo` arrays is very similar (see TABLE 5.2).

<table>
<tr><th>A displayInfo array</th><th>An editInfo array</th><th>An exportInfo array</th></tr>
<tr><td><pre>[
  {
    slot: 'textName,
    label: "Name",
     size: 15,
    type: 'string
    path: 'name,},

  {
    slot: 'textAge,
    label: "Age",
    size: 3,
    type: 'integer,
    path: 'age
  },
],</pre></td><td><pre>[
  {
    slot: 'textName,
    label: "Name",
    size: 15,
    type: 'string,
  },

  {
    slot: 'textAge,
    label: "Age",
    size: 3,
    type: 'integer,
  },
]</pre></td><td><pre>[
  {
    slot: 'textName,
    label: "Name",
    path: 'name,
  },

  {
    slot: 'textAge,
    label: "Age",
    path: 'age,
  }
]</pre></td></tr>
</table>

TABLE 5.2 Comparing a `displayInfo`, `editInfo`, and `exportInfo` array.

A good way to share that information is via the `defaultDefinitions` frame. This frame contains frames that can be used within any of the three info arrays. Here is an example:

```
defaultDefinitions: {
   name: {
      slot:'textName,
      label:"Name",
      size: 5,
      type:'string,
      path:'name,
   },
   age: {
      slot:'textAge,
      label:"Age",
      size:3,
      type:'integer,
      path:'age,
   }
}
```

The items in the `defaultDefinitions` frame can then be used by specifying their slot names within one of the three arrays. For instance, the `displayInfo` array could now contain:

```
[ 'name, 'age]
```

Likewise, the `editInfo` array could be slimmed down to this:

```
[ 'name, 'age]
```

Notice that a frame in an `info` array that has a slot named `slot` specifies an entry in `defaultDefinitions` with additional (or overridden) slots. For example, if you wanted the `age` slot to be uneditable, you could override that value in your `editInfo` simply by doing this instead:

```
[ 'name, {slot: 'age, uneditable: true}]
```

Common Questions about Newton Connection

How Do I Debug?

Unfortunately, if your NewtonScript code running within NCK causes an exception, NCK will only tell you that an error occurred. In addition, `Print` statements don't work because there is no place to send the output.

Fortunately, there is a special version of NCK for developers named "NCK for Developers". In this version, there is a Listener window in which you, the developer, can execute NewtonScript statements within NCK, and where you can read `Print` statements. In the case of an exception, debugging information is printed to the Listener window.

This version of NCK is available on PIE Developer CDs, available to Newton Associates and Newton Partners (see "Apple's Associates and Partners Program for Newton" on page 440).

In My Import Routine, How Do I Convert a String to an Integer?

The `StringToNumber` function converts a string to a real number (just as in the NewtonScript interpreter on the Newton). However, many other global functions that exist on the Newton aren't available within NCK. In particular, `RIntToL` is not present.

The `Floor` function is available, however. Therefore, to convert, use:

```
i := StringToNumber(text);
if i then// StringToNumber can return nil
   i := Floor(i);
```

Why Don't Changes to My Meta Data Take Effect?

There are a number of steps between editing a meta data frame in NTK and seeing the change in NCK.

The two most common errors programmers make are:

1. *Not updating the meta data soup.* Make sure that you have updated the `version` slot in the meta data frame (otherwise, your code will not update the existing meta data frame in the meta data soup). In addition, make sure that you run your application (since it is in your `viewSetupFormScript` that you probably update the meta data soup).

 *Note:* While developing your meta data frame, you might consider not checking the version number in the code that installs your meta data frame; if an existing meta data frame exists, remove it, and then install the new one. In this way, you don't have to update the version slot each time you modify the meta data frame.

Another possibility is to set the version slot to Time(). This way, each time you build, the version number will automatically increase.

A third thing to consider is to create a small application which installs that meta data frame in its InstallScript. That way, you won't have to explicitly run the application in order to get the meta data soup updated. In addition, the application will build and download quickly. Once you've finished writing your meta data frame, you can copy it into your full application.

---

2. *Not synchronizing.* After you update your meta data soup, you must synchronize with NCK so that NCK obtains the new meta data frame.

What Does It Mean When My Directory Information Is Corrupted?

This message occurs when your meta data frame is incorrect. One reason might be that you are missing a required slot (like `BrowserExport`). Another reason might be that you do not have all the required slots for a particular operation (for example, you have an `editInfo` array, but no `EditImport` function). A final reason might be an inconsistency between slots. For instance, you might have an entry in the `displayInfo` array which is a symbol, but that symbol isn't present in the `defaultDefinitions` frame.

When Does NCK Do Validation in the Detail Window?

In the Detail window, NCK will verify the value typed when the user tabs to a new field or clicks in a new field. NCK will verify, for instance, for an `'integer` slot that the input value doesn't contain characters. However, a bug in NCK 2.0 means that verification doesn't actually occur if the user uses the arrows to switch to a different soup entry, or if the user clicks the New Entry button. Therefore, your `EditImport` must do validity checking on its own.

Newton Connection Checklist

This chapter has provided an introduction to adding Newton Connection support to your application. You should be able to take the information here and in three passes implement the various parts of Newton Connection in your code.

To be sure, there are many more capabilities in Newton Connection. You should, nevertheless find that by following the step-by-step guidelines presented here that you have all you need for the basic support.

Finally, last of all we provide a checklist version of all steps involved in implementing Newton Connection in an application.

The Checklist

1. Add a meta data frame to the Meta Data soup.
2. Create a meta data frame named "`nckMetaData`" as a slot in your application base template.

Displaying Soup Entries

3. Add the `soup`, `name`, and `version` slots to your `nckMetaData` frame.
4. Add the `DisplayInfo` array to your `nckMetaData` frame. This array deals with the characteristics of the Application Browser window. Fill in each frame in the `DisplayInfo` array.
5. Add the `BrowserExport` function to your `nckMetaData` frame. This converts a soup entry to displayable text.

User Editing of Entries

6. Add the `editInfo` array to your `nckMetaData` frame. This allows the user to edit soup entries.
7. Add the `EditImport` function to your `nckMetaData` frame. This converts a text frame back into a soup entry.
8. Add the `EditExport` function to your `nckMetaData` frame, if necessary. Remember, you only add this if you have different text frames for the Browser and Detail windows.

Importing and Exporting of Data from the Desktop

9. Add the `exportInfo` array to your `nckMetaData` frame. This controls what can be imported and exported to a text frame.
10. Add the `TabExport` function to your `nckMetaData` frame, if necessary. This returns either a `nil` value or a text frame for export. If this function is not present, `BrowserExport` will be used instead.
11. Add the `TabImport` function to your `nckMetaData` frame. This takes a text frame and converts it into a soup entry.
12. Add the `NativeExport` function to your `nckMetaData` frame. Newton Connection calls this function for each entry that is exported from the soup.
13. Add the `EntryValidation` function to your `nckMetaData` frame. This function is called for every `NativeImport` and its job is to check for the existence of required slots and their values. You do all your data and entry verification here.

14. Add the `NativeImport` function to your `nckMetaData` frame. This function takes a frame that comes from a native format and saves it as a soup entry.

Centralizing and Sharing Data

15. Centralize all your validation code. Have `TabImport` and `EditImport` call `EntryValidation`.
16. Create a `defaultDefinitions` frame to share information in your `displayInfo`, `editInfo`, and `exportInfo` arrays.

Chapter 6

Intelligent Assistance

To me, there is something superbly symbolic in the fact that an astronaut, sent up as ***assistant*** *to a series of computers, found that he worked more accurately and more* ***intelligently*** *than they.*

——Adlai E Stevenson

One of the features in the Newton Operating System that has received the most hype is Intelligent Assistance. Given all the limelight, it is certainly worth knowing what Newton Intelligent Assistance (IA) is and more importantly what it can and cannot do. To this end, we first discuss Intelligent Assistance as a general feature of a Newton application, the user interface required for its implementation, and then get down to the nuts and bolts of the implementation.

The User Interface of Intelligent Assistance

On the Newton, several different things are meant when we speak about Intelligent Assistance. Most importantly, there are two notably different types of intelligent assistance that your application should provide:

Intentional Assistance	The user actively requests assistance by tapping the assist button on the Newton.
Brownie Assistance	When users attempt to do something, your application anticipates their needs as much as possible.

Brownie Assistance

Let us look at the second type of assistance first. Brownie Assistance is similar to having a small elf that looks after you and does things for you unasked: *You leave out a dish of brownies and your dishes get washed, floors mopped, taxes filed, and so on.* On the Newton, Brownie Assistance occurs for a user when she or he is attempting to do a task and the Newton application assists with the task by trying to complete as much of the task as possible. For example as shown in FIGURE 6.1, if I am looking at Mary Gunther's name card and then tap "fax" in the action button. I should find that Mary's information is already provided for me in the fax slip.

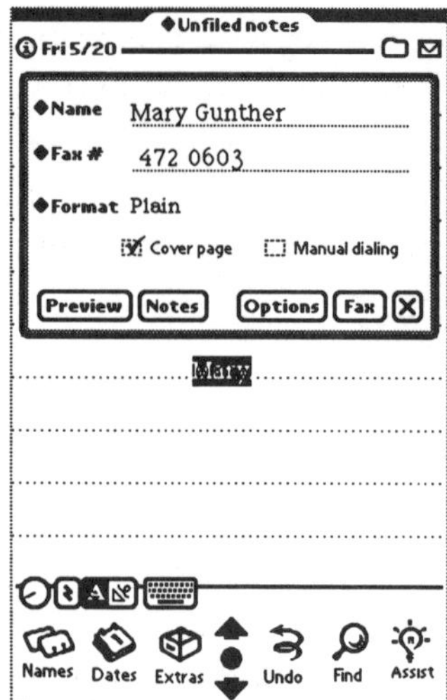

FIGURE 6.1 Brownie Assistance on the Newton.

Another type of Brownie assistance occurs when the program itself provides information in a more usable form. For example, you as the programmer could provide a `protoLabelInputLine` which is also a dynamic picker for input instead of requiring the use of an input line without a picker (see FIGURE 6.2 for a comparison). Using a picker, the user can customize the picker with certain common choices and rely on the input line portion for the rarer entries.

FIGURE 6.2 Adding a picker to an input line to enhance ease of use.

The trick in implementing Brownie assistance is a combination of relying on Newton's common data storage model and using programmatically sophisticated user input data tools. By relying on common data storage you have access to soups from other programs. For example, you can pull names from the names soup for faxes, dates from the calendar for notes about meetings, and information about your user from the preferences soup. Likewise, when you use sophisticated user input tools like modifiable pickers, your users find tasks far easier to complete.

While we will not be covering the implementations of these types of assistance in detail, it is well worth your time to add these aspects to your applications. A well-designed application that relies on others' soups and has minimized the need for a user to write is a far more useful tool.

Intentional Assistance

The second type of assistance, Intentional Assistance, occurs when the user taps the "Assist" button and wants assistance with a task (see FIGURE 6.3). When the user taps this button, a new slip is presented similar to the one in FIGURE 6.4. At this point, the user can do one of two things. She or he can write in a phrase similar to "Fax Bob", "Plan trip", or "Do expense" in the input line. Alternatively, the user can tap the Please picker and will be presented with types of words the Newton can help with (see FIGURE 6.4). In either case, when the "Do" button is tapped the Newton will attempt to interpret the phrase and assist with the task.

FIGURE 6.3 The Newton Intelligent Assistance button.

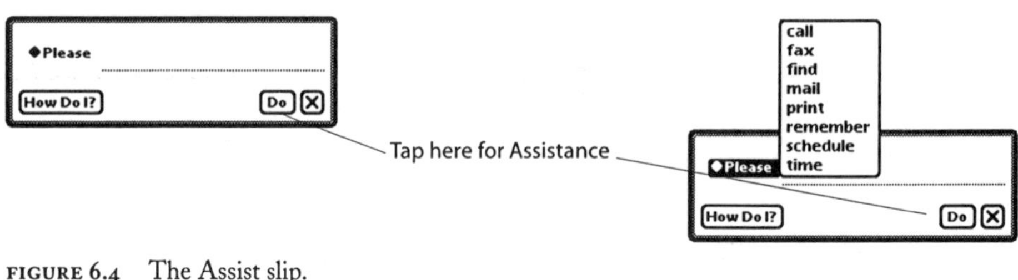

FIGURE 6.4 The Assist slip.

When you implement IA in your application, you will be adding your own main verb to this list in the IA picker. For example, a travel application might use the verb "fly," and a spreadsheet application might use "calculate" as verbs in this list (see FIGURE 6.5).

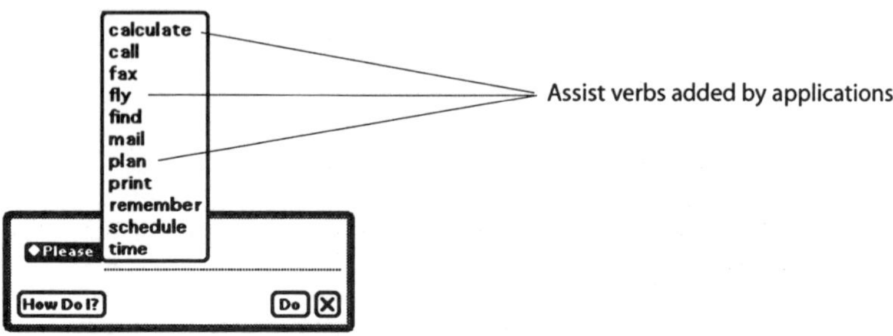

FIGURE 6.5 A Please picker with custom application verbs added.

If the user has used a verb in the list or written in another verb that the Newton recognized, then the operating system will open the appropriate application to the correct view and complete the task. For example, in our travel application, when the user taps "fly" the Newton would open the application to the view containing flight information and present a slip for the user to finish filling in the flight information (FIGURE 6.6).

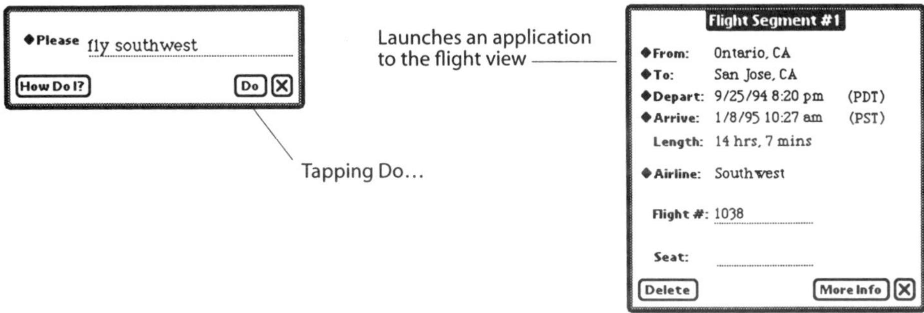

FIGURE 6.6 IA opening an application from a Please picker.

On the other hand, if the user enters a phrase in the IA slip that is unrecognized, then the Newton responds with the message shown in FIGURE 6.7.

FIGURE 6.7 When a phrase is not recognized by IA.

An Overview of Intelligent Assistance

Before we walk through the mechanics of implementing IA in a Newton application, it is worth defining a few key terms:

`ParseUtter`	The global function called on the text in the assist slip when the user taps Do.
Task template	A frame that an application registers with IA. It contains target templates, action templates, and a `PostParse` method.

Target template	A frame defining nouns that IA will recognize.
Action template	A frame defining verbs that IA will recognize.
`DoPostParse`	The method within an application base template which completes the IA task.
`PostParse`	The method within an application's task template which calls `DoPostParse`.
Result frame	A clone of the Task template augmented with additional slots. It is returned by `ParseUtter` after a phrase has been interpreted.

How It All Works

This is what happens when the user enters an IA request using the "Assist" button and/or the assist slip. The Newton OS takes the string the user has entered and tries to match its contents against the nouns and verbs registered by various applications and against special built-in types (the Newton knows how to recognize other words like dates, times, or names). Once each word (or phrase consisting of multiple words) has been categorized (it may match nothing), then IA uses the matched nouns and verbs to determine which application's task template most closely matches the string. The application that wins the contest gets sent the `PostParse` message by IA. FIGURE 6.8 shows a graphic representation of how this entire process works.

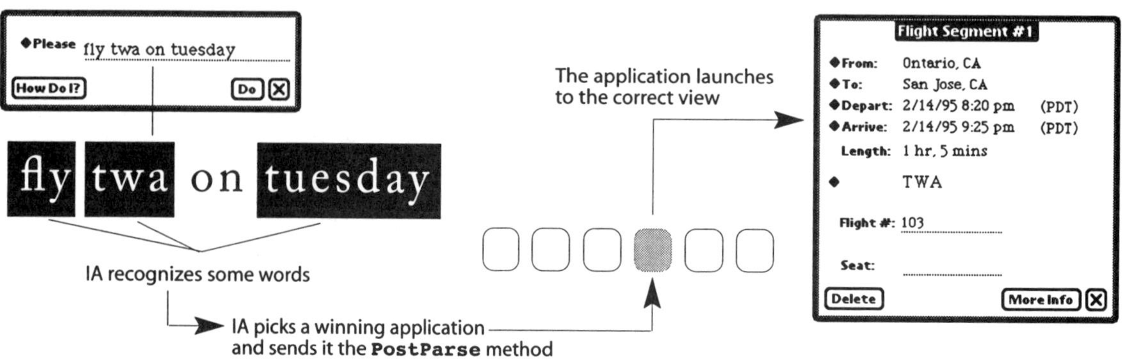

FIGURE 6.8 IA parsing string and sending it off to an application.

What the Application Has to Do for IA

The application needs to handle several different tasks:

- Register itself with IA so that it will be considered for phrase matching.
- Select the words (both verbs and nouns) it wants to recognize.
- Handle the mechanics of closing the IA slip and opening the application.
- Fill in the appropriate information in the relevant view.

The first two tasks are handled in your application's `InstallScript` and in the task template. An application's `DoPostParse` routine is responsible for these last two items. These items, in turn, can be broken down into further discrete tasks:

- Closing the IA slip.
- Opening the application.
- Creating and displaying a confirmation slip for the user.
- Filling in all of the slip's relevant information based on the words in the assist slip.
- Closing the application if the user cancels the confirmation slip (if the application wasn't already open).
- Implementing IA.

Of course, you also have to clean up after yourself when you are through. These tasks, as usual, happen in the `RemoveScript` of your application.

Implementing IA —The Checklist

It is easiest to add support for IA in stages, so that is what we will do. First, we will get a minimal action and task template in place. At this point, we can then see how IA hands a string off to our application. Once that material is tested, we will refine the IA templates, adding a target template and a real `DoPostParse` routine. Here is a basic checklist for IA's steps.

Get Minimal IA Working

1. Create an action template.
2. Create a task template.
3. Create a `DoPostParse` routine.
4. Install the task template.
5. Remove the task template.

Extend the Task Template

6. Create the target template.
7. Choose built-in targets.
8. Add targets to the task template.
9. Choose an appropriate primary action.

Further Refinement

10. Extend the `DoPostParse` method to extract necessary information.
11. Add a confirmation slip.

Getting Minimal IA Working

The first thing to do is to get enough of IA implemented in your application so that you can test it. After you have these basic components in place, then you will build up a more complete set of IA functions.

1. Create an Action Template

An action template is a frame that describes an action. Typically you will place this template in your project data file. Action templates contain these three slots:

`value`	This slot is not currently used by IA. By convention it contains a string or symbol describing the action template.

`isa` This should contain the symbol `'dyna_user_action`.

`lexicon` This contains an array of strings. Because any one of the strings will match this action template, make them synonyms. The first word in the lexicon is also used in the please picker.

Here is a sample action template:

```
{
   value: "my flying actions",
   isa: 'dyna_user_action,
   lexicon: ["fly", "make reservation"]
}
```

With this action template, IA would match either the string "fly" or the string "make reservation". The first word in the lexicon, "fly", will also appear in the Please picker as shown in FIGURE 6.9.

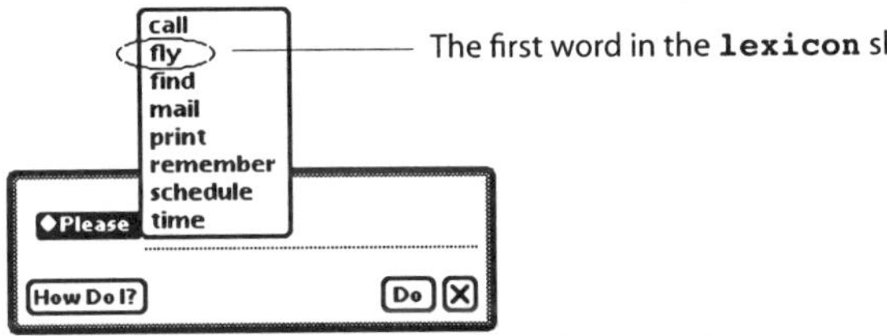

FIGURE 6.9 Adding a word to the Please picker.

Action templates are commonly defined as constants, thus they can appear in the Project Data file as:

```
DefConst('kFlyAction,
   {
      value: "my flying actions",
      isa: 'dyna_user_action,
      lexicon: ["fly", "make reservation"]
   }
);
```

2. Create a Task Template

A task template is a frame that describes the structure of an application's IA phrases and a method which gets executed when an IA phrase is matched. This is normally stored as a slot in your application base template. For the moment, use the task template described next as a recipe; a later section will describe the unfamiliar slots in more detail. The slots you should know about at this point, however, are the following:

`value` — This slot is not currently used by IA. By convention it contains a string or symbol describing the task template.

`primary_act` — This contains the action template you defined in the last step.

`signature` — For now, it contains an array with one element, the action template.

Here is a complete task template:

```
{
   value: "my flying template",
   isa: 'task_template,
   primary_act: kFlyAction,
   preconditions: ['action],
   signature: [kFlyAction],
   postParse: func()
      GetRoot().(kAppSymbol):DoPostParse(self),
}
```

As we said earlier, unlike your action template, your task template is normally defined in a slot in your application base template (see FIGURE 6.10). A good name for the slot is `taskTemplate`.

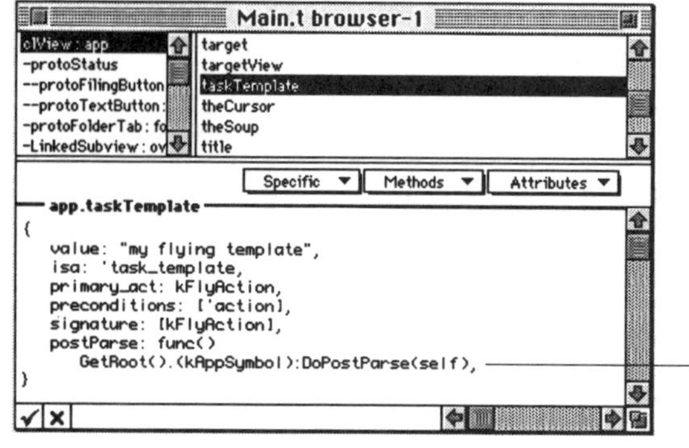

FIGURE 6.10 A `taskTemplate` in an application.

3. *Create a* `DoPostParse` *Routine*

When the user invokes Intelligent Assistance with an action (as specified in your task template), the `DoPostParse` message will be sent to your application base view (via the `PostParse` method of the task template). For now, all your `DoPostParse` method does is beep, simply showing that it has executed. You create your `DoPostParse` method as a slot in your application base template as shown in FIGURE 6.11.

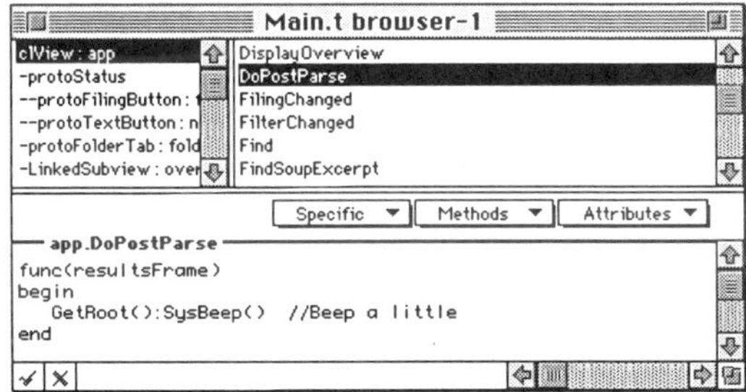

FIGURE 6.11 A `DoPostParse` method in an application.

4. *Install the Task Template*

Your must register your task template in order for IA to be aware of it. This is normally done in the application's `InstallScript` since IA can be invoked even when the application is closed.

```
InstallScript := func(partFrame)
begin
   … // other stuff
   local theTemplate :=
      partFrame.theForm.taskTemplate;
   partFrame.IATask :=
      RegTaskTemplate(theTemplate)
end
```

`RegTaskTemplate` takes a task template, does a `TotalClone` on it and then returns that cloned frame (or `nil` if registration was unsuccessful). Because unregistration requires the cloned frame, it must be stored somewhere until `RemoveScript` time. The only place available is in the `partFrame` itself.

5. Remove Task Template

Now, you have to remove what you installed:

```
RemoveScript := func(partFrame)
begin
   ...
   if partFrame.IATask then
      UnRegTaskTemplate(partFrame.IATask)
end
```

`UnRegTaskTemplate` takes the return result of `RegTaskTemplate`. A common bug when implementing IA is forgetting to unregister. Thus, when you modify the task template and reinstall, you have both the old task template and the new one. In this case, IA won't be able to decide which one to use since they'll both match the input.

Test

At this point, you can test IA. Here are some things to do:

1. Open the Assist slip and check that the first string from the lexicon is in the Please picker.
2. Try IA with each of the verbs in the lexicon. Each should cause a beep. If you get tired of writing in each verb in the assist slip, you can also use the `ParseUtter` global function from the Inspector (see FIGURE 6.12). Notice that when you do this, the Newton will open the Assist slip and fill it in with the string you passed to `ParseUtter`. Last of all it will also beep at you.
3. Remove your application from the Newton and make sure that the Please picker no longer contains your verb.

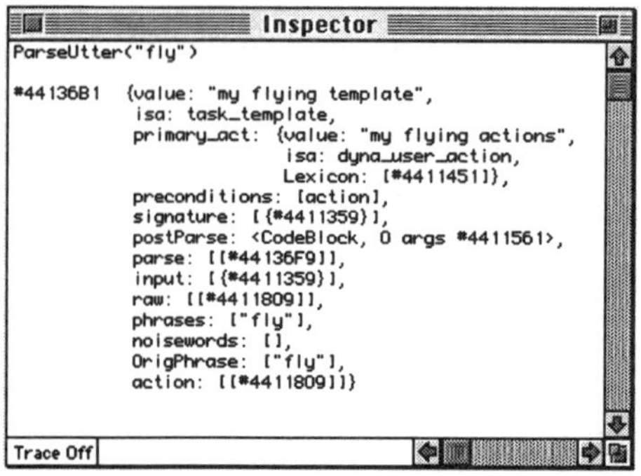

```
Inspector
ParseUtter("fly")

#44136B1  {value: "my flying template",
           isa: task_template,
           primary_act: {value: "my flying actions",
                         isa: dyna_user_action,
                         Lexicon: [#4411451]},
           preconditions: [action],
           signature: [{#4411359}],
           postParse: <CodeBlock, 0 args #4411561>,
           parse: [[#44136F9]],
           input: [{#4411359}],
           raw: [[#4411809]],
           phrases: ["fly"],
           noisewords: [],
           OrigPhrase: ["fly"],
           action: [[#4411809]]}
Trace Off
```

FIGURE 6.12 Doing a `ParseUtter` in the Inspector.

Extending the Task Template

Now you are in a position to extend the task template to match nouns as well as verbs. Once you have added nouns, IA will be capable of recognizing all the sentences you might be interested in matching.

6. Create the Target Template

A target template is a frame that specifies a noun (or set of nouns) to match. It is found as a constant in your project data file and it contains three slots:

`value`	This slot is not currently used by IA. By convention it contains a string or symbol describing the target template.
`isa`	This should contain the symbol `'dyna_user_obj`.
`lexicon`	This contains an array of strings made up of noun words. Any one of the nouns in the array matches this target template.

Here is a sample target template that will match any of the three airline strings in the lexicon:

```
{
    value: "my airlines",
    isa:   'dyna_user_obj,
    lexicon: ["Southwest", "American", "US Air"]
}
```

Target templates are usually defined as constants, thus they can appear in the Project Data file as shown in FIGURE 6.13.

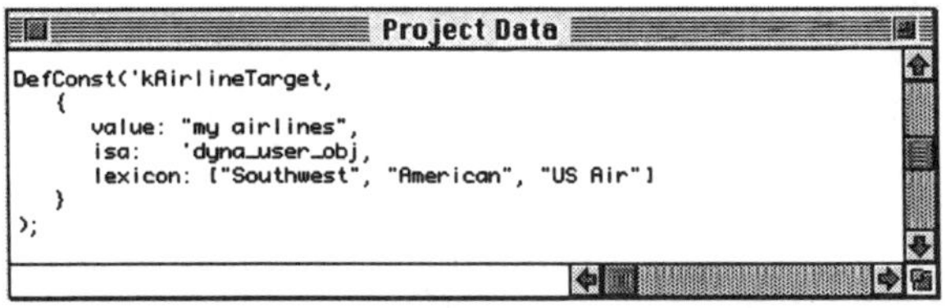

```
Project Data
DefConst('kAirlineTarget,
    {
        value: "my airlines",
        isa:   'dyna_user_obj,
        lexicon: ["Southwest", "American", "US Air"]
    }
);
```

FIGURE 6.13 Defining a constant target template in the Project Data file.

7. Choosing Built-In Targets

Applications can add new nouns to those recognized by IA only by specifying a list of strings. Intelligent Assistance, however, also has some built-in types that make it more flexible than this. Four of these types match certain patterns, and two, in particular, are recognized using the "Names" soup.

These built-in types are also referenced using symbols:

`'date`	This matches a date. Examples are "12/1/94" and "Monday".
`'time`	This matches a time in 12- or 24-hour format. The minutes are optional. Examples are "3:15", "3:15 AM", "15:15", "5", and "23".
`'phone`	This matches a phone number. Examples are "555-1212", "5551212", "(408) 555-1212", "4085551212", "+44 (0)81 974 5523", and "0130 815 199".
`'parsed_number`	This matches a floating point number or an integer that isn't matched as a `'date` (0 to 23) or as a `'phone` (7 or 10 digits).
`'person`	This matches a person found in the "Names" soup. Examples are "Royce", "Walthrop", and "Royce Walthrop".
`'company`	This matches a company found in the Names soup. Examples are "Newton", "Cafe", "Newton Cafe", and "Conglomerated Credit".

Unfortunately, there is no way to extend these types. Thus, you cannot have IA look in other soups besides the Names soup.

How IA Matches

Each application stores a description of the structure of the strings for which they are looking. Here are two examples an application might use:

```
call_verb phone home_noun person work_noun company
pay_verb person company parsed_number date
```

A complete match is not necessary. For instance, using the first sequence above, "call Bob home" will match, as will "call Bob work". After all, it is unlikely that a user will have both "work" and "home" in one sentence. Other matches would include "call 555-1212" or "call Newton Cafe".

The structure that an application uses is represented by the elements in the `signature` slot in the task template. For example, the preceding example might have a `signature` slot matching it as shown in FIGURE 6.14.

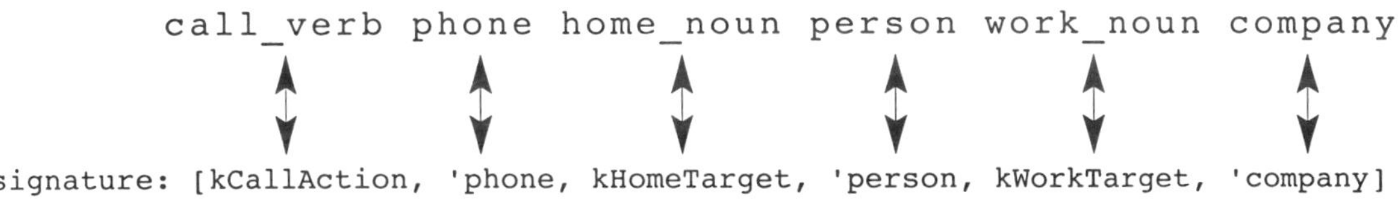

FIGURE 6.14 Relationship between the structure of strings and a `signature` slot.

IA matches phrases using a bottom-up approach. First, it uses the lexicons from all installed action and target templates, along with the built-in targets to identify the phrases in an input string. Only after IA has identified all the phrases (which will be placed in a result frame), does it begin to look at application task templates to find the one that matches best.

If an action is part of the string, it matches the task template with that action. If the user did not provide an action, IA determines whether a set of targets uniquely identifies a task template. If no unique identification occurs, then IA creates a list of those task templates that do match targets and provides a reduced Please picker. For example, if we simply type "southwest" in the assist slip we get the slip shown in FIGURE 6.15 prompting for more information.

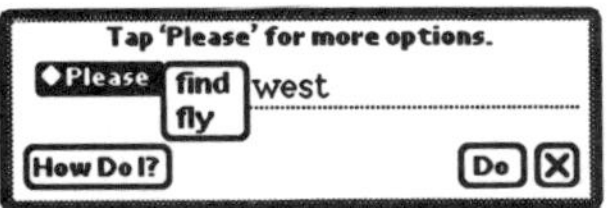

FIGURE 6.15 A reduced Please picker.

8. Add Targets to the Task Template

Before you can understand how to add these targets, two more slots in the task template need definition:

`signature`	This slot in the task template is an array containing an action template, target templates, and symbols for built-in targets. The order of these array entries does not matter, since IA ignores the order of user input.
`preconditions`	This slot in the task template is an array parallel to the signature slot. Each entry in this array contains symbols of your own choosing. Thus, the number of entries in this array should be equal to the number of entries in the `signature` array and the symbolic names should correspond to their counterparts in the `signature` array.

Having defined these two new slots, you will now replace your old task template with a much fuller one:

new code to handle targets

```
{
   value: "my flying template",
   isa: 'task_template,
   primary_act: kFlyAction,
   preconditions: ['action, 'airline, 'when],
   signature: [kFlyAction, kAirlineTarget, 'date],
   postParse: func()
      GetRoot().(kAppSymbol):DoPostParse(self),
}
```

9. Choose an Appropriate Primary Action

Remember that the value in the `primary_action` slot of the task template will determine what shows up in IA's Please picker. It is possible to make this an action template different from the action template in the `signature` array, but it is normally the same.

Test

At this point, you can test to determine if IA is correctly matching your task template. Here are some tests worth trying:

1. Try parsing a unique target, without any accompanying action (e.g., "Southwest").
2. Try parsing a non-unique target (e.g., "Friday") and verify that your action appears in the reduced Please picker.
3. Try a combination of targets (e.g., "Southwest Friday").

Further Refinement

Now that you have complete task, action, and target templates, it is time to handle the remaining tasks involved in supporting IA. Once your application has been chosen as the best match, you need to actually do something with the phrase the user entered in the Please picker. These activities are all handled in the `DoPostParse` method that is called by the `PostParse` method in the task template.

The Result Frame

When a task template is matched, IA creates a result frame by cloning that task template and adding some extra slots to it. This result frame is passed as a parameter to `DoPostParse`.

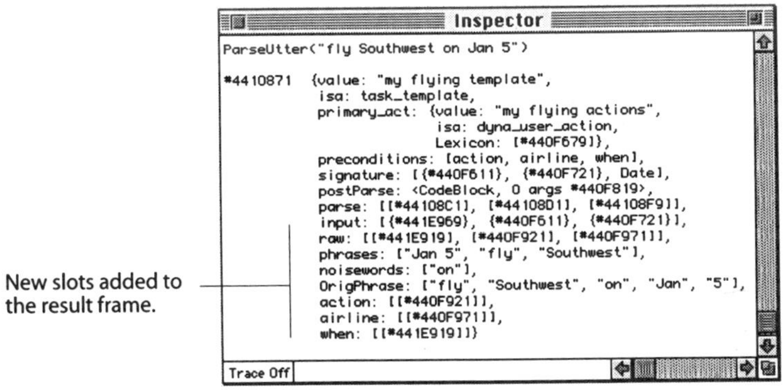

```
Inspector
ParseUtter("fly Southwest on Jan 5")

#4410871  {value: "my flying template",
           isa: task_template,
           primary_act: {value: "my flying actions",
                         isa: dyna_user_action,
                         Lexicon: [#440F679]},
           preconditions: [action, airline, when],
           signature: [{#440F611}, {#440F721}, Date],
           postParse: <CodeBlock, 0 args #440F819>,
           parse: [[#44108C1], [#44108D1], [#44108F9]],
           input: [{#441E969}, {#440F611}, {#440F721}],
           raw: [[#441E919], [#440F921], [#440F971]],
           phrases: ["Jan 5", "fly", "Southwest"],
           noisewords: ["on"],
           OrigPhrase: ["fly", "Southwest", "on", "Jan", "5"],
           action: [[#440F921]],
           airline: [[#440F971]],
           when: [[#441E919]]}
Trace Off
```

New slots added to the result frame.

FIGURE 6.16 Doing a `ParseUtter` in the Inspector.

You can see this result frame by doing a `ParseUtter` from the Inspector. For example, executing `ParseUtter("fly Southwest on Jan 5")` returns the result frame shown in FIGURE 6.16.

The added slots are fairly easy to understand, once you know what their contents are describing.

`phrases`	`["fly", "Southwest", "Jan 5"]`. This contains the phrases/words that were actually matched.
`input`	`[kFlyAction, kAirlineTarget, whenFrame]`. This contains an array of matches. This array is parallel to the `phrases` array, but the contents are action or target templates. For built-in targets, the entry is a frame containing a `value` slot.
`parse`	`[[kFlyAction], [kAirlineTarget], [whenFrame]]`. This contains the same elements as the `input` slot, but each element in the array is itself an array.
`raw`	This contains an array of arrays. Each array contains information on each word that IA tried to match in the string.
`noisewords`	`["on"]`. This contains those phrases which were not matched.
`origphrase`	`["fly","Southwest","on","Jan 5"]`. This contains the original string, broken up into an array of string/words. Since your application receives the original input in the `DoPostParse` routine, you can add further intelligence to how you interpret that phrase (e.g., look in other soups, add sophisticated natural language processing).
`action`	`[[kFlyAction]]`. This is the action template that was matched by IA.
`airline`	`[[kAirlineTarget]]`. This is the target template that was matched by IA.

`when` `[[whenFrame]]`. This contains the built-in type that a word was matched to, in this case a date. The `whenFrame` has a `value` slot containing the matched string.

Additionally, for every IA match from the `signature` array, a slot is created whose symbol comes from the corresponding entry in the `preconditions` array. This new slot contains an action template, target template, or frame with a `value` slot (just like the `input` array). This frame is stored in an array within an array.

For built-in targets other than `'person` or `'company`, the `value` slot contains the matching phrase. Thus, it is easy to obtain the date (or time, etc.) as a string. For persons and companies, or action and target templates, however, there is no `value` slot. Looking at the `airline` slot in the results frame, for instance, doesn't tell you which airline was provided, just that one was. Thus, to deal with person, company, or your own action and target templates, you must iterate through the input array trying to find the person, company, action or target template. Once you find it, you can then obtain the actual phrase from the corresponding entry in the `phrases` array.

The IsA Function

The Newton has an `IsA` function that determines whether a template is of a particular type. You can use this function to determine whether a template in the input array is a built-in target. The syntax is:

```
IsA(template, typeSymbol)
```

This function returns a boolean; here are examples of its usage:

```
IsA(resultsFrame.input[1], 'date)
IsA(resultsFrame.when, 'date)
```

*10. Extend the **DoPostParse** Method to Extract Necessary Information*

Now that you see some of the code you might add to your `DoPostParse`, let us look at a sample `DoPostParse` that reads information from a `resultsFrame`. This method is an updated version of your earlier `DoPostParse` that just beeped. Remember this is still a slot in your application base template:

```
DoPostParse : func(resultsFrame)
begin
   local date;
   local airline := "no airline specified";

   // try to read phrases associated with targets
   foreach i, template in resultsFrame.input do
   begin
      if resultsFrame.airline exists and
         template = resultsFrame.airline[0][0] then
         airline := resultsFrame.phrases[i]
      else if IsA(template, 'date) then
         date := StringToDate(template.value);
   end;
   Print("fly");
   Print(airline);
   if date then
      Print(ShortDateStr(date, 0))
   else
      print("no date specified");
end;
```

Although the date could have been read directly from the `when` slot in the `resultsFrame`, it was necessary to loop through each element of the input array anyway. Because of this, it seems easiest to have just one way to read the targets.

11. Add a Confirmation Slip

Your confirmation slip (which will be a template in your application), needs to display the information read by the `DoPostParse` method and provide confirmation and cancel buttons. If the user confirms, you must carry out the task. It may also be necessary to open the application (unless it is already open), and to close the assist slip. Here is some additional code to add to your `DoPostParse` routine which will open the application if necessary, send a message to the confirmation slip, and close the assistant:

```
DoPostParse: func(resultsFrame)
begin
   local wasOpen := call kViewIsOpenFunc
      with (self);
   if not wasOpen then
      self:Open();
```

```
        // get information from results frame
        confirmSlip:DoIt(infoFromResultsFrame, wasOpen);
        GetRoot().assistant:Close();
    end;
```

The `DoIt` method of the `confirmSlip` opens the `confirmSlip`, sets a slot to keep track of whether the application was initially open, and determines whether the confirm slip ends up confirmed or canceled:

```
confirmSlip.DoIt: func(info, wasOpen)
begin
    self.wasConfirmed := nil;
    self.wasOpen := wasOpen;
    :Open();
    // set values of children based on info
end;
```

The `viewQuitScript` of the `confirmSlip` must close the application if the application was initially closed, or if the user cancels the action. If the user confirms, the script must carry out the necessary task:

```
confirmSlip.viewQuitScript: func()
begin
    if not wasConfirmed then begin
        if not wasOpen then
            GetRoot().(kAppSymbol):Close()
    end else begin
        // read info from children and carry out task
    end;
end;
```

The `confirmSlip` should contain a confirm button that carries out the task specified in the `confirmSlip`. It then closes the `confirmSlip` after setting the inherited slot `wasConfirmed` to `true`:

```
confirmButton.buttonClickScript: func()
begin
    wasConfirmed := true;
    base:Close();// close confirm slip
end;
```

Common Questions about IA

When I call `RegTaskTemplate`, why does the Newton run out of memory?

If the `postParse` routine is created at runtime (rather than at compile time), the function object has a non-nil lexical environment. When the task template is `TotalCloned`, it tries to clone the lexical environment, which has references to the ROM.

To solve this problem, make sure your `postParse` routine is created at compile time. An easy way to do this is to define your `postParse` within an evaluate slot in NTK. In this chapter, we've recommended that your task template be defined in an evaluate slot in your application base template, and that your `postParse` routine is defined within that frame.

For more information on this problem, see "Function Objects with No Lexical Environment or Message Context" on page 238.

Why don't new versions of my task template do anything?

If your `RemoveScript` doesn't unregister the task template with `UnRegTaskTemplate`, there will be two task templates installed: the old one (which wasn't unregistered) and the new one (which was just registered). When IA looks through the registered task templates to find one that matches, both will match; because IA can't decide which one matches best, it won't match either.

The fix is to make sure your `RemoveScript` is executing `UnRegTaskTemplate`. You'll have to reset your Newton to remove the old task template.

Intelligent Assistance Checklist

This chapter discussed the user interface of IA in a typical application. Next we gave you a brief overview of the important terms in IA and how it works in general. We then showed you the easiest way to implement IA, which is in stages. We did this in three passes. First, we got a minimal IA working and then we tested it. Next, we refined it by adding a target template and some built-in targets as well.

Last of all, we added a fully fleshed out version with a working `DoPostParse` method and a confirmation slip.

The Checklist

Here is the IA checklist for easy reference for implementing IA in your application.

Getting Minimal IA Working

1. Create an action template.
2. Create a task template.
3. Create a `DoPostParse` routine.
4. Install the task template.
5. Remove the task template.

Extending the Task Template

6. Create the target template.
7. Choose built-in targets.
8. Add targets to the task template.
9. Choose an appropriate primary action.

Further Refinement

10. Extend the `DoPostParse` method to extract necessary information.
11. Add a confirmation slip.

Chapter 7

Advanced Debugging

Look for a tough wedge for a tough log.

—*Publilius Syrus*

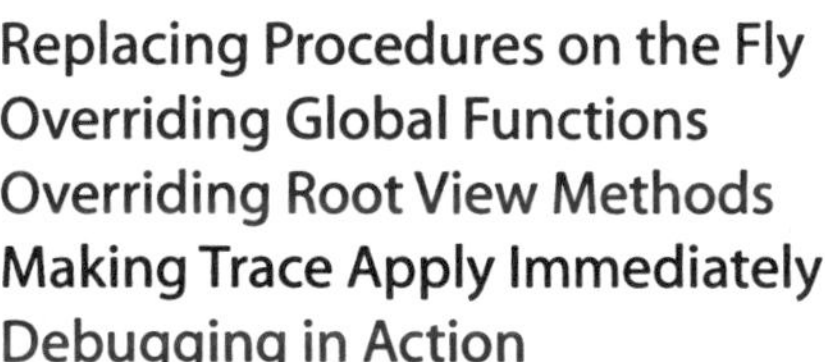

This chapter discusses several little-known debugging techniques which should be included in the arsenal of any Newton programmer.

Replacing Procedures on the Fly

You are probably familiar with the common debug cycle on the Newton:

- You compile and download.
- You run the application looking for an error.
- You then debug the error using the Inspector or ViewFrame (or a combination of both).

When a bug is found in a method, the straightforward approach is to fix it, and then start the whole process over again: compile, download, run the application, and finally test to make sure the fix is correct. The good news about this approach is that this compile/download/run cycle is much quicker than on many platforms (e.g., Macintosh Programmer's Workshop). The bad news, as any programmer will tell you, is that the cycle can never be quick enough.

A speedier approach is actually available. You can use the Inspector to replace the method that has the error while still running the application. For example, let's say you have a template named "myButton" which has a method named `Method1`:

```
func(a, b, c)
begin
   local x := 1;
   local y := 1;
   local z := 1;

   if a then begin
      y := y * 2;
      z := :Method2(a, b, c, x, y);
   end;
   return x * y * z;
end
```

Imagine that you have a problem with `Method1`, and you would like to print its parameters. You can do this by executing the following in the Inspector:

```
Debug("myButton").Method1 := func(a, b, c)
begin
   Print(a);
   Print(b);
   Print(c);
   inherited:Method1(a, b, c);
end
```

When you execute this method, it is compiled from the Inspector and stored in the NewtonScript heap of the Newton. The `Method1` slot is now part of the myButton view, and now whenever the Method1 message is sent to myButton, the new overridden version will execute instead. Since part of the new method is to print out the parameters, it will do so and then call the old version.

Notice that you could even completely replace Method1:

```
Debug("myButton").Method1 := func(a, b, c)
begin
   local x := 3;
   local y := 3;
   local z := 5;

   if a then begin
      BreakLoop();
      y := y * 2;
      z := :Method2(a, b, c, x, y);
   end;
   return x * y * z;
end;
```

The one requirement for this technique to work is that the view to which you are adding a method must exist. Thus, you cannot add code to a closed view (unless, of course, it is declared to a view that is still open).

You can also use this approach to add code incrementally to a view—one method at a time—thus making debugging much simpler. The only tricky part is remembering to copy the finished product back into your project once you're satisfied that it is correct. The best code in the world won't help your application if it is left sitting in the Inspector window.

Overriding Global Functions

There are times when you will find it useful to replace a global function. For example, you may want to know when a particular function is called, or what its parameters are. As an example of this, let's say you want to find out when (or even whether) a particular application calls BroadcastSoupChange and what soup name it passes when it does.

Here is how you perform this neat little trick. First, you create a global function that references the old global function (usually, you'll want to augment the behavior of a function, not completely replace it). Remember that all global functions are stored in the functions frame, which is itself a global variable. From the Inspector, execute this code:

```
functions.OldBroadcastSoupChange :=
   functions.BroadcastSoupChange;
```

Second, create a new function that takes the same number of parameters and that calls the old function:

```
functions.BroadcastSoupChange := func(s)
begin
   Print("BroadcastSoupChange called with" && s);
   OldBroadcastSoupChange(s);
end;
```

Overriding Root View Methods

Another technique that can be very useful is being able to override root view methods for debugging purposes. Methods like `Notify`, `FindSoupExcerpt`, `Open`, and `Close` are all root view methods that at times might be useful to override for debugging purposes.

The technique is very similar to the one for replacing methods in your own application. Simply add a slot to the root view that overrides the method found in the root template. All you have to do is execute the following code from the Inspector:

```
GetRoot().MethodToOverride := func(params)
begin
   whatever you want to do
   inherited:MethodToOverride(params);
end;
```

Making `trace` Apply Immediately

The `trace` global variable controls tracing and it would probably be very useful if you could precisely control when it turns on and off. And as many programmers have learned, simply setting it to `true` (or `'functions`) for your whole program only causes enormous amounts of information to be printed to the Inspector. Given this bulky bitstream, many programmers would like to be able to set the variable to `true` only around a particular section of code and then be able to examine just that output.

While you would like it to be, it's not as easy as simply surrounding your code with the following:

```
trace := true;
code to trace
trace := nil;
```

The reason this doesn't work is that the NewtonScript interpreter only occasionally looks at the value of the `trace` variable. Once you think through the logic of it, you will realize that this is actually a good thing; if the interpreter checked the value on every instruction, the speed of the interpreter would be severely slowed for all Newtons. Since slowing the Newton down is not a choice that anyone wants, the NewtonScript interpreter currently only checks the value of this variable when a function is called (function calls are expensive enough that one check of a global variable has no substantial impact). Of course, this does leave you with the problem of how to trace around just a little bit of code.

Happily there is a nice solution to this problem. To cause tracing to happen around a particular section of code, all you have to do is make sure to call a function immediately after setting the `trace` variable:

```
trace := true;
Apply(func() nil, []);
code to trace
trace := nil;
Apply(func() nil, []);
```

Simply by executing an empty function using `Apply`, we've caused the interpreter to check the `trace` variable and start tracing before the code of interest is executed. Similarly, as soon as the code of interest completes, we turn tracing off.

There is one small proviso that accompanies this trick and that is that you cannot depend on it to work in the future versions. The particular manner in which the `trace` global variable is checked may well be different some day.

Debugging in Action

In this section, we provide some demonstrations of debugging Newton problems. The techniques from this chapter, along with other debugging techniques, will be used to solve the problems.

A Problem with Naming

Problem Description

We've got an application in a floating window that displays the user's name. If the user edits the name in Prefs, the floating window doesn't update the name (even though it provides a `soupChanged` method and registers in the `soupNotify` array). FIGURE 7.1 shows the problem.

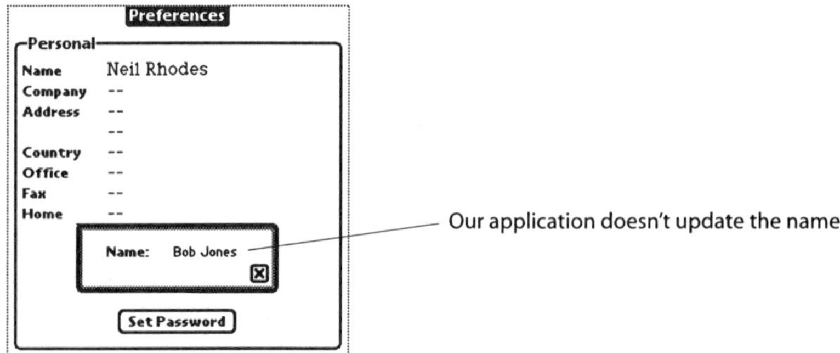

FIGURE 7.1 Application showing user name after changing the name in Preferences.

Tracking Down the Problem

First we open our application and check the `soupNotify` array from the Inspector to make sure we're really there, and that we're registered for the proper soup:

```
soupNotify
#4411101  ["Notes", paperRoll, "Names", cardFile,
   "Calendar", calendar, all, finddrawer,
   "System", preferenceRoll,
   "System", DisplayName:Calliope]
```

Our application symbol

Our application is in the array with the proper soup name (the `userConfiguration` is stored as an entry in the System soup). Our application symbol seems to be correct.

Maybe the problem is in our `soupChanged` method. Let's add a Print statement to see whether that method is being reached. Instead of recompiling, let's modify the method on the fly from the Inspector:

```
printDepth:=0   // so as not to print the root view
#0        0
```

```
theFloater := Debug("floater")

#44113F9   {_Parent: {#4410059},
            _proto: {#600230CD},
            viewCObject: 0x110A3F4,
            name: {#44110E9},
            viewclipper: 17867778,
            base: <1>,
            viewFlags: 577,
            viewBounds: {#4411B41}}

theFloater.soupChanged := func(soupName) begin
   Print("Got to soupChanged");
   inherited:soupChanged(soupName);
end
#441A389   <CodeBlock, 1 args #441A389>
```

Now, we'll edit the name in the Preferences application. After changing the name, nothing prints to the Inspector. After closing the name expando, nothing prints. Now we reason: *Maybe the Preferences application has a watchdog timer, and only writes to the soup and calls* `BroadcastSoupChange` *after a time delay.* Okay let us test that… After waiting 5 minutes, nothing prints, so we can conclude that there is no watchdog timer. Now we imagine: *Maybe the application only writes to the soup and calls* `BroadcastSoupChange` *when the application is closed.* We test that and still find that nothing prints after closing the application.

What now? *Time to check to make sure the application symbol in the Project Settings dialog matches what's in the* `soupNotify` *array.* It seems to.

We'll override `BroadcastSoupChange` to see when (if ever) it is getting called. From the Inspector:

```
functions.OldBroadcastSoupChange :=
   functions.BroadcastSoupChange;
functions.BroadcastSoupChange := func(s)
begin
   Print("BroadcastSoupChange(" & s & ")");
   OldBroadcastSoupChange(s);
end;
#4416661   <CodeBlock, 1 args #4416661>
```

After doing the same editing in the Preferences application (changing the name, closing the name expando, waiting five minutes, and closing the application), we find that the "got to BroadcastSoupChange" message never prints in the Inspector. Now we reason: *Maybe we didn't override* `BroadcastSoupChange` *correctly.* We open the Names application and make a change, and still no message prints. We make a note in the Notepad, no message prints. We add an appointment to the Calendar, no message prints.

Out of desperation, we insert a storage card. Lo and behold, the following prints in the Inspector:

```
"BroadcastSoupChange(Calendar)"
"BroadcastSoupChange(Repeat Meetings)"
"BroadcastSoupChange(Calendar Notes)"
"BroadcastSoupChange(Repeat Notes)"
"BroadcastSoupChange(To do)"
"BroadcastSoupChange(Names)"
"BroadcastSoupChange(Notes)"
"BroadcastSoupChange(Names)"
"BroadcastSoupChange(Personal Groups)"
```

From this information we can conclude that the built-in applications Names, Dates, Notepad, and Prefs don't call `BroadcastSoupChange` when a change is made. Thus, we'll have to rethink our application. One solution to the problem would be to make it an auto-close application so that when the Prefs application opens, our application closes. In that way, the user won't see both at once.

To restore the Newton to normal, and get rid of all the new code we had added to find the problem, we can do either of the following:

- Reset
- Execute this code in the Inspector:

```
RemoveSlot(GetRoot().|DisplayName:Calliope|,
   'soupChanged)
#440FCB1 {_Parent: {#4415E21},
          _proto: {#600248DD},
          viewCObject: 0x110A910,
          name: {#440F989},
          viewclipper: 17862822,
          base: <1>,
          viewFlags: 577,
          viewBounds: {#440FD21}}

functions.BroadcastSoupChange :=
   functions.OldBroadcastSoupChange
#30E439   <CodeBlock, 1 args #30E439>
RemoveSlot(functions, 'OldBroadcastSoupChange);
   nil // so whole functions frame isn't printed
#2        NIL
```

We should still keep the `soupChanged` method in the application, however. Some other application may change the name and call `BroadcastSoupChange`, a future version of Prefs may call `BroadcastSoupChange`.

It follows then that if we're going to keep the method, we'd better test it. From the Inspector we can do that with this code:

```
BroadcastSoupChange(ROM_SystemSoupName);
```

Indeed, our application redraws the new user name.

Another Problem

Here is another problem to try. It will help build your skills at dissecting and manipulating views.

Problem Description

We've just started writing an application that will be displaying rows of information. The application base view resizes itself to fit the screen size; likewise `rowHolder` (a subview) resizes itself based on the size of the application. The number of rows is determined dynamically based on the height of the `rowHolder`.

Our current problem is that tapping on the application to open it causes the Newton to reset.

FIGURE 7.2 shows information from the row proto. FIGURE 7.3 shows the main layout, along with the slots which were set. After both of these figures we give you a text version of the whole program to look at for problems.

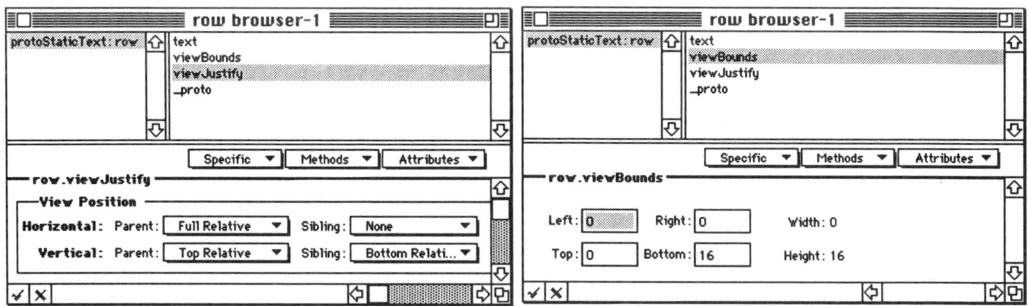

FIGURE 7.2 Settings of the row proto.

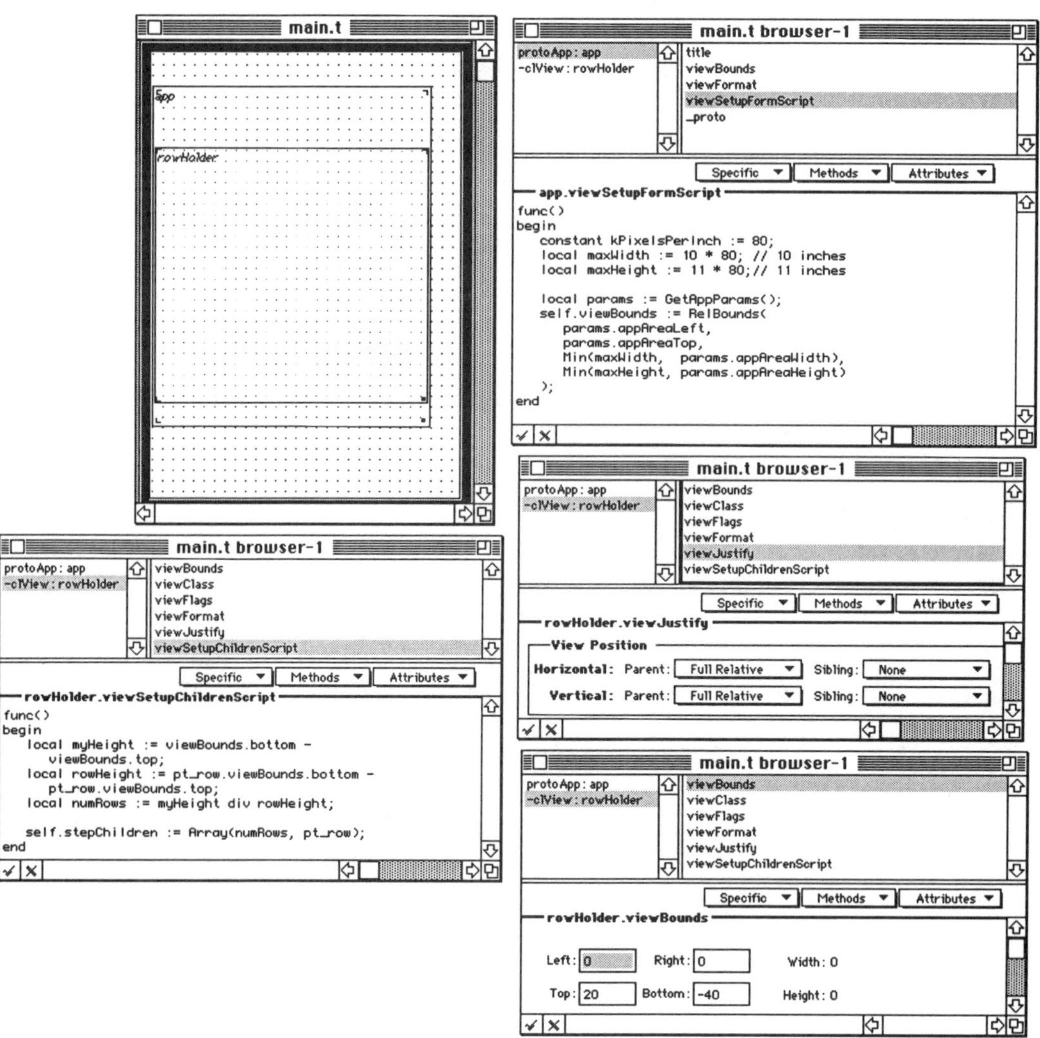

FIGURE 7.3 Settings of the main layout.

Here is a text dump of the entire project:

```
// ---- File main.t ----

app := {
   title: "Application",
```

```
    viewBounds: {left: 4, top: 26, right: 216, bottom: 278},
    viewFormat: 83951953,

    viewSetupFormScript:
    func()
    begin
        constant kPixelsPerInch := 80;
        local maxWidth := 10 * 80; // 10 inches
        local maxHeight := 11 * 80;// 11 inches
    local params := GetAppParams();
    self.viewBounds := RelBounds(
        params.appAreaLeft,
        params.appAreaTop,
        Min(maxWidth,  params.appAreaWidth),
        Min(maxHeight, params.appAreaHeight)
    );
    end,

    _proto: protoApp,
    debug: "app"
};

rowHolder := /* child of app */ {
    viewBounds: {top: 20, left: 0, right: 0, bottom: -40},
    viewFlags: 1,
    viewFormat: nil,
    viewJustify: 240,

    viewSetupChildrenScript:
    func()
    begin
    local myHeight := viewBounds.bottom -
        viewBounds.top;
    local rowHeight := pt_row.viewBounds.bottom -
        pt_row.viewBounds.top;
    local numRows := myHeight div rowHeight;
    self.stepChildren := Array(numRows, pt_row);
    end,

    viewclass: 74,
    debug: "rowHolder"
};

// ---- Beginning of User Protos ----

// ---- File row ----

row := {
    text: "A row",
    viewBounds: {top: 0, left: 0, right: 0, bottom: 16},
    viewJustify: 8396850,
    _proto: protoStaticText,
    debug: "row"
};
```

Tracking Down the Problem

Since the Newton resets before throwing an exception, setting `breakOnThrows` won't be of any use. Instead, tracing to keep a log of what happened before the reset will probably be more helpful.

From the Inspector, set `trace`. To begin with, we'll only trace functions. If that doesn't provide enough information, we'll switch to tracing all:

```
trace := 'functions
```

Now, we tap on the application. The Inspector starts printing out a great deal. The Newton resets while the Inspector is printing:

```
Sending TrackHilite(18070461) to #4414091
    => TRUE
    Sending buttonClickScript() to #4414091
        Calling HasVar(cardSocket)
        => NIL
        Calling =(MultipleRows:Calliope, card)
        => NIL
        Calling =(NIL, NIL)
        => TRUE
        Calling HasVar(MultipleRows:Calliope)
        => TRUE
        Calling getVar(MultipleRows:Calliope)
        => #440C8D9
        Calling =(MultipleRows:Calliope, copperfield)
        => NIL
        Sending Open() to #440C8D9
            Calling *(10, 80)
            => 800
        => 800
        Calling *(11, 80)
        => 880
    => 880
    Calling GetAppParams()
    => #4416601
    Calling min(800, 240)
    => 240
    Calling min(880, 320)
    => 320
    Calling RelBounds(0, 0, 240, 320)
    => #4416621
    Calling not(NIL)
    => TRUE
    Calling FontHeight(#163)
    => 14
    Calling StrFontWidth(#60027E1D, #163)
    => 64
    Calling +(64, 6)
    => 70
    Calling SetBounds(0, -2, 70, 14)
```

```
=> #4416749
Sending calcTime() to #4416819
    Calling Time()
    => 47920947
    Calling div(47920947, 60)
    => 798682
=> 798682
Calling *(798682, 60)
=> 47920920
Calling -(47920947, 47920920)
=> 27
Calling div(27, 5)
=> 5
Calling >(5, 6)
=> NIL
Calling div(798682, 12)
=> 66556
Calling *(66556, 12)
=> 798672
Calling -(798682, 798672)
=> 10
Calling GetAppParams()
=> #44168F1
Sending parent() to #44167B1
=> #440C8D9
Calling -(320, 0)
=> 320
Calling >(320, 320)
=> NIL
Calling -(-40, 20)
=> -60
Calling -(16, 0)
=> 16
Calling div(-60, 16)
=> -3
Calling Array(-3, #600284DD)
```

The very last function that executed was `Array`, with a first parameter of -3. That is, it is trying to create an array with -3 elements. In a better world, `Array` would have thrown an exception for this case; in our not-so-perfect world, the Newton resets.

 *Note:* If `Array` had been written in NewtonScript, rather than in C++, a reset probably wouldn't have occurred. NewtonScript has such extensive runtime checking that errors usually throw exceptions, rather than causing things like bus errors. We must remember, however, that the underlying operating system is written in C++; errors at that level can indeed cause major errors and thereby force the whole system to reset.

So the big mystery question is "Who called `Array`?" A quick look at our trace shows that the base view's `viewSetupFormScript` executed (notice the calls to `GetAppParams` and `RelBounds`). Indeed, we seem to be in the `rowHolder`'s `viewSetupChildrenScript` when we crash and burn. Further, this fits with logic (the children of the base view are created soon after the base view's `viewSetupFormScript`), and with the code from our application (the `viewSetupChildrenScript` has two subtractions, a division, and a call to `Array`).

Now the question can be refined: "Why are we calling `Array` with -3?" Looking earlier in the trace, we see that we divided -60 by 16 in this line of code:

```
local numRows := myHeight div rowHeight;
```

We do this because the value of `myHeight` is equal to -60. Having this valuable clue we can learn even more. Looking back at the trace farther, we can see that there was a call to subtract:

```
Calling -(-40, 20)
    => -60
```

Now, if we compare this to the actual code, we can see where the correspondence occurs:

```
local myHeight := viewBounds.bottom -
   viewBounds.top;
```

Thus the values that we are using, `viewBounds.bottom` is -40 and `viewBounds.top` is 20, are what give us the `-60` value. But does this make sense? At first glance, no. But remember that the `rowHolder` is using parent full justification. Therefore these numbers signify that the `rowHolder` is inset 20 pixels from the top of its parent and 40 pixels from the bottom.

The problem, then, is that `viewBounds` is interpreted with respect to `viewJustify`. To deal with this situation, we will have to use a different approach. We will use a call to `LocalBox` which will return the width and height of a view. By changing the calculation of `myHeight` to use `LocalBox`, all the problems are solved:

```
local myHeight := :LocalBox().bottom;
```

We're still looking at the `viewBounds` of `pt_row`, but since we know the kind of justification it uses, and since we can't call `LocalBox` (because the row view hasn't been created yet), we'll leave the calculation of `rowHeight` alone.

Chapter 8

Advanced NewtonScript

A moment's insight is sometimes
worth a life's experience.

—Oliver Wendell Holmes, Sr.

Some of the most sophisticated aspects of NewtonScript can be difficult to implement correctly unless you understand these features in detail. Particularly important features to so comprehend are function objects, frame maps, and exceptions.

Learning to use function objects is essential, because any time you are dealing with functions in NewtonScript, you are actually dealing with function objects. Learning how frame maps get created and using them efficiently will allow you to greatly optimize the memory and space usage in your program. Last, but not least, learning to program with exceptions (as many people will tell you) will give you far more robust code. All in all, learning about each of these features of NewtonScript will help clarify both the structure of the language and your code.

Function Objects

A function object is more than just a sequence of instructions. It also provides context information so that the object has access to all the environment that was available at the time it was created.

Function Objects in NewtonScript

In NewtonScript, function objects (sometimes called closures) are first-class objects that can be manipulated and stored just like other values. For example, you can store a function object in:

- A local variable
- An array
- A slot in a frame
- A soup entry.

You can use function objects in a variety of ways as well: You can pass a function object as a parameter, `Clone` it, or `DeepClone` it. This consistency between function objects and other values makes NewtonScript very flexible.

While these aspects of function objects are usually well understood by NewtonScript programmers, there is another aspect to a NewtonScript function object that is less clear:

- When a function object is created, by executing a `func` statement, it saves the environment that exists at that time.

Because the function object saves the environment, it can have access to local variables, parameters, and inherited variable lookup that existed at its creation time.

The term "function object," rather than just "function," is used to emphasize the fact that one `func` statement can give rise to many different function objects. What differs between these function objects is the environment that exists at the time the `func` statement is executed.

Function Environment

To understand how you can use this aspect of a function in your applications, let us first review the environment of most NewtonScript functions. Usually, you create a function object at compile time. They are slots in a template, edited using a slot browser, or they are functions created in a Project Data file. In either case, the environment that exists is the environment of NTK. Thus, there are no local variables, parameters or inherited variables available when that function object is created.

Now, look at a different way to create a function object. In this case, the environment in which it is created will matter. This will be a runtime function object. Please note that, by definition, these will be nested functions (ones created inside other functions).

Here is a small example:

```
outerFunction := func(aParameter)
begin
   local aVariable := 3;

   local nestedFunction := func(nestedParameter)
   begin
      local nestedlocal := 5;

      Print(aParameter);
      Print(aVariable);
      Print(nestedParameter);
      Print(nestedFunction);
      Print(nestedlocal);

   end;
   ...
```

The function object `nestedFunction` is created as `outerFunction` is executing. At the runtime point when `nestedFunction` is actually created, the environment includes a local variable, `aVariable`, and a parameter, `aParameter`. Since `nestedFunction` has access to that environment, it can access both its own parameters and locals, as well as those of the function in which it is nested.

How Are Function Objects Used—Call/Perform/Apply

Before we actually talk about how you can use function objects in your programming, we need to address the manner in which you call a function object. There are four ways you can do this:

1. With a compile time argument list (`Call`):

   ```
   Call functionObject with (argumentList)
   ```

2. With a runtime argument list (`Apply`):

   `Apply(functionObject, argumentArray)`

3. By sending a message with a compile time argument list (`:` and `:?`):

   `frameExpression:message(argumentList)`

4. By sending a message with a runtime argument list (`Perform`):

   `Perform(frameExpression, messageSymbol, argumentArray)`

Here is some sample code that uses each of the four ways to call function objects:

```
local Pow := func(num, iterations)
begin
   for i := 1 to iterations do
      total := total * num;
   return total;
end;

Call Pow with (2, 3)
8

local argArray:= [2, 3];
Apply(Pow, argArray)
8

local account := {
   balance: 0,
   Deposit: func(amount)
      return balance := balance + amount,
};

account:Deposit(50)
50
Perform(account, 'Deposit, [75])
125
Print(account)
{balance: 125,
            Deposit: <CodeBlock, 1 args #4419361>}
```

Abstract Data Types

One use of function objects is to implement abstract data types. These are types that can only be modified procedurally; thus their actual data remains hidden. Though it might appear so, frames with methods don't provide the same functionality. In a frame, the data values in the slots are visible and can be modified even when not using the appropriate methods.

For example, consider the following account generator:

```
MakeAccount := func()
begin
   local balance := 0;
   local Deposit := func(amount) begin
      return balance := balance + amount;
   end;
   return Deposit;
end;
```

Calling `MakeAccount` returns a function object:

```
myAccount := call MakeAccount with ()
```

Notice that this function object (the one returned from `MakeAccount`) references the `balance` local variable from `MakeAccount`. Even though `MakeAccount` is no longer executing, since the nested function, `Deposit`, references `balance`, the `balance` variable continues to exist. Thus, calling `myAccount` modifies the hidden variable `balance`:

Executes this code

```
call myAccount with (50)
50
call myAccount with (75)
125
```

Notice also that one function object can return multiple function objects, each of which references shared data. For instance, suppose you want both Deposit and Clear capabilities in your account:

```
MakeAccount := func()
begin
   local balance := 0;
   local Deposit := func(amount) begin
      return balance := balance + amount;
   end;
   local Clear := func() begin
      balance := 0;
   end;
   return [Deposit, Clear];
end;
```

Because `MakeAccount` needs to return two values (two function objects), it returns them in an array:

```
myAccount := call MakeAccount with ();
myOtherAccount := call MakeAccount with ();

call myAccount[0] with (50)
50
call myOtherAccount[0] with (40)
40
call myAccount[0] with (75)
125
call myAccount[1] with ()
0
call myOtherAccount[1] with ()
0
```

Using an array for the two function objects is somewhat inconvenient, however, since the numbers 0 and 1 don't describe the `Deposit` and `Clear` functions in a very useful manner. To fix this problem, we will rewrite `MakeAccount` to return the two function objects in a frame rather than in an array. This way, the function objects can be referenced by name, rather than by array location.

```
MakeAccount := func()
begin
   local balance := 0;
   local d := func(amount) begin
      return balance := balance + amount;
   end;
   local c := func() begin
      balance := 0;
   end;
   return {
      Deposit: d,
      Clear: c,
   };
end;
```

Now we can call `MakeAccount` with our two values, `myAccount` and `myOtherAccount`.

```
myAccount := call MakeAccount with ();
myOtherAccount := call MakeAccount with ();


call myAccount.Deposit with (50)
50
call myAccount.Deposit with (75)
125
call myOtherAccount.Deposit with (10)
```

```
10
call myOtherAccount.Deposit with (60)
70
call myAccount.Clear  with ()
0
call myOtherAccount.Clear  with ()
0
```

Remember, however, that we are using the above frame only as a way to store two named values. We are not using object programming, for example, as no messages are being sent. You can tell this most clearly by looking at the graphic diagram in FIGURE 8.1 which shows the relation between the preceding two function objects, their code, and environments.

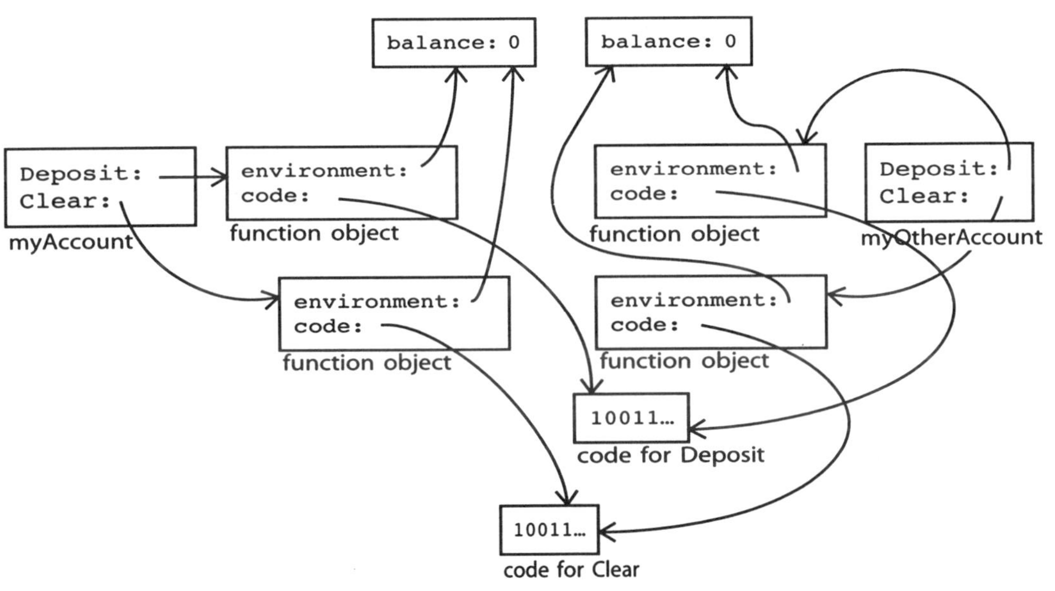

FIGURE 8.1 Function objects, their code, and environments.

Admittedly, this use of function objects to create abstract data types is not common. There are cases, however, where function objects are necessary in your Newton programming. One of the most common examples occurs when an application is supporting Date Find.

Using Function Objects to Support Date Find

Here is an excerpt of code from `DateFind`:

```
func(comparison, time, ...)
begin
   cursor := Query(..., {
      type: 'index,
      validTest: func(e)
      begin
         if comparison = 'dateBefore then
            return e.date < time
         else
            return e.date > time;
      end
   });
   return cursor to caller
end;
```

Even though the `validTest` function is embedded in the cursor, it is called from the Find results slip to display the found entries. Notice how `DateFind`'s `comparison` and `time` parameters are used by the nested function `validTest`. Keep in mind that the `validTest` is called after the `DateFind` function has finished executing.

Here is another example of how you might use function objects. Imagine that you want to count the number of entries in a particular query. In such a case, you might use a function object to count the number of entries in a cursor using `MapCursor`:

```
CountEntries := func(cursor)
begin
   local total := 0;
   MapCursor(cursor, func(e)
            begin
               total := total + 1;
               return nil;
            end);
   return total;
end
```

The function object passed to `MapCursor` increments a variable in `CountEntries`. Notice that the function object returns `nil`, and thus `MapCursor` will end up returning an empty array.

Stack Frames/Activation Records

In most programming languages, when a function is entered, an activation record (also called a stack frame) is pushed on the stack. This activation record contains the parameters to the function and local variables. When the function returns, the activation record is popped from the stack.

For NewtonScript, some allowance needs to be made for variables which are closed over; that is, variables which are accessed by nested functions. Since nested functions may need to access variables from outer functions even after the outer function has exited, a stack-based system which always pops outer variables from the stack could not help but fail.

One possible implementation could be to allocate activation records in dynamic memory (the heap). Like all other allocated memory, when no more references are made to the memory, it can be garbage collected. Thus, the activation record is not freed when a function object exits if any nested functions still exist. Only when all nested functions are freed is the activation record available for garbage collection.

Another implementation involves storing part of an activation record on the stack and part on the heap (only those variables referenced by nested functions need be on the heap). Happily, your code does not need to rely on any particular method and thus you need not worry about which is the actual implementation.

Message Context

A function object has access to more than local variables and parameters from enclosing functions. It also has inheritance lookup based on the value of self at the time the function object was created. This means that a function object has access to all variables, including inherited slots, that are available to the code that created the function object.

This inheritance lookup is implemented by storing a message context as part of a function object. This message context contains the value of self at the time a function object is created. Calling a function object restores self to the value stored in its message context.

The major difference between calling a function object and sending a message is this:

- **Sending a message sets the value of self to the frame where the message is sent.**

Thus, sending a message causes the message context of the function object to be ignored.

There are times, however, when you need to use inheritance in a function that has not been executed in response to a message send. A common case of this in Newton programming is found in the implementation of filing. Filing is usually implemented by creating a cursor that contains a `validTest`. The cursor is usually created when the application opens:

```
app.viewSetupFormScript := func()
begin
   …
   self.theCursor := Query(…, {type: 'index,
      validTest: func(e)
      begin
         return labelsFilter = '_all or
            labelsFilter = e.labels;
      end,
   });
   …
end
```

The cursor saves the `validTest` function object and calls it every time the cursor is moved. Now, look at what happens when the `validTest` is called:

- First, the `labelsFilter` variable is looked up as a local,
- Next, using inheritance the variable is searched for *based on the value of self at the time the* `validTest` *function object was created.*

Since self was the application view when the `validTest` was created, self is set to the application base view when the `validTest` is called.

Sending a Message Changes the Message Context

Thus, the major difference between sending a message and calling a function *has to do with the value of self.* When a message is sent, self is set to the frame that was sent the message. When a function is called directly, self is based on the message context of the function object.

Function Objects with No Lexical Environment or Message Context

When a top-level function object is created at compile-time–either as a slot in a template editor or in the top-level of the Project Data file–the lexical environment and message context are empty (technically, the lexical environment and message context exist, but are removed after the function is created). This is important for you to remember when you are creating stand-alone function objects (ones that have no references to your package). Examples of such function objects are those copied to a soup, or the one used as a `postParse` routine for Intelligent Assistance. If you use a function object that has a nonempty message context here, the whole receiver will be copied into the soup (or into memory in the case of Intelligent Assistance). This will be a really bad idea, in most cases. Thus, for function objects that need to execute independently of your package, make sure they are created at compile time, rather than runtime.

Summary

In NewtonScript, function objects are first-class objects which have access to the environment that exists at the time they are created. They have access to variables in enclosing functions, as well as to inherited slots based on the value of self at the time the function object was created. Because of these characteristics, you can do particular types of things with function objects that cannot be done in most other languages.

Frame Maps

The difference between arrays and frames is an issue of access. In an array, an integer index value specifies the location of each element. In a frame, slot symbols specify the location of each element. Further, the mapping from slot symbol to location is done using a *frame map*.

A frame map is basically a table; each element of the table is a symbol. The index in the table specifies the location of the corresponding element in the frame. If you look at FIGURE 8.2, you can see this graphically represented.

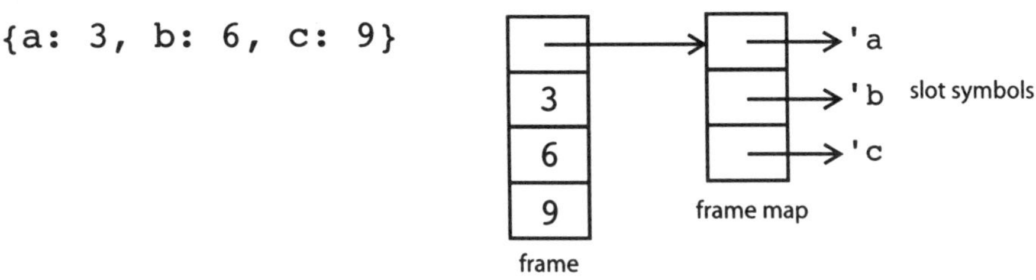

FIGURE 8.2 A frame and its frame map.

For objects with the same slots, the frame map can be shared as you can see in FIGURE 8.3, which shows two objects that share the same frame map. How does NewtonScript know when to share frame maps? As you might imagine, it would be quite time consuming if the NewtonScript interpreter checked to see whether frame maps could be shared every time an object were created or a slot were created or removed in an object.

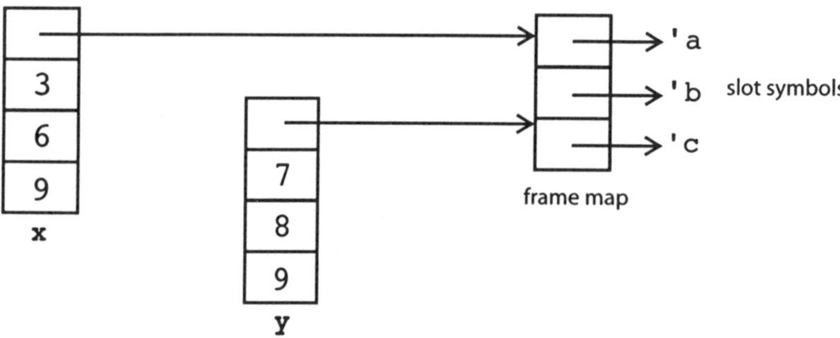

FIGURE 8.3 Two objects sharing a frame map.

Instead, the NewtonScript interpreter shares frame maps for those frames it knows are shared. In other words:

- Frames share frame maps when they are created by the same frame constructor.

A frame constructor is the technical name for the `{}` syntax which creates a frame. For example, here is a frame constructor which creates a frame with three slots:

```
{
    a: 3,
    b: 5,
    c: 10,
}
```

When NTK compiles the above frame constructor, it stores the frame map in your package. When the code is executed, it creates a frame in the NewtonScript heap which contains a reference to the frame map in the package.

Now, here is where things get interesting. Notice that even if two different frame constructors have the same slots in the same order, they do not share frame maps. Thus, in the following code, `x` and `y` **do not** share frame maps:

```
x := {
   a: 3,
   b: 6,
   c: 9,
};
y := {
   a: 7,
   b: 8,
   c: 9,
};
```

So what is the fix to this? How can you make two frames share the same frame constructor? For example, how can you create the `x` and `y` frame with the same frame constructor, and thereby ensure that they will share the same frame map? Here is one way:

```
local CreateFrame := func(p1, p2, p3)
begin
   return {a: p1, b: p2, c: p3};
end;
x := call CreateFrame with (3, 6, 9);
y := call CreateFrame with (7, 8, 9);
```

There is another way to share frame maps that is even easier: frame maps are also shared (as much as possible) when objects are copied. For example, cloning a frame with `Clone` does not create a new frame map; on the contrary, the two frames share the same frame map:

```
x := {a: 3, b: 6, d: 9};
y := Clone(x);
y.a := 7;
y.b := 8;
y.c := 9;
```

Frame Maps and Cloning

Different cloning functions will have different effects on the use of frame maps. Here are the four most often used clone functions and a description of what they do with frame maps.

`Clone`	This method does not copy the frame map.
`DeepClone`	This method does not copy frames maps either; it does, however, copy magic pointers.

`TotalClone`	This method copies the frame map and not the magic pointers.
`EnsureInternal`	This is the smartest method by far. If copies as necessary, including frame maps but not magic pointers (because magic pointers are already internal in the system ROM).

Frame Maps and Memory Use

When a slot is added to a frame which shares a frame map, the frame map cannot be modified directly; instead there is a form of frame map inheritance where a frame map can reference another frame map. In this way, adding one slot to a frame doesn't require duplicating its frame map.

Frame maps are, for the most part, transparent. Since they act to save space, you can notice their effect most when you do things which prevent frame map sharing. For instance, if you create bounds frames with the following code, your package is bigger than it should be:

```
x := {top: 10, bottom: 30, left: 6, right: 56};
```

If you use the global routine `SetBounds`, instead, the frame map for the frame will be in the Newton ROM rather than in your package.

As another example, we've seen code in applications that creates frames in the following way:

```
MakeFrame: func()
begin
   x := {};

   if ... then
      x.a := ...
   else
      x.a := ...;

   if ... then
      x.b := ...
   else
      x.b := ...;

   if ... then
      x.c := ...
   else
      x.c := ...
   ...
```

```
    return x;
end;
```

If `MakeFrame` is called 100 times, it will create 100 frames which have completely separate frame maps. What's worse, each of the frame maps is stored in the NewtonScript heap, rather than in the package. Since each of the slots is added dynamically, the frame map must be constructed dynamically.

A slight change to the function can suddenly make all 100 of those frames maps shared, a whopping memory savings (only 1 copy will be in the package):

```
MakeFrame: func()
begin
    x := {a: nil, b: nil, c: nil, …};

    if … then
        x.a := …
    else
        x.a := …;

    if … then
        x.b := …
    else
        x.b := …;

    if … then
        x.c := …
    else
        x.c := …
    …
    return x;
end;
```

Soups and Frame Maps

For soups, the requirements for sharing frame maps are more relaxed. Since each entry in a soup must be completely self-contained (not referring to anything outside the soup), it isn't possible for the soup entries to refer to frame maps in the package, or in the Newton ROM. When an entry is added to a soup (or modified with `EntryChange`), conceptually, the entry's frame map (along with any other frame maps in enclosed frames) is copied into the soup. In reality, however, before copying a particular frame map, the soup's master frame map list is consulted to see whether there is already a frame map that has the same slot names in the same order. If so, that frame map is shared; otherwise, a new entry in the master frame map list is created. Thus, the two soup entries on the left in FIGURE 8.4 do share the same frame map, while the two entries on the right do not.

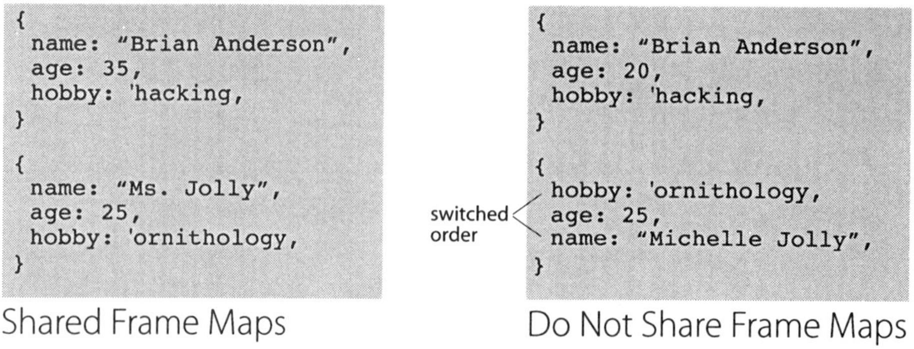

```
{
 name: "Brian Anderson",
 age: 35,
 hobby: 'hacking,
}

{
 name: "Ms. Jolly",
 age: 25,
 hobby: 'ornithology,
}
```

Shared Frame Maps

```
{
 name: "Brian Anderson",
 age: 20,
 hobby: 'hacking,
}

{
 hobby: 'ornithology,
 age: 25,
 name: "Michelle Jolly",
}
```

switched order

Do Not Share Frame Maps

FIGURE 8.4 Sharing frame maps for soup entries.

For a soup then, frame map sharing occurs if two frames have the same slots in the same order. This might argue then for not sharing frame constructors for frames that will be soup entries. However, by sharing a frame constructor, you guarantee that the slots and order are the same. So, even for soup entries, it is a good idea to have one function in your application which is responsible for creating the frame.

This sharing of frame maps in a soup may explain something quite puzzling in the Names soup. When an entry is created in the Names soup, it has slots with `nil` values for fields which haven't been filled in rather than having no slot at all. In the most extreme case, an entry in the Names soup with only a Company filled in looks like:

```
{
    sortOn: "",
    cardType: 0,
    phones: [],
    email: NIL,
    company: "Apple Computer",
    address: NIL,
    address2: NIL,
    city: NIL,
    region: NIL,
    country: NIL,
    postal_code: NIL,
    bday: NIL,
    _uniqueID: 195
}
```

There are eight slots which haven't been filled in and yet are taking up space in the soup. We had expected something more like:

```
{
    sortOn: "",
    cardType: 0,
    phones: [],
    company: "Apple Computer",
    _uniqueID: 195

}
```

However, by making sure that the currently unused slots exist, the Names application is ensuring that all of the entries in the Names soup share one frame map. If each entry had the minimal number of slots, then adding a new slot (due perhaps to the user editing the entry in the Names application) would cause a new frame map.

What's worse is that the possible number of frame maps is huge because the order of slots in the frame matters. Thus, if for one entry, a `city` is added, and then a `country`, and on the next entry a `country` is added, and then the `city`, two unshared frame maps would exist in the soup.

For fun, we calculated that just 9 possibly different slots could lead to a whopping 362,880 different slot orderings and thus different frame maps.

Exceptions

An Overview of Exceptions

Exceptions are a way of dealing with errors. An exception is a report of an abnormal condition. It terminates the execution of the current function and of the function that called it, and so on up the call chain until an exception handler is found to respond to the error. The exception handler is in turn usually responsible for cleanup and for reporting the error/exception to the user. Exceptions are an alternative to returning error codes.

If you look at FIGURE 8.5, you can see how this whole mechanism operates. Here we have function A calling B calling C calling D calling E. Function B has the exception handler and function E throws the exception. At this point, function E terminates followed likewise by E and C. Function B's normal flow is stopped and execution continues within B's exception handler.

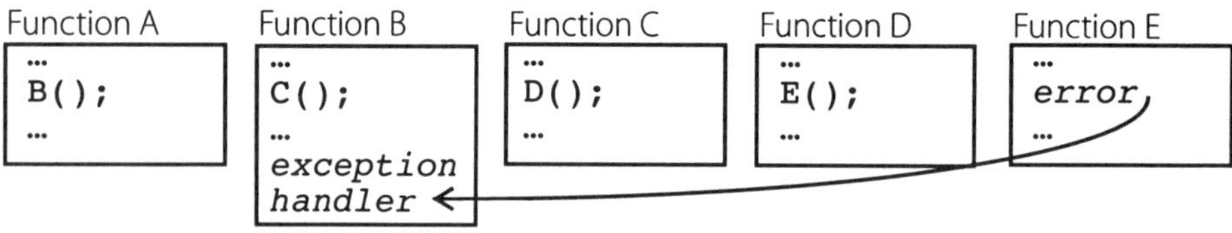

FIGURE 8.5 Calling chain when E throws an exception.

The Advantages of Using Exceptions

In your programming, there are a couple of clear advantages to using exceptions as opposed to returning error codes. The first is that functions that do not generate exceptions (like C and D), but which call functions that do, don't need to do anything special. Without any extra work the exception will be propagated to calling functions (like B). On the other hand, if you relied only on error codes, C and D would need to return error codes as well, since they call the function with the error in it.

The second advantage is the simplification of functions. Since the execution of a function terminates as soon as an exception occurs, you can be sure that the statements that occur prior to the exception are successfully executed. Conversely, functions which use error codes commonly look like this:

```
X := func()
begin
   error := A();
   if error = noEr then begin
      error := B();
      if error = noErr then begin
         error := C();
      end;
   end;
   return error;
end;
```

It is easy to see how the logic of the functions can get lost among the error checking. Instead, with exceptions, X could be neatly written as:

```
X := func()
begin
   A();
   B();
   C();
end;
```

Working with Exceptions

Now that you have a fairly good idea why programming with exceptions makes sense, let's look at the type of code that would profit from its use. Code that typically follows this pattern will be of interest:

```
set some state
code which could cause an exception
restore the state
```

Now, we will look at the types of states that benefit from exception handling. First, it is worth reviewing those states you might be familiar with from other types of machines. After that, we can look at the Newton.

Exception Handling in Other Environments

On desktop machines using traditional languages, exception handlers are common. Here are some examples of common states that require exception handlers:

- Allocate some memory
  Do something with the memory that could cause an exception
  Deallocate the memory

- Lock some memory
  Do something with the memory that could cause an exception
  Unlock the memory

- Open a file
  Write to the file
  Close the file

- Create a temporary file
  Use the temporary file
  Delete the file

Each of these requires an exception handler. When a resource has been acquired, it must be given up in the event that an exception occurs.

Exception Handling in the Newton Environment

Because of NewtonScript's superior handling of memory, the need for exception handlers is minimized. Since NewtonScript has garbage collection, explicitly deallocating memory is not necessary. Likewise, the Newton has no memory locking, making unlocking unnecessary as well. Another nice point is the Newton's lack of files; if you don't have a file open, you don't have to close it.

As you can see then, none of the typical states on other platforms is a problem on the Newton. *So what is?* Commonly you'll only need exception handling when calling `EntryChange` and the `PutAway` method used for beaming and mailing. A rarer situation involves the creation of a temporary soup on the Newton; in that case you would use exception handling to ensure its deletion.

The Structure of Exceptions

Now, let us look at how you would write an exception handler using proper NewtonScript syntax. Next, we will look at the hierarchical relationship of exceptions.

Exception Handling Syntax

The actual syntax for exception handlers in NewtonScript is:

```
try
   code which could generate an exception
onException exceptionSymbol do
   code to handle this type of exception
onException exceptionSymbol2 do
   code to handle this type of exception
```

To propagate an exception from within an exception handler, use the `Rethrow` function:

```
Rethrow();
```

The Hierarchy of Exceptions

There is a hierarchy of exceptions (FIGURE 8.6 shows a small subset of the whole tribe). This hierarchy exists so that you can write exception handlers for a specific exception, a group, or for all of them.

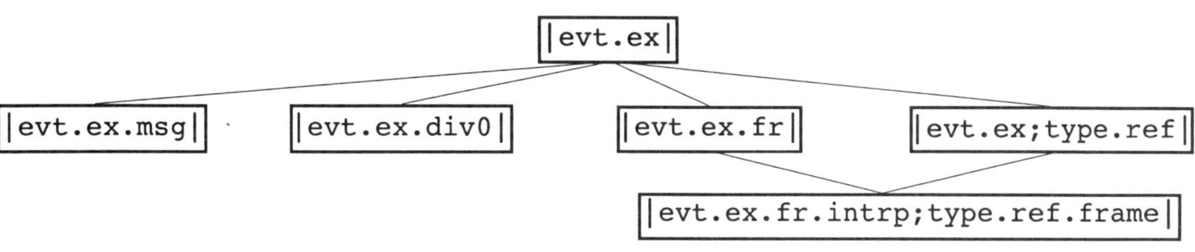

FIGURE 8.6 A small subset of the exception hierarchy.

Every exception symbol contains an `evt.ex` part (since the symbols contain embedded periods, they must be surrounded by | |). Portions after the . part signify lower levels in the hierarchy. For example, the exception `|evt.ex.fr.intrp|` is a kind of `|evt.ex.fr|` exception. You use semicolons (as shown in FIGURE 8.6) to separate multiple exception types.

So, if you wish to catch all exceptions use:

```
onException |evt.ex|
```

You can also use more specific exception types. For instance, to catch only `type.ref` exceptions, use:

```
onException |type.ref|
```

It is also possible to cascade exception handlers. In this case, the first one that matches the current exception is the one that gets called. For example:

```
try
   code
onException |evt.ex.div0|
   deal with divide-by-zero error
onException |type.ref|
   deal with a type error
onException |evt.ex|
   deal with all other errors
```

Using Exceptions in Newton Applications

When doing communications programming using endpoints, it is common to use exception handlers. Other than that, the use of exception handlers is fairly rare. There are, however, two typical problems that you might encounter on the Newton that could benefit from the use of exceptions. The first involves using `Entry-Change` and the second involves the `PutAway` method. Let's explore each in turn.

Exception Handling and EntryChange

`EntryChange` is the mechanism used on the Newton to allow an application to save a modified entry back into its soup. Although the view system will provide a reasonable notification to the user (see FIGURE 8.7), shortly you will see that more is needed.

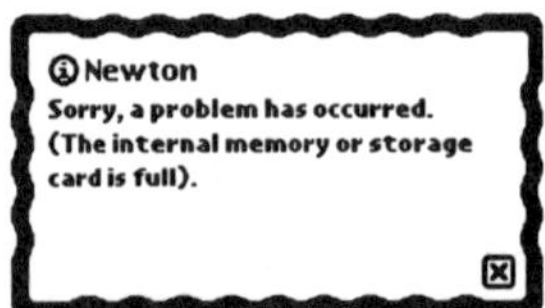

FIGURE 8.7 Default view system notification when out of storage space.

Here is why. Look at these code snippets that handle saving a modified entry when the user scrolls up:

```
viewScrollUpScript: func()
begin
   ...
   :SaveModifiedEntry(currentEntry);
   :DisplayPreviousEntry();
end;
SaveModifiedEntry: func()
begin
   ...
   EntryChange(currentEntry);
   ...
end;
```

If `EntryChange` throws an exception, the notification shown in FIGURE 8.7 will be presented by the view system. A problem remains, however. The user is stuck viewing the current entry and any effort to save it (like closing the application), causes `SaveModifiedEntry` to throw an exception (due to `EntryChange`). Thus, no progress can be made.

The solution is to change the behavior of `SaveModifiedEntry`. Rather than having it swing between successfully saving or throwing an exception, modify its behavior to successfully save or to notify the user of the error. While we're at it, we can rename the function to more accurately describe its new responsibilities:

```
viewScrollUpScript: func()
begin
   ...
   :HandleModifiedEntry(currentEntry);
   :DisplayPreviousEntry();
end;

HandleModifiedEntry: func()
begin
   ...
   try
      EntryChange(currentEntry)
   onException |evt.ex.fr.store| do
   begin
      :Notify(kNotifyAlert,
         EnsureInternal(kAppName),
         EnsureInternal("There is not enough space to
         save the modified item"));
      EntryUndoChanges(currentEntry);
   end;
   ...
end;
```

The call to `EntryUndoChanges` is necessary; that is how we wipe out modifications made to the entry. Otherwise, the soup holds the unmodified entry, while the entry cache contains a modified entry. The right course of action is to have the view reflect the fact that we were unable to save the entry. Thus, we must revert back to the saved entry and then display that version.

 *Caution:* The union soup method `AddToDefaultStore` does not throw an exception when a store is full, although the documentation says that it does. In reality, in the 1.3 Newton OS (and previous versions), `AddToDefaultStore` shows that it fails by returning `nil` and by calling `Notify` to alert the user. This failure to follow documented behavior can create problems for applications that do not check to ensure that `AddToDefaultStore` succeeds.

You might want to consider writing a wrapper around `AddToDefaultStore` that will throw an exception in the event that `AddToDefaultStore` returns `nil`:

```
DefConst('kMyAddToDefaultStore, func(unionSoup,
frame)
begin
   result := unionSoup:AddToDefaultStore(frame);
   if not result then
      Throw('|evt.ex.fr.store|, '{error: 10617});
   return result;
end);
```

Exception Handling and PutAway

The other place where exception handling is commonly needed is in the `PutAway` method that is used for beaming and mailing. Remember that `PutAway` calls `kRegisterCardSoupFunc`, adds an entry to a soup, and then calls `kUnRegisterCardSoupFunc`. Since adding an entry may throw an exception, an exception handler is needed to call `kUnRegisterCardSoupFunc`. Here is a stripped-down `PutAway` method that shows the problem:

```
PutAway := func(item)
begin
local soup;
  soup := call kRegisterCardSoupFunc with
    (kSoupName, kSoupIndexes, kAppSymbol, kAppObject);
  call kMyAddToDefaultStore with (soup, item.body);
  call kUnregisterCardSoupFunc with (kSoupName);
end
```

The Sample Code with Exception Handling

Now, look at how to fix the problem with exception handling:

```
PutAway := func(item)
begin
  local soup;

  soup := call kRegisterCardSoupFunc with
    (kSoupName, kSoupIndexes, kAppSymbol, kAppObject);
  try
    call kMyAddToDefaultStore with (soup, item.body);
  onException |evt.ex| do begin
    call kUnregisterCardSoupFunc with (kSoupName);
    ReThrow();
  end;
  call kUnregisterCardSoupFunc with (kSoupName);
end
```

Hopefully, you can now see that exception handling is a helpful tool in programming. First, we covered exception handling in general. Next, we turned to its structure in NewtonScript and last to where in your application you might profit by using exception handlers. At this point all that remains is for you to incorporate this new technique into your bag of programming tricks.

Summary

The three important NewtonScript concepts that we discussed in this chapter are function objects, frame maps, and exception handling. Correctly understanding each of these concepts should greatly benefit you in your applications.

Chapter 9

Performance Tuning

Act quickly, think slowly.

—*Greek Proverb*

Creating Efficient Code
Measuring NewtonScript Performance
Speeding Up Your NewtonScript
Using and Allocating Memory Efficiently
View Performance
Summary

There are a number of ways you can affect Newton application performance. The most important of these areas are:

- *Writing good NewtonScript.* Writing in a manner that takes advantage of both NewtonScript and the C++ that underlies the operating system can give you much smaller and quicker code.
- *Using the Newton view system correctly.* It is essential to understand how the view system operates. Otherwise, you will be con-

stantly performing view operations that are time consuming and unnecessary.

- *Designing your Newton data storage (soups) correctly.* The difference between an application with a good soup design and one without can be quite dramatic. It can be the difference between a slug no one will use and a responsive and useful application.

In this chapter, we discuss the first two areas. We will looking at optimizing soup design in Chapter 10 as well. Before we turn to the discussion of NewtonScript, there are a few points worth noting about performance tuning in general.

Creating Efficient Code

Certainly, it would be better to write code that is clean and efficient from the outset. How does one do that, however? By studying examples, of course.

Some Resources

There are two books that contain particularly good discussions of optimizing code. The first is devoted solely to optimization, and the second covers all aspects of programming:

- Jon Bentley, *Writing Efficient Programs*, Prentice Hall, 1982, ISBN 0-13-970244-X.
- Steve McConnell, *Code Complete*, Microsoft Press, 1993 ISBN 1-55615-484-4.

Both books correctly point out that optimization should be done *after-the-fact*. You measure the performance of an application, determine where the application is spending most of its time, and then concentrate on optimizing those areas.

Measuring NewtonScript Performance

When you want to optimize the performance of NewtonScript within your application, you need to focus on three important areas:

- Writing code that is generally efficient.
- Taking advantage of the speediest ways of doing things within NewtonScript.
- Using memory correctly. This involves minimizing your use of frame heap memory by maximizing the amount of code you keep in your package. Just as importantly, you need to have code that avoids unnecessary memory allocation.

Measuring Performance of a Newton Application

For the Newton, using NTK 1.01, the only real way to measure performance is to call `Ticks()` before and after a piece of code. By doing so, you can see how long the code took to execute. Because `Ticks()` returns a number in sixtieths of a second, however, you may want to use it repetitively in a loop if you have to measure a quick operation. For this sort of thing, we usually use:

```
call func()
begin
   local t := Ticks();
   // code to test
   return Ticks() - t;
end with ();
```

The code to be timed is embedded in a function. This way the code can include loops (remember, the Inspector doesn't allow loops at the top level).

Here's an example of how you would execute this within the Inspector:

```
call func()
begin
   local t := Ticks();

   local total := 0;
   for i := 1 to 1000 do begin
   end;

   return Ticks() - t;
end with ();

#1D        8
```

```
call func()
begin
   local t := Ticks();

   local total := 0;
   for i := 1 to 1000 do begin
      total := total + i;
   end;

   return Ticks() - t;
end with ();
```

#38 14

Notice that there are two timings here. The first timing measured 1000 iterations of an empty loop, while the second measured 1000 iterations of a loop with an integer addition in each iteration. As you can see, the difference between the two is 6 ticks, or 1/10 of a second. From this simple little timing, we can now estimate the speed of adding two local integer variables to be roughly 1/10,000 of a second.

Speeding Up Your NewtonScript

There are a number of small things you can do to increase the performance of your NewtonScript. Taken together, they can make a noticeable difference in the speed of your application.

Note: In this book, we always use the `return` statement to return a value from a function. If an explicit `return` isn't provided, the value of a function is the value of the last statement. For example:

```
func(a, b)
begin
   return a * b;
end
```

can be written as:

```
func(a, b)
begin
   a * b;
end
```

Leaving out the `return` statement reduces (by one byte) the amount of code which is generated, and increases the speed (only very slightly). However, we find it easier to read with an explicit `return` statement.

Use Built-in Functions in Preference to Writing Your Own

Since NewtonScript is an interpreted byte-code (at the time this book was written, NTK didn't provide a NewtonScript compiler), the best way to increase NewtonScript speed is to use raw NewtonScript as little as possible.

For example, if you want to sort an array, you should not write your own sorting routine:

```
call kMySortFunc with (anArray)
```

Instead, you are far better off calling the built-in `Sort` function. Since many of the built-in global functions are written in C++, they run much faster than your custom NewtonScript code. Using a built-in function also has the additional advantage of reducing the size of your package.

This knowledge should shape your writing of other Newton Script code as well. For example, some functions take other functions as parameters. `Sort`, which is one of them, takes a function as the second parameter. Now, let us look at some possible ways you might write this function. Assume you have an array of frames, each of which has a `height` slot, and you wish to sort the frames in increasing order of height. Certainly, one way to do this is:

```
Sort(
   myArray,
   func(a, b) return a.height -  b.height,
   nil);
```

Notice, however, that the third parameter of `Sort` is a symbol specifying the slot on which to sort. Therefore, you could easily rewrite this example as:

```
Sort(myArray, func(a, b) return a - b, 'height);
```

This should be slightly quicker, since the two accesses of the `height` slot are now made in C++ rather than in NewtonScript. Timing these two variations yields just this result. With a sample of 500 elements the first case took 884 ticks, whereas the second took 810.

What would be much better, however, is to avoid NewtonScript entirely (even `return a-b`). You can pull off this trick with `Sort` by passing the symbol `'|<|` (which specifies a numeric comparison) as the second parameter:

```
Sort(myArray, '|<|, 'height)
```

This last example is substantially faster than the first two, taking only 42 ticks (a whopping 20 times increase in speed).

If you use built-in NewtonScript functions in preference to writing your own, you can expect performance improvements as well as a reduction in the size of your package.

Always Declare Your Local Variables

Local variables should always be declared using the `local` keyword. One good reason for this rule is code readability:

- If you define your locals using the keyword, then a reader of your code can easily distinguish between local variables and globals or inherited slots.

Additionally, at some point in the future, NewtonScript may start issuing warnings or errors for undeclared local variable usage. You might as well be prepared.

Ignoring all of these reasons, however, still leaves you with the most compelling argument of all—speed. When the NewtonScript compiler sees the use of a declared local variable, it generates code that accesses the variable via a numeric index. On the other hand, the code to access variables not declared locally requires a dynamic lookup based on the variable name. This dynamic lookup is slower.

The situation is even worse when you are assigning to a variable for the first time. If a variable isn't declared local, the assignment requires a three-part lookup, first with a lookup as a local, then using both proto and parent inheritance, and finally as a global. It is only if the variable isn't found at this point that the interpreter creates the variable as a local.

The following example demonstrates just what sort of speed sacrifice you make by omitting the `local` keyword. When this code gets called with an array of 1000 elements it takes 66 ticks:

```
SumArray := func(anArray)
begin
   total := 0;
   foreach elem in anArray do
      total := total + elem;
   return total;
end;
```

By simply declaring the local variables, the time is more than cut in half (31 ticks):

```
SumArray := func(anArray)
begin
   local total := 0;
```

```
   foreach elem in anArray do
      total := total + elem;
   return total;
end;
```

Loop Variables and Local Declarations

Happily, you don't have to declare every local variable. Loop variables are automatically declared as locals for you. Thus, the following variables in bold do not need to be declared as locals:

```
for i := 1 to 100 do
   ...
foreach elem in array do
   ...
foreach slotSymbol, value in frame do
   ...
```

Use `foreach` *Correctly*

It is also worth knowing which type of loop is the fastest for which jobs. For instance, it is quicker to iterate through the elements of an array using `foreach` than it is with an explicit `for` loop. Here are two examples:

```
SumArray1 := func(anArray)
begin
   local total := 0;
   foreach elem in anArray do
      total := total + elem;
   return total;
end;

SumArray2 := func(anArray)
begin
   local total := 0;
   for i := 0 to Length(anArray) - 1 do
      total := total + anArray[i];
   return total;
end;
```

With `foreach`, you have a specialized NewtonScript construct that the interpreter knows how to handle quickly. When we timed these two samples, `SumArray1` took 18 ticks, while `SumArray2` took 32 ticks (for a 1000-element array). `foreach` is also easier to use, since you can't accidentally get the starting or ending conditions wrong.

Even with NewtonScript constructs like `foreach`, however, measurement remains important. There are times when using it is slower. For example:

```
SetArray1 := func(anArray)
begin
   foreach index, elem in anArray do
      anArray[index] := index;
end;

SetArray2 := func(anArray)
begin
   for index := 0 to Length(anArray) - 1 do
      anArray[index] := index;
end;
```

Although `SetArray1` is easier to read, the interpreter sets the value of `elem` (which is unused) as well as the value of `index` each time through the loop. `SetArray1` took 42 ticks when called with a 1000-element array. `SetArray2` took only 16 ticks.

Function Calls Are Expensive

Compared to some other languages, the cost of calling a function is expensive in NewtonScript. Here is what we mean: The Newton can make almost 1000 function calls per second (if the functions do nothing, and no lookup is necessary to call them). Thus, the following code took between 60 and 70 ticks to execute:

```
g := func()
begin
   local f := func()
      begin
      end;

   Print(Ticks());
   for i := 1 to 1000 do
      call f with ();
   Print(Ticks());
end
```

You may not want to make your functions as small as you would in other languages, since fewer larger functions will execute quicker than many small ones. Sending a message is somewhat more expensive, as well. This is mostly due to the lookup cost for the method in the frame that received the message.

Many Newton programmers implement utility functions as methods in the base view. For example, a program which calculates sales tax might have a `CalcSalesTax` method in the base template:

```
func(amt, state)
begin
   ...
   return tax;
end
```

Throughout the application, this method would be called with:

```
GetRoot().(kAppSymbol):CalcSalesTax(amount, state)
```

Each call to the utility routine requires a message send. A somewhat quicker approach would be to create the utility routine as a constant routine in Project Data:

```
DefConst('kCalcSalesTax,
   func(amt, state)
   begin
      ...
      return tax;
   end)
```

The function could then be called with:

```
call kCalcSalesTax with (amount, state)
```

Cache Expensive Calculations

Currently, the NewtonScript compiler does no common subexpression elimination optimization. Thus, every expression in your code will be evaluated. This means that if you refer to an inherited variable many times in a method, the lookup for that slot must also occur many times.

To deal with this problem, you might want to consider caching the results of operations. For instance, instead of this type of code:

```
Method := func(s)
begin
   for i := 1 to 100 do
      s := s + inheritedVar;
   return s;
end;
```

you can use:

```
Method := func(s)
begin
   local cachedVar := inheritedVar;
   for i := 1 to 100 do
      s := s + cachedVar;
   return s;
end;
```

The latter code avoids 999 inherited variable lookups.

Along similar lines, if you are accessing nested frames repeatedly, consider caching the result of the dot lookup. So, rather than:

```
for i := 1 to 100 do
   a.b.c.d[i] := i;
```

use:

```
local arr := a.b.c.d;
for i := 1 to 100 do
   arr[i] := i;
```

The latter code took 6 ticks while the former code took 7 ticks.

Eliminate Loop Invariants

A common compiler optimization is to move expressions whose values don't change outside of a loop. Unfortunately, NewtonScript doesn't currently do this optimization for you—so you must do it. Rather than writing your code like this:

```
for i := 1 to 100 do
   :Method(x + y + 53 * z, 2 * i);
```

you should use this instead:

```
local value := x + y + 53 * z;
for i := 1 to 100 do
   :Method(value, 2 * i);
```

By making this change, the calculation of `x + y + 53 * z` is done only once, instead of 100 times.

Another optimization one could make to the above code is to remove the multiplication:

```
local value := x + y + 53 * z;
for i := 2 to 200 by 2 do
   :Method(value, i);
```

Using and Allocating Memory Efficiently

Efficient handling of memory within a NewtonScript application involves both attention to memory allocation and consideration of whether memory is allocated at compile-time (within your package) or at runtime (in the frame heap). First, let's deal with the more simple issue of how you measure your memory use.

Measuring Memory Use

To measure space in a program, your best bet is to call `Stats()`. This method returns the total amount of memory available. Since `Stats` doesn't take into account memory that needs to be garbage collected, however, you should always call `GC()` first.

Here is the type of code we use to see how much memory a function takes:

```
call func()
begin
   GC();
   local m := Stats();

   // code to test
   ...

   GC();
   return m - Stats();
end with ();
```

Here is an example of it in use:

```
call func()
begin
   GC();
   local m := Stats();

   local x := [1, 2, 3, 4, 5, 6, 7, 8];

   GC();
   return m - Stats();
end with ();

Free: 71352, Largest: 71352
Free: 71304, Largest: 71304
#C0       48
```

The preceding example shows you that this eight-element array takes 48 bytes.

Efficiently Allocating Memory

Allocating memory takes time. In addition, it uses a very precious resource on the Newton, RAM memory. For those reasons, it is best to treat such allocations like walking a tightrope above the jaws of a snapping alligator—something a sane person avoids whenever possible. Here are some ways to slip past those teeth.

Using Constant Frames and Arrays

If you are using an array or frame that will not change at runtime, make sure that it is constant. For example, there is a substantial difference between:

```
myFrame := {a: 3, b: 4}
```

and:

```
kMyFrame := '{a: 3, b: 4}
```

The former is technically a frame constructor that allocates memory for the frame at run time from the frame heap. The second frame, on the other hand, is created at compile time and is stored as part of your package. What could be better than such a nice trade of RAM for ROM?

There is also a syntax you should use for defining a constant array:

```
'[1, 2, 3]
```

You can likewise use `DefConst` to create computed constants:

```
DefConst('kMyFrame, {a: 3, b: 4 * 5});
```

It is also worth knowing that in the current implementation of NewtonScript, strings literals are constants. Thus, the following code creates a string that is stored in the package, rather than in the frame heap:

```
s := "abcd";
```

Sharing Frame Maps

As shown in FIGURE 9.1, frames created with the same frame constructor share the same frame map. (A frame map is the mapping from slot symbol to location within the frame.) You should ensure wherever possible that you share frame maps.

```
M := func(p1, p2, p3, p4)
begin
   return
      {a: p1, b: p2,
       c: p3, d: p4};
end;

call M with (1, true, 3, 4);
call M with (2, nil, 0, 9);
```

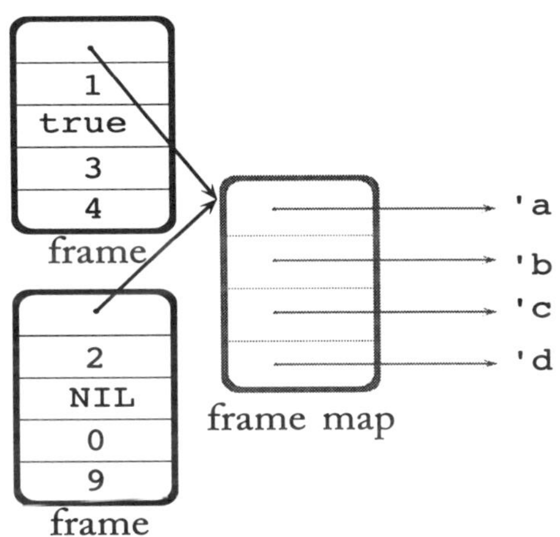

FIGURE 9.1 NewtonScript frame maps.

For example, if you need to create a bounds frame at run-time, don't use:

```
{left: 10, top: 5, right: 30, bottom: 100}
```

You should instead call `SetBounds` (or `RelBounds`):

```
SetBounds(10, 5, 30, 100)
```

With `SetBounds`, the frame map of the rectangle will be in the Newton's ROM, rather than in your package.

On a similar note, you should ensure that your frame constructor contains all the slots your frame will ever need; do not add the new slots dynamically. Adding a new slot not only requires allocating memory for the slot value (an unavoidable four bytes), but also requires allocating memory for the frame map entry. By making sure the initial frame constructor contains the necessary slots, the frame map can be in the package, rather than in the frame heap.

FIGURE 9.2 and FIGURE 9.3 contrast creating slots in the frame constructor versus dynamically creating slots. In FIGURE 9.2, the frame map is in the package. In FIGURE 9.3, each slot is created dynamically, resulting in a frame map in RAM.

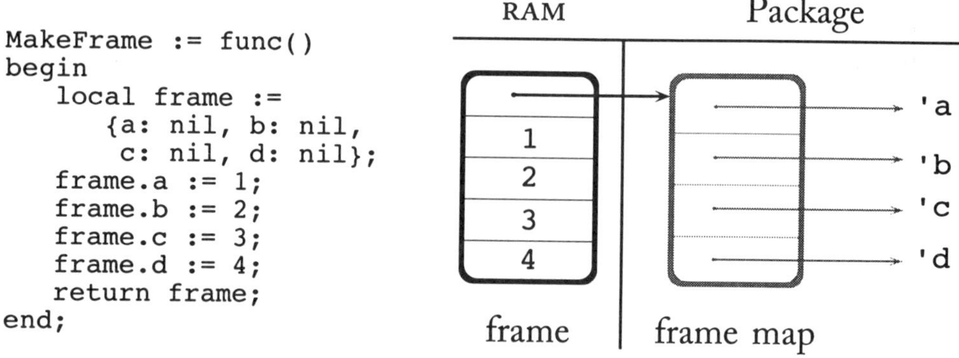

FIGURE 9.2 Creating slots using a frame constructor.

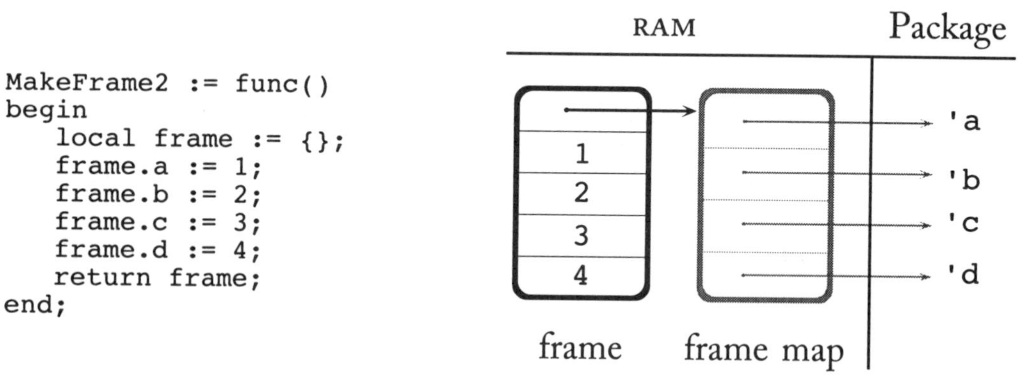

FIGURE 9.3 Creating each slot dynamically.

The first routine uses 32 bytes of the frame heap; the second uses 64 bytes. Notice also that the first routine is not just smaller, it is faster; calling each routine 100 times takes only 17 ticks for the first routine, while the second takes 32 ticks (since memory allocations for each slot and for each entry in the frame map are time consuming).

Don't Blindly Use **`EnsureInternal`**, **`TotalClone`**, or **`DeepClone`**

Remember that NewtonScript is specifically designed to share memory. When you use `EnsureInternal`, `TotalClone`, or `DeepClone`, you are specifically undoing this good thing about NewtonScript—you are making copies so that memory is not shared. Don't blindly litter your code with calls to these routines, rather make sure you have a good reason for each use.

Watch Out for Operators that Allocate Memory

It is important to allocate only the necessary amount of memory in your routines. Look at the following function, for example, which returns the reverse of a string:

```
Reverse := func(s)
begin
   local reversedString := "";
   local stringLength := StrLen(s);
   for i := 0 to stringLength - 1 do
      reversedString := s[i] & reversedString;
    return reversedString;
end;
```

When this `Reverse` routine is called with a string of length 20, it makes 20 memory allocations, since each call to the `&` operator creates a new string. Although the first 19 of these allocations can be reclaimed (since the intermediate string concatenations are not used), they definitely take time.

Here is a better way to reverse a string that avoids these unnecessary allocations:

```
Reverse2 := func(s)
begin
   s := Clone(s);
   local stringLength := StrLen(s);
   for i := 0 to stringLength div 2 do
   begin
      local tmp := s[i];
      s[i] :=  s[stringLength - 1 - i ];
      s[stringLength - 1 - i ] := tmp;
   end;
   return s;
end;
```

Here, a different approach is taken. The original string is duplicated (one memory allocation) and then pairs of characters are swapped. These types of optimizations will often gain you a two times speedup as is evidenced from a time measurement of the two routines. Using `Reverse` on a 200-character string took 40 ticks, while `Reverse2` took 10 ticks.

Allocate Once Rather than Many Times

As a general rule, you should allocate only once rather than multiple times. For example, if you want to create an array with 1000 integers in it, don't use:

```
a := [];
for i := 1 to 1000 do
   AddArraySlot(a, i);
```

Instead, you should use code like this:

```
a := Array(1000, nil);
for i := 0 to 999 do
   a[i] := i;
```

The first method does 1001 memory allocations (and takes 67 ticks), while the second does only one (and takes only 27 ticks).

Smart String Routines

Yet another way to reverse is to use the built-in smart string routines. These routines allocate a large amount of memory for a string. When you use this technique, your string can be appended to, but as the length changes, no new memory allocations are necessary (unless the new size exceeds the original memory allocation). Nicely enough, once the operations are complete, the excess memory (beyond the length of the string) can be returned to the system.

Here is an example of reversing a string using the smart string routines:

```
Reverse3 := func(s)
begin
   local stringLength := StrLen(s);
      // SmartStart takes byte count, not char count
   local reversedString :=
      SmartStart(2 * stringLength + 2);
   for i := 0 to StrLen(s) - 1 do
```

```
            SmartConcat(reversedString, i,
               s[stringLength - 1 - i]);
         SmartStop(reversedString, stringLength);
         return reversedString;
      end;
```

With a 200-character string, `Reverse3` took 22 ticks.

View Performance

You can do a number of things to affect the performance of your application's views. Certainly, the best place to start with these enhancements is to understand how the view system decides to draw and re-draw views. In this way, your application will be designed to do as little of this as necessary.

Fill Patterns

In many cases, the decision to use a fill pattern for a view is based on the look you are trying to achieve. It is true that it will always be faster to use a fill pattern rather than to try and draw a pattern in your `viewDrawScript`. This is because the view system has optimized pattern drawing.

What about fill patterns that have no visual effect? What is their impact on the view system? For example, if you have a protoInputLine on a white background, it will certainly display the same whether you give it no fill pattern or a white fill pattern. Is the drawing time the same in both cases, however?

Actually, specifying a fill pattern will cause the drawing of the view to be somewhat slower (since drawing the fill pattern is more code to execute). Consider, however, what happens when that view is dirtied (either in your code with a `Dirty()` call, or by the user writing a word in a protoInputLine).When a view with a fill pattern is dirtied, only that view must be redrawn (see FIGURE 9.4). On the other hand, when a view with no fill pattern is dirtied, each of its ancestors must be redrawn, all the way up the chain until one is found with a fill pattern (see FIGURE 9.5). The Newton view system does this because the background needs to be set appropriately for the view that was dirtied.

A good rule of thumb for judging when to use a fill pattern is:

- All views that are explicitly dirtied should have a fill pattern, whenever possible.

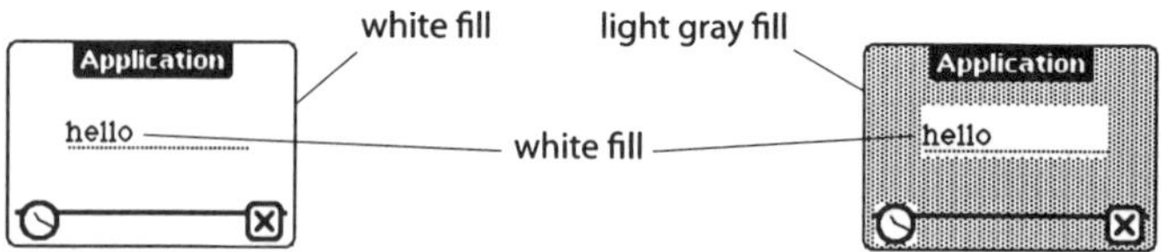

FIGURE 9.4 Two different applications with protoInputLines with a fill pattern. When the protoInputLine is dirtied, only the protoInputLine must redraw.

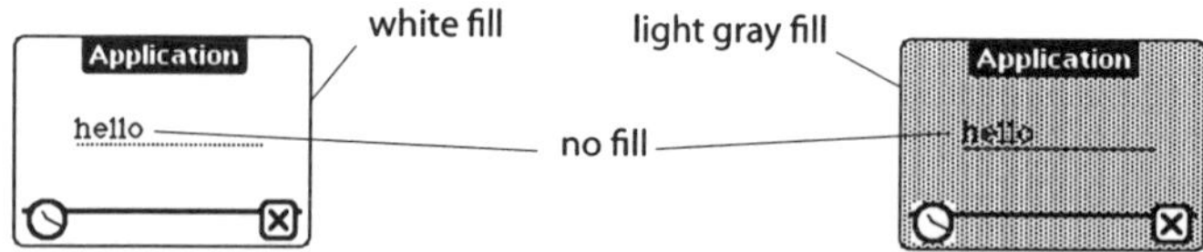

FIGURE 9.5 Two different applications with protoInputLines with no fill pattern. When the protoInputLine is dirtied, the protoInputLine and protoApp must both redraw.

The View System Update Mechanism

To get the best performance, it is crucial to know how and when the view system updates views. Here are some of the general rules.

When Views Are Dirtied

A view is dirtied when:

- It is sent the `Dirty` message. This message may be sent directly by your code, or indirectly when you use `SetValue`. For instance:

  ```
  SetValue(aStaticTextView, 'text, "new text")
  ```

  dirties `aStaticTextView` so that the new text will redraw. The view system may also send this message when the user does something to dirty a view, like write text into a protoInputLine.
- It is opened or shown.
- An obscured portion of the view is no longer obscured. For example, if a notify slip obscures part of a static text, the static text will

be dirtied when the notify slip is closed. Similarly, if a dragable keyboard obscures part of a view, the view is dirtied as soon as the keyboard is dragged away from it.

When Views Are Drawn

A view is not drawn immediately after it is dirtied. Instead, like the Macintosh or Windows, Newton redraws during its main event loop. Further, the drawing is done to an offscreen bitmap. During the drawing operations, the screen is not refreshed from the bitmap, but is only refreshed after the drawing operations are completed. In this way, multiple views dirtied during one operation can be redrawn to appear on-screen all at once, rather than staggering in one at a time.

Note: The details of how the view system handles dirtying and updating are subject to change in future Newton system software. Nevertheless, an understanding of how the view system works today can be very useful.

How the View System Knows Which Views to Draw

The view system does not keep track of a (potentially large) list of views that need to be redrawn; instead it keeps track of only two variables:

- *The update rectangle.* This is a rectangle in global coordinates specifying what portion of the screen must be updated.
- *The update view.* This view is the starting point for navigating the view hierarchy to redraw.

How the Update Rectangle and Update View Are Maintained

For simplicity's sake, assume for the moment that the update view is `nil` (that is, no view currently needs updating). When a view is dirtied, the view system sets the update rectangle to the bounds of the dirty view (outset, if necessary, by any frame). After this, the view system checks to see whether the view has a fill pattern. If it does, then the update view is also set to the dirty view (see FIGURE 9.6).

Notice, however, what happens if the view does not have a fill pattern. In this case, the update view is set to the closest ancestor of the dirty view that does have a fill pattern. Now, all those ancestors must be redrawn as well (see FIGURE 9.7).

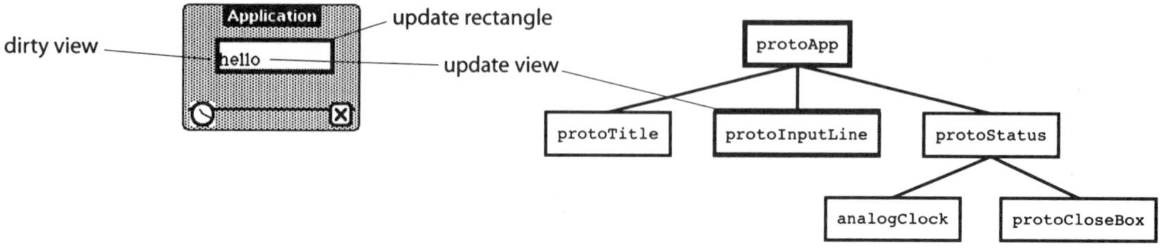

FIGURE 9.6 The update rectangle and update view after dirtying a filled view.

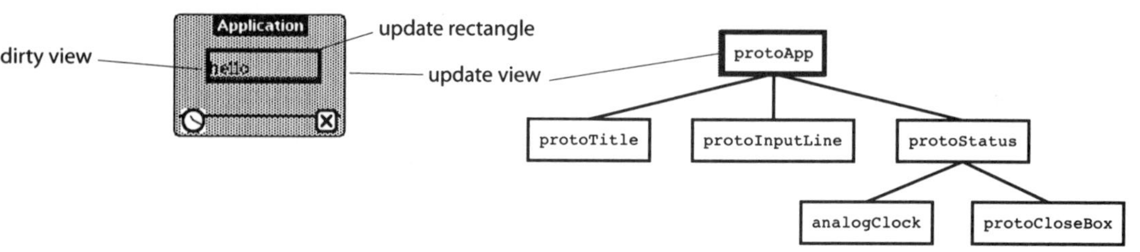

FIGURE 9.7 The update rectangle and update view after dirtying a view without a fill pattern.

In cases where there is already an update view and update rectangle, then the update rectangle is set to the smallest rectangle that will enclose both the old update rectangle and the update rectangle of the dirty view. Similarly, the update view is set to the lowest common ancestor of both the old update view and the update view for the new dirty view.

FIGURE 9.8 shows a protoInputLine with a fill pattern that has been dirtied. The resulting update view is the protoInputLine itself, while the update rectangle is the bounds of the protoInputLine. Now, things change. Before the dirty view is updated (that is, before the update event in the event loop), the analogClock gets dirtied (perhaps, the time displayed in the clock changed).

The analogClock gets has a fill pattern, and would have been made the update view if there weren't already an update view in place. Since there is an update view, however, the new update view is the lowest ancestor of the two views (the analogClock and the update view). In this case, the ancestor is the protoApp.

Similarly, the new update rectangle is the smallest rectangle enclosing the old update rectangle and the update rectangle of the newest dirty view. This means, of course, that when the view system redraws, it will need to redraw every view within the protoApp, since they all intersect the new update rectangle.

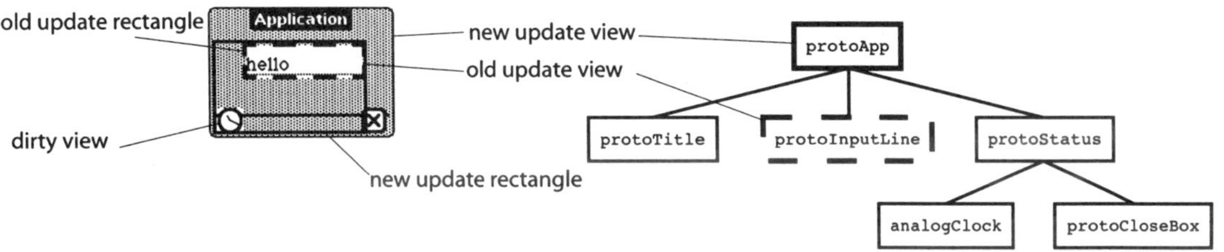

FIGURE 9.8 The update rectangle and update view modified after a second view is dirtied.

How the Update Rectangle and Update View Are Used to Draw

In the event loop, when the view system decides it is time to draw, it uses the following pseudocode:

```
UpdateAView(aView)
begin
   if aView intersects updateRectangle then begin
      Draw aView
      foreach childView of aView
         UpdateAView(childView)
   end
end
```

The system calls `UpdateAView` with the update view. This, in turn, recursively traverses all the descendants of the update view, drawing them if they intersect the update rectangle.

The amount of space used to keep track of what needs to be redrawn is minimal: a rectangle and a reference to a view. However, in comparison to the Macintosh, for instance, the Newton will indeed often be drawing more views than necessary. The Macintosh has better performance because it keeps an update region, which describes which bits of the screen need to be updated. Because you cannot rely on the speedier design found on Macintosh, it is up to you to design your views so that they are redrawn in the most efficient manner within the Newton system. This means putting them into well-constructed view hierarchies.

Using Hierarchies of Views

Arranging your views in proper hierarchies can speed up drawing. For example, assume you have a 10 x 5 grid of nonfilled views in a parent view as shown in FIGURE 9.9. If one of them is dirtied, the parent view and the dirty view will need to be redrawn. Examination of the `UpdateAView` pseudocode reveals, however, that all 50 children of the parent view will also be examined to see whether they intersect the update rectangle.

By putting each column into its own enclosing view you can speed up this update process. Such enclosing views should be invisible to the user (use a `clView` with no fill or frame). FIGURE 9.10 shows a resulting hierarchy with this type of design.

Now, when a view is dirtied, the same update view and update rectangle will result. Further, when the `UpdateAView` pseudocode operates on each of the five children of the parent view, four of them will not even intersect the update rectangle. Only the views in the same column as the dirty view will need to be checked for intersection. Thus, as FIGURE 9.10 shows, instead of checking 50 views for intersections, we're down to a mere 15 that need to be checked (the 5 new enclosing views, plus the 10 children in the dirty view's column).

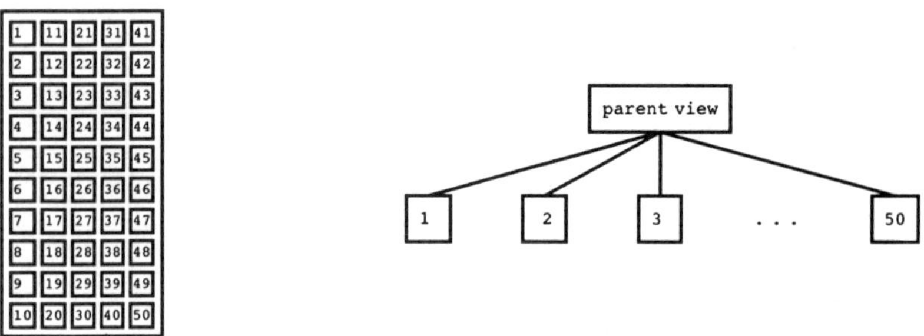

FIGURE 9.9 A parent view with 50 direct child views.

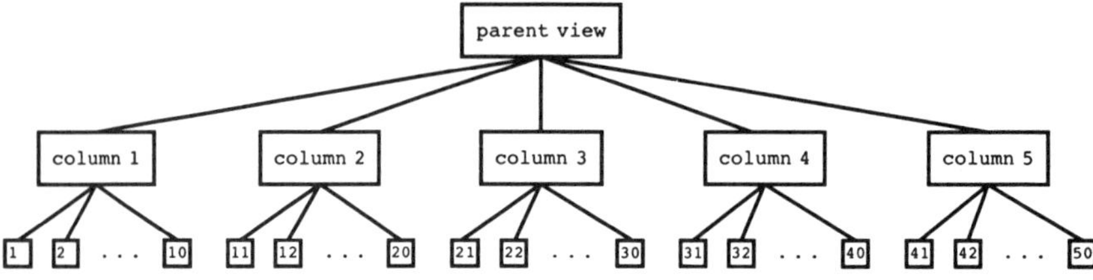

FIGURE 9.10 A parent view with 5 children, each with 10 views.

In addition to speeding up updates, arranging your views in hierarchies can quicken responses to stylus taps. The view system follows the same recursive search through the hierarchy on a stylus tap that it does for updating. Thus, without enclosing views for each column, the view system must check each view to see whether the tap is within its bounds. With enclosing views, the view system can quickly rule out whole columns at a time.

Using `RefreshViews`

As should be obvious by now, dirtying more than one view at a time can cause many unnecessary redraws of views. This is especially true if the two dirty views are far apart on-screen. Indeed, you might find it faster in some cases to redraw the first view before dirtying remaining views. This is done with the `RefreshViews()` call. `RefreshViews()` redraws the screen immediately; and after it is finished, the update view is set back to `nil`.

This circumstance of dirtying more than one view at a time happens more often than you might realize. For instance, FIGURE 9.11 shows a protoLabelPicker in which selecting a new item from the picker causes a protoStaticText view to change as well. Here is the sequence of events: When an item from the protoLabelPicker is chosen, the picker is dismissed, dirtying the area underneath the picker. Then, the protoLabelPicker's `labelActionScript` calls `SetValue` on the static text, dirtying the view:

```
func(index)
begin
   SetValue(statText, 'text, labelCommands[index]);
end;
```

Since two views have been dirtied, the update rectangle is the bounding box enclosing both of them. Unfortunately, in this case, that encloses most views on the screen.

You can be very clever here, however. You can modify the `labelActionScript` (which is called after the popup view is closed) so that it refreshes the views under the picker *before* dirtying the static text view. Here is some code that does this:

```
func(index)
begin
   RefreshViews();
   SetValue(statText, 'text, labelCommands[index]);
end;
```

Now, only the static text must be redrawn after the picker is dismissed.

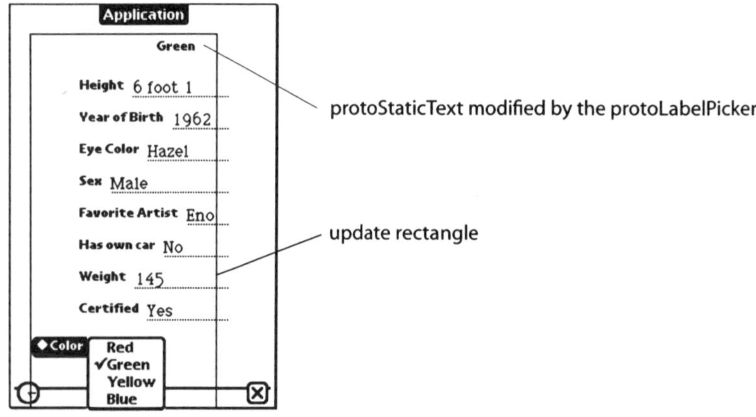

FIGURE 9.11 A protoLabelPicker which modifies a protoStaticText.

Summary

You can see that performance tuning of a Newton application involves a grab bag of techniques. It also does not hurt to start off with a well-thought-out coding style. Nevertheless, you will still need to pay special attention to certain portions of your code where most of the time is spent. Happily, by optimizing these few key areas you can both increase the responsiveness of your program and use memory much more efficiently. Remember this bag of tricks includes relying on built-in functions rather than creating your own custom NewtonScript; defining your locals; using the right loop for the job; and defining constants where possible.

You can tighten up your memory use by paying attention to important details such as frame maps. You should also never blindly use the cloning functions to accomplish a task that can be handled in another way. Likewise, being aware of how operators use memory can help you determine when to change to a more efficient method.

Last of all, your application views should be placed within logical view hierarchies. A well-considered design can go a long way to giving your application a more responsive feel to stylus taps and snappier performance in general.

Chapter 10

Advanced Soup Design

Many things difficult to design prove easy to performance.

—Samuel Johnson

We recently had a student in a Newton programming class tell us about his early days as a Newton programmer. He and his partner had come up with a great idea for a Newton application. They had already written much of the application when the fun started. It began when they designed their application's data soup. After filling the application up with some sample data, they found to their horror that what had been a great idea was now an application so slow that no one would use it. One partner called the other and announced the bad news and the prognosis for their future (Roughly, he said: "We're dead").

Being persistent (and having no alternative) they went back to the drawing board and eventually realized that their initial soup design was flawed. The application was for use by doctors to see which patients had undergone particular operations.* In the initial soup design, an entry had been based on a patient record and contained all the information about that particular person, including what operations the person had undergone. Using this design, operation searches by a doctor that were supposed to be useful were instead abysmally slow.

After some thinking, they realized they were using the wrong design and index. Even though there were many patients, doctors needed to know about only a few operations. Their old searches were slow because each patient record had to be searched for a particular operation that would rarely be present.

They redesigned the soup to be structured around operations instead of patient names and suddenly they had a working application. Now, they could quickly search though the indexed operations (of which there were few) for the patients who had received that particular procedure. Searches were now snappy and they were back in business. But not without a lot of pain† along the way.

It is our hope that this chapter contains enough information about designing soups to spare you such anguish, but enough moralizing for now. The essential ingredients in the design of a Newton soup include:

- Creating the right searching parameters.
- Having indexes on the right slots or combinations of slots.
- Designing soup entries that take into account application requirements.

First, we will look at how soup indexes work, how searches occur, and what type of indexes to use for various search requirements. Next, we will look at how to best structure soup slots to give you efficient indexes. Last of all, we will look at how you take application requirements and turn them into a good soup design.

---

* The application description has been altered, though the gist of the problem remains the same.

† Although the doctors would probably have labeled it discomfort.

Using Indexes Effectively

Using an index efficiently requires that you understand how they work and how the structure of an index affects the performance of data searches.

How Indexes Work

An index is like a binary tree with key values in nodes of the tree. Searching for a particular key value is done by starting at the root node, and then comparing the search key with the value stored in the node. If the key is less, searching continues at the left child; if the key is greater, searching continues at the right child. Although indexes are slightly more complex than this (they are more like a B-tree data structure in which there are actually more than two children for each node), the concept remains essentially the same. See FIGURE 10.1 for a typical index example.

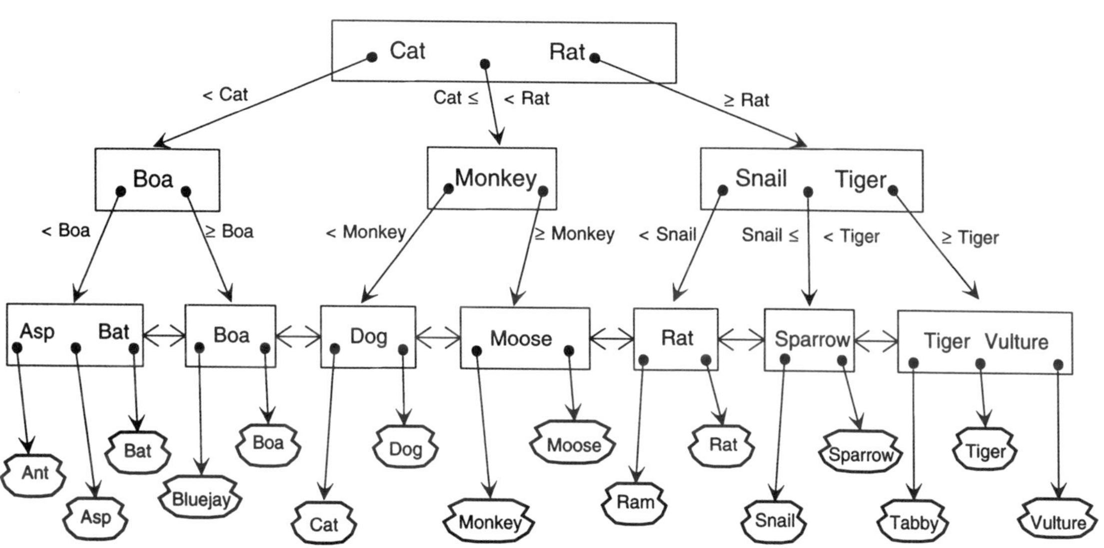

FIGURE 10.1 Structure of an index.

Notice in the index in FIGURE 10.1 that the nodes at the leaf of the tree reference the actual entries in the soup and contain forward and backward pointers. This enables traversal to the next and previous node in index order.

Indexes on the Newton aren't implemented using exactly this data structure; the concept is similar, however.

Searching through Every Entry

Given this structure for indexes, what is the most appropriate method for traversing the soup? The answer, of course, depends on your search criteria. For example, imagine you are using an index to traverse a soup and want to visit a subset of contiguous entries. The simplest approach would be to go through each entry checking to see whether it matches your criteria. On the Newton, this simple approach will result if you construct your soup query using only a `validTest` (see FIGURE 10.2).

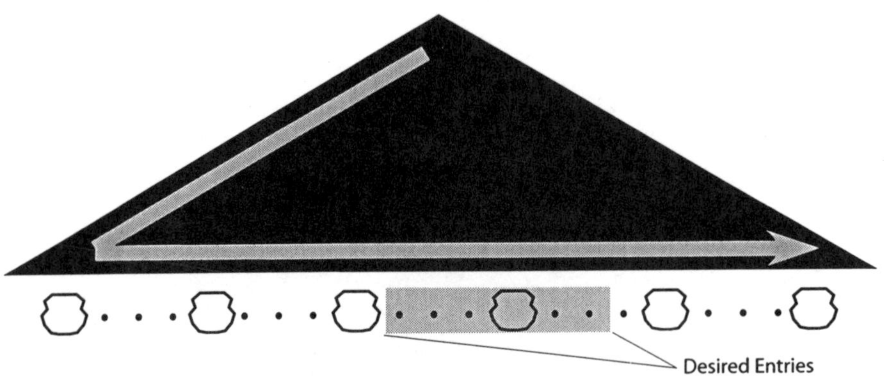

FIGURE 10.2 The entries visited when using only a `validTest`.

Starting at the Right Place

A more complicated approach involves starting your search at a more precise point. This is accomplished by skipping as many entries as possible by using a `startKey`. If you place a `startKey` in your query, some of the leaf nodes at the left of the tree can be missed altogether (see FIGURE 10.3). The cursor uses the `startKey` to skip rapidly through nodes in the tree to the desired beginning node. This is where the saving ends, however, because all nodes from there to the end of the tree must be visited.

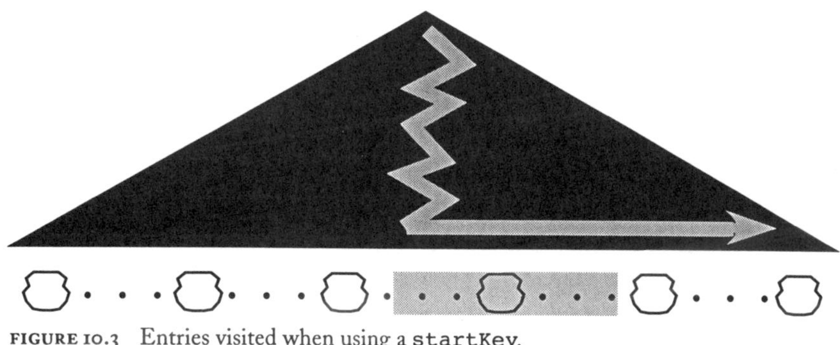

FIGURE 10.3 Entries visited when using a `startKey`.

Ending at the Right Place as Well

If you add an `endTest` as well as a `startKey`, then the leaf nodes at the right of the tree can also be skipped in the search (see FIGURE 10.4). The cursor uses `endTest` to determine when the last of the matching nodes has been reached and thus searching stops.

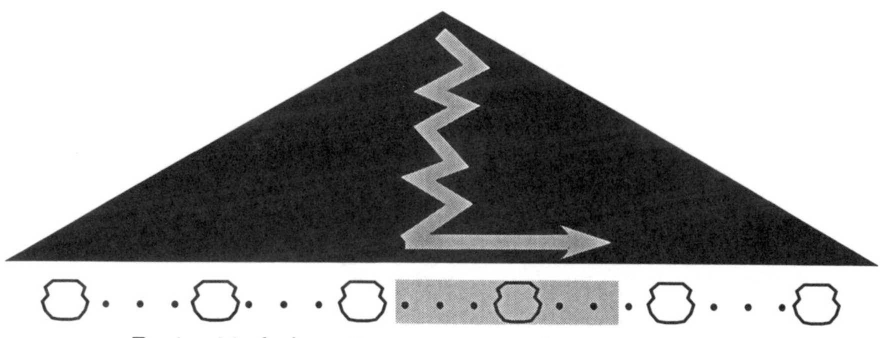

FIGURE 10.4 Entries visited when using a `startKey` and `endTest`.

When to Use an Index

There are two clear cases in your soup design where you should use an index:

- When you need to access entries in sorted order.
- When you need to search quickly for a particular entry or set of entries that share a common index value or range of index values.

There are other cases where you might need to search, but using an index will be of no help. This occurs if the set of valid entries does not share a common index value or range of index values. For example, imagine trying to find all entries whose age multiplied by weight exceeds their height multiplied by shoe size. No index will help find exactly those entries (unless you have the weirdly constructed and very specialized index on the integer slot `ageTimesWeightMinusHeightTimesShoeSize`). Barring such unusual types of requirements, you can do a great deal to improve soup performance by carefully choosing which slots in a soup will be indexed. Just remember the following as a general rule:

- *Index any slot that you will use for sorting or searching.*

Unfortunately, indexes are not free, so it will not make sense to index on every slot in a soup. Indexes take up space in the soup and make adding, deleting, and modifying entries slower.

When to Use `startKey` Only

You should use a `startKey` if you can rule out a sequence of initial entries based on an indexed slot value. You need to be aware that a `startKey` only affects where the cursor is initially placed in the soup; no subsequent operations are affected. In particular, if you move backward in the soup (using `Move` or `Prev`), you can move to entries before the `startKey`. Of course, you can always use a `Reset` to take you back to the entry matching the `startKey`. Thus, you should always use a `startKey` any time there are some initial entries that can be ignored and when you won't be moving backward through the cursor.

A common case crying out for a `startKey` would be counting all names from the Names soup that start with the letter 'K' or greater. Here is some code that does this:

```
q := Query(GetUnionSoup(ROM_cardfilesoupname),
    {  type:        'index,
       indexPath:   'sortOn,
       startKey:    "K",
});
```

```
total := 0;
e := q:Entry();
while e do begin
   total := total + 1;
   e := q:Next();
end;
Print(total);
```

When to Use `endTest` *Only*

You should provide an `endTest` when a sequence of final entries can be ruled out of the search. For example, to support "Find Dates Before" if there is an index on a `date` slot, you could use the following query:

```
DateFind := func(theDate, …)
begin
   …
   q := Query(soup, {type: 'index,
      indexPath: 'date,
      endTest: func(e) return e.date > theDate);
   …
end;
```

When to Use Both `startKey` *and* `endTest`

You can use both a `startKey` and an `endTest` if your entries are located in a specific range of entries based on an indexed slot value. One example might be counting all names from the Names soup that start with 'K' through 'P'. Here is code that uses both a `startKey` and an `endTest` to perform this type of count:

```
q := Query(GetUnionSoup(ROM_cardfilesoupname),
{type: 'index,
   indexPath: 'sortOn,
   startKey: "K",
   endTest: func(e)
      begin
         local char := e.sortOn[0];

         if char >= $K and char <= $P then
            return nil
         else if char >= $k and char <= $p then
```

```
                    return nil
                else
                    return true;
            end,
    });

    total := 0;
    e := q:Entry();
    while e do begin
       total := total + 1;
       e := q:Next();
    end;
    Print(total);
```

Notice that the `endTest` checks both for a soup entry which is greater than desired (going off the right-hand side) and for a soup entry which is less than desired (going off the left-hand side). Remember that the `startKey` is ignored by `Next` and `Prev` and only `endTest` will prevent `Prev` from going before the K's (although the code we've used which tests the cursor doesn't actually ever use `Prev`).

Dynamic Indexes

Normally, you make all the indexes you need when you create your soup (`RegisterCardSoup` takes an array of indexes that it uses whenever it creates a new soup). In specialized circumstances, however, you may want to add an index dynamically by using the soup method `AddIndex`. For temporary situations in which you will remove the index when you're through, this is fairly straightforward:

- Use `AddIndex` to add the index, do your operations with the index, and then use `RemoveIndex`.

Both `AddIndex` and `RemoveIndex` work with union soups.

For example, imagine you want to use a temporary index as part of printing a number of addresses. You could create an index on the postal code so that addresses come out in postal code order. If the operation is very rare, however, it will not make sense to maintain the postal code index permanently, (remember that indexes take time and space). Thus, a temporary index might be just the right thing.

Here is the code to accomplish this type of task:

```
myAddressSoup:AddIndex({structure: 'slot,
   path: 'postalCode, type: 'string});
q := Query(myAddressSoup, {type: 'index,
   indexPath: 'postalCode});
e := q:Entry();
while e do begin
   PrintAnEntry(e);
   e := q:Next();
end;
myAddressSoup:RemoveIndex('postalCode);
```

There is a potential problem in this situation, however. Because union soups are dynamic, tricky things might result if a store is removed *after* the index is added, but *before* the index is deleted. In this case, the store will have an extra index.

Navigating with a Cursor

It is useful to know how the cursor operations work with indexes. Here are the standard cursor methods and their relationship to indexes and queries:

`Prev()`	Moves forward one entry in index order.
`Next()`	Moves backward one entry in index order.
`Move(`*`num`*`)`	Moves *num* entries forward (or backward if negative) in index order. It is only slightly faster than repeated calls to `Next` (or `Prev`).
`Reset()`	Moves to the first entry (takes into account the `startKey`). This operation uses the index, so it is fast.
`Goto(`*`entry`*`)`	Moves to the specified entry (if it is part of the cursor). This operation is fast.
`GotoKey(`*`keyValue`*`)`	Moves to the first entry (in index order) whose indexed value is greater than or equal to `keyValue`. Since this uses the index, it is fast.

Referring to Entries Uniquely

Even though entries, soups, and stores all have unique IDs, you need to keep track of all three in order to identify an entry uniquely. This is because the entry ID is unique only within a soup on a given store. Thus, an entry in the same soup on another store could have the same ID. Currently, entry IDs are assigned sequentially (this is subject to change). Soups and stores, however, are assigned random numbers when they are initialized (but don't count on this either).

Now, you might ask yourself if you really need to keep track of all three numbers. Given the existence of union soups, it may be clear that a store ID is necessary along with an entry ID (a union soup can easily have two different entries with the same entry ID), but why would you ever need the soup ID number?

Imagine you want to save information about the current entry so that when the application is reopened it can display that entry. First off, you know that the application needs to save the information in the `prefs` soup. But what information needs to be saved? Obviously the entry ID and the store ID. But is that enough? The answer is "No," and here is the dreaded situation that proves it:

> A user with an external store on a card runs the application, closes it (the entry ID and store ID are saved), and then takes the card to a different Newton where she or he deletes the soup and then runs the application on the new Newton. This new Newton will create a soup (with the same name) on the external store. Now, the user creates some entries on that store (one of which duplicates the entry ID of the original saved entry). Now, the user pulls the card and returns it to the original Newton. When the application runs, it will find a soup of the correct name existing on the store and with a matching store ID. In addition, it will find an entry with the correct saved entry ID. Unfortunately, it will not be the original entry.

The addition of the soup ID allows an application to handle this very situation.

Caution: *Caution:* A bug in Newton Connection Kit 2.0 causes unique IDs for soup entries to change after a restore.

Sorting on More Than One Slot

In some cases, it would be useful to sort on more than one key. For instance, opening the Names application in the overview mode is currently very slow. It happens whenever there are a lot of names and you are looking at a particular folder. This is because there is no index on the `labels` slot (where the folder value is stored). You might ask, why doesn't the Names application use such an index? While it is true that it would be able to retrieve entries quickly in a given folder if it had such an index, those entries would no longer be displayed alphabetically on the name in this instance. What the Names application really needs is the ability to filter *first* on the `labels` slot, and then sort on the `name` slot.

Soups in general do not allow you to sort on more than one slot. Happily, there are ways to simulate such a sort. Let us look at some Names soup entries to see how you can perform this trick. The Names soup currently stores entries in the following form:

```
{sortOn: "Cooper", labels: 'Business, …}
{sortOn: "Jones", labels: 'Personal, …}
{sortOn: "Leyland", labels: 'Business, …}
{sortOn: "Oakley", labels: nil, …}
{sortOn: "Smith", labels: 'Personal, …}
```

Thus, to display the entries from the Business folder, all the entries must be visited.

If the Names soup also had an indexed slot `labeledSortOn` containing not just the name but the folder as well, then displaying entries could be much quicker. Standard entries would look like this instead:

```
{labeledSortOn: ":Oakley",
   sortOn: "Oakley", labels: nil, …}
{labeledSortOn: "Business:Cooper",
   sortOn:"Cooper", labels: 'Business, …}
{labeledSortOn: "Business:Leyland",
   sortOn:"Leyland", labels: 'Business, …}
{labeledSortOn: "Personal:Jones",
   sortOn: "Jones", labels: 'Personal, …}
{labeledSortOn: "Personal:Smith",
   sortOn: "Smith", labels: 'Personal, …}
```

To display the entries from the Business folder, a `Query` with a `labeledSortOn` index could be done with a `startKey` of "Business:" and an `endTest` to check that the `labeledSortOn` slot didn't begin with "Business:". In this particular example, displaying the Business folder would require visiting only three entries (the two in the Business folder, plus one more to determine that there are no more Business entries).

If you remember from our earlier discussion on indexes, this is the fastest searching method available (as opposed to the slowest method that Names currently uses). Thus, you end up with a much quicker display in the overview. The disadvantage, of course, is that the `labeledSortOn` slot must be updated when either the name or `labels` slot changes.

The "All Folders" display would be done with a Query using the original `sortOn` index.

Creating a Synthetic Slot

The technique of creating a synthetic slot (which contains a value based on one or more other slots) can be used to handle sorting and searching on either simple or complex criteria. We mentioned earlier in this chapter the silly example of displaying only those entries whose age multiplied by weight exceeds their height multiplied by shoe size. If each soup entry had a slot which contained the result of that comparison (perhaps an integer slot named `ageTimesWeightMinusHeightTimesShoeSize`), then an appropriate `startKey` (of `0`) could quickly lead to those entries.

Unfortunately, this technique isn't useful for ad hoc queries, unless you create a slot in your soup entries and an index for that slot on the fly.

Indexing on Two Slots

If you have a slot that contains only a few distinct values, it is quite possible to simulate an index on two slots. Let's continue with the same Names soup example. Here again are some standard Names soup entries:

```
{sortOn: "Cooper", labels: 'Business, …}
{sortOn: "Jones", labels: 'Personal, …}
{sortOn: "Leyland", labels: 'Business, …}
{sortOn: "Oakley", labels: 'nil, …}
{sortOn: "Smith", labels: 'Personal, …}
```

Let us see what happens when we add a new slot whose symbol is the value of the `labels` slot, and which contains the same value as the `sortOn` slot. Now, the new entries will look like this:

```
{business: "Cooper",
   sortOn: "Cooper", labels: 'Business, …}
{business: "Leyland",
   sortOn: "Leyland", labels: 'Business, …}
{personal: "Jones",
   sortOn: "Jones", labels: 'Personal, …}
{personal: "Smith",
   sortOn: "Smith", labels: 'Personal, …}
{unfiled: "Oakley",
   sortOn: "Oakley", labels: 'nil, …}
```

If there were an index on `business`, `personal`, `unfiled`, etc., then finding all entries in the business folder (sorted by name) would be quite easy. A `Query` with an `indexPath` of `'business` would iterate through each item in the business folder sorted by name (if a slot that is indexed isn't present, or contains `nil`, it is skipped by that indexed `Query`).

Again, there is some maintenance involved. If the folder changes (that is, the value of the `labels` slot changes), then a slot must be removed and a new slot added. In addition, indexes must be created and deleted as folders are created and deleted.

Alternatives to Soups

Perhaps the answer to your design problems is to not use a soup at all. For instance, read-only data is a perfect example of data screaming for non-soup storage. The alternatives to soups include, storing data in arrays or frames or as binary objects within your package. Further, these can be more compact because you may choose to have your package stored in compressed form. In addition, retrieval directly from your data structure is usually quicker than reading from a soup.

There is also the possibility of combining structures; you could store your initial package data in an array, frame, or as a binary object and keep any new or modified data in a soup. Thus, you could have the flexibility of soups when you need it, while having the speed of alternate data structures when you don't.

Designing Your Soup

When designing the format of your soup(s) and the indexes that you'll maintain, make sure to consider *fully* the particular needs of your application. If nothing else, the story related at the beginning of this chapter should have convinced you of this.

So, to help you get an idea of how to best create a soup design, we will look at an example application and work through several possible designs. This application is for cataloging CD collections and for each CD, we'll need to keep track of the song titles, the title of the CD, and the artist(s). The application requirements are straightforward:

- Display all CD titles alphabetically.
- Know the titles of songs on a particular CD.
- Know the artist for a particular CD.
- Know what CD a particular song is on.
- Know what CDs an artist has done.

TABLE 10.1 shows three sample CDs, each of which contains the data we will be using. We will also be referring to other CDs by these same artists in some of our examples.

Naive Soup Design

A naive approach would be to design this soup with each CD as an entry. The title would be kept as a string, and the artist and song titles stored in arrays. The index would be on the `title` slot. You can't create an index on the song titles array, although you could create indexes on each element of the array. Creating an index on each element wouldn't be particularly useful however; you really want an index on all song titles, which you can't create. For the same reason, you couldn't create an index on the artists array. Given the CDs in TABLE 10.1, this is how each entry in this soup design would look:

```
{
   title:"Abbey Road",
   artists: ["Beatles"],
   songTitles: [
      "Come Together",
      "Something",
```

```
        "Maxwell's Silver Hammer",
        "Oh! Darling",
        "Octopus's Garden",
        "I Want You (She's So Heavy)",
        "Here Comes the Sun",
        "Because",
        "You Never Give Me Your Money",
        "Sun King",
        "Mean Mr. Mustard",
        "Polythene Pam",
        "She Came in through the Bathroom Window",
        "Golden Slumbers",
        "Carry that Weight",
        "The End",
        "Her Majesty",
    ],
}

{
    title:"So"
    artists: ["Peter Gabriel"],
    songTitles: [
        "Red Rain",
        "Sledgehammer",
        "Don't Give Up",
        "That Voice Again",
        "In Your Eyes",
        "Big Time",
        "We Do What We're Told (milgram's 37)",
        "This Is The Picture (excellent birds)",
    ],
}

{
    title:"Wrong Way Up"
    artists: ["Brian Eno", "John Cale"],
    songTitles: [
        "Lay My Love",
        "One Word",
        "In the Backroom",
        "Empty Frame",
        "Cordoba",
        "Spinning Away",
        "Footsteps",
        "Been There Done That",
        "Crime in the Desert",
        "The River",
    ],
}
```

With this design it is worth noting that the format is very simple; there is a 1 to 1 correspondence between CDs and entries in a soup.

Title: Abbey Road *Artist: Beatles*	*Title: Wrong Way Up* *Artists: Brian Eno, John Cale*	*Title: So* *Artist: Peter Gabriel*
Song Titles	**Song Titles**	**Song Titles**
Come Together	Lay My Love	Red Rain
Something	One Word	Sledgehammer
Maxwell's Silver Hammer	On the Backroom	Don't Give Up
Oh! Darling	Empty Frame	That Voice Again
Octopus's Garden	Cordoba	In Your Eyes
I Want You (She's So Heavy)	Spinning Away	Mercy Street
Here Comes the Sun	Footsteps	Big Time
Because	Been There Done That	We Do What We're Told (milgram's 37)
You Never Give Me Your Money	Crime in the Desert	This Is The Picture (excellent birds)
Sun King	The River	
Mean Mr. Mustard		
Polythene Pam		
She Came in through the Bathroom Window		
Golden Slumbers		
Carry that Weight		
The End		
Her Majesty		

TABLE 10.1 CD collection.

So, let's go ahead and see how well this design meets our various application requirements.

Display All CD Titles Alphabetically

Clearly this is accomplished very easily. All you have to do is use the `title` index in a `Query` to obtain the titles in alphabetical order. This would take time proportional to the number of CDs.

Know the Titles of Songs on a Particular CD

All you have to do is display the entries from the `songTitles` array. This would take time logarithmically proportional to the number of entries (to find the entry).

Know Who the Artist Is for a Particular CD

Here, you would display the `artists` slot from an entry. This would take time logarithmically proportional to the number of entries (to find the entry).

Know What CD a Particular Song Is On

This is very hard because you would have to use a `validTest` that searches through all entries. It would, of course, be very slow. This would take time proportional to the number of CDs.

Know What CDs an Artist Has Done

This is likewise difficult and slow. Once again you would have to resort to the use of a `validTest`. This would take time proportional to the number of CDs.

Space Required

This design would require space proportional to the number of `cds` (the entry and `title`) plus the number of song titles (the `songTitles`) plus the number of cd-artist pairs (the `artists`).

Better Soup Design

By using some extra space, you could deal more appropriately with the last two requirements. By just creating some additional entries, you can handle the problems that couldn't be solved using indexes. You do this by simply adding soup entries that keep track of what CDs a particular artist has done:

```
{
   artist: "Beatles",
   cdTitles: ["Abbey Road", "Let it Be", …],
},

{
   artist: "Peter Gabriel",
   cdTitles:["So", "Us", …],
}

{
   artist: "Brian Eno",
   cdTitles: ["Wrong Way Up", "Neroli", "Discreet Music", …],
}

{
   artist: "John Cale",
   cdTitles: ["Wrong Way Up", "Paris 1919", …],
}
```

Notice that the new entries have an `artist` slot while the old entries have an `artists` (the plural) slot. By having an index on this new `artist` slot, you can easily display what CDs a particular artist has done. You can also choose to store these entries in a different soup or in the same soup if you prefer. Since the slots that are indexed have different symbols, the indexes won't interfere with one another. The `artist` index will only index those entries that contain such a slot. Similarly, the `title` index will only index those entries that have a `title` slot.

Finally, in order to handle the mapping from song title to CD, you would add some other types of entries. You would also add an index on the `songTitle` slot:

```
{
   songTitle:   "Lay My Love",
   cdTitle:     "Abbey Road",
},

{
   songTitle:   "Come Together",
   cdTitle:     "Abbey Road",
},
```

```
// and so on...

{
   songTitle:  "Crime in the Desert",
   cdTitle:    "Wrong Way Up",
},

{
   songTitle:  "The River",
   cdTitle:    "Wrong Way Up",
},
```

All the operations in the naive design which took time proportional to the number of cds now take time logarithmically proportional (a substantial time savings). Notice, however, that even though all the requirements have been satisfied, there are numerous disadvantages as well. One particular disadvantage to using nonhomogenous soups like this (where each entry is not the same type of item) is with Newton Connection. It expects that all the entries in a soup can be displayed in one overview. Thus, if you'll be supporting Newton Connection, you are better off creating different soups for the different types of entries, not storing different types of entries in one soup. You should also notice that this design has a great amount of storage redundancy. For each CD:

- The song titles are stored twice.
- The CD titles are stored three times.
- The artists are stored twice.

Perhaps there is a better way that does not use up such vast amounts of space....

An Even Better Soup Design

By using ideas from relational databases, you could create a design even better than the last one. TABLE 10.2 through TABLE 10.6 will give you an idea of how the data will be organized. Notice that the keys are values that uniquely identify a particular CD, artist, or song.

CD KEY *(symbol)*	TITLE *(string)*
'|abbey road|	"Abbey Road"
'so	"So"
'|wrong way up|	"Wrong Way Up"

TABLE 10.2 CD information entries.

ARTIST KEY *(symbol)*	ARTIST NAME *(string)*
'beatles	"Beatles"
'|peter gabriel|	"Peter Gabriel"
'|brian eno|	"Brian Eno"
'|john cale|	"John Cale"

TABLE 10.3 Artist information entries.

SONG KEY *(symbol)*	SONG TITLE *(string)*
'|lay my love|	"Lay My Love"
'|come together|	"Come Together"
'|the river|	"The River"
'footsteps	"Footsteps"

TABLE 10.4 Song information entries.

CD KEY *(symbol)*	ARTIST KEY *(symbol)*
'|abbey road|	'beatles
'so	'|peter gabriel|
'us	'|peter gabriel|
'|wrong way up|	'|brian eno|
'|wrong way up|	'|john cale|

TABLE 10.5 CD-artist relationship entries.

CD KEY *(symbol)*	SONG KEY *(symbol)*
`'|abbey road|`	`'|lay my love|`
`'|abbey road|`	`'|come together|`
`'|wrong way up|`	`'|the river|`
`'|wrong way up|`	`'|spinning Away|`
`'|so|`	`'|in your eyes|`

TABLE 10.6 CD-song relationship entries.

Remember, that if we were to use strings, we would have the same storage problems. We could, on the other hand, use unique integer values. Another possibility is to use unique symbols (perhaps a string converted to a symbol), since the string representing a symbol is stored only once in a soup. The advantage of symbols is that they are easy to read; the disadvantage is that the characters of the symbol must be stored in the soup, taking up more space. Another problem with using symbols occurs when a title is not unique (for instance, two bands use the same title for their CD), or if titles contain non-ASCII characters (since symbols are 7-bit ASCII). In that case, we could add a unique ID to the end of the title.

Below are some sample entries for each of the five types of relational entries we have in our application soup. The first entries are sample CD-*info* entries and they would have an index on both the `cdKey` and `cdTitle`:

```
{
   cdKey:   '|abbey road|,
   cdTitle: "Abbey Road",
},

{
   cdKey:   'so,
   cdTitle: "So",
},

{
   cdKey:   '|wrong way up|,
   cdTitle: "Wrong Way Up",
},
```

The second type of entries represents *artist-info* entries. They would also need an index on both the `artistKey` and the `artistTitle` slots:

```
{
   artistKey:  'beatles,
   artistTitle:"Beatles",
},

{
   artistKey:  '|peter gabriel|,
   artistTitle:"Peter Gabriel",
},

{
   artistKey:  '|brian eno|,
   artistTitle:"Brian Eno",
},

{
   artistKey:  '|john cale|,
   artistTitle:"John Cale",
},
```

The third type of entries are *song-info* entries and they would need indexes for both the songKey and songTitle:

```
{
   songKey:    '|lay my love|,
   songTitle:  "Lay My Love",
},

{
   songKey:    '|come together|,
   songTitle:  "Come Together",
},

...

{
   songKey:    '|crime in the desert|,
   songTitle:  "Crime in the Desert",
},

{
   songKey:    '|the river|,
   songTitle:  "The River",
},
```

The fourth set of entries handles the CD-*artist* relations. For these, you would need an index on artist and cd:

```
{
   artist:  'beatles,
   cd:      '|abbey road|,
},

{
   artist:  'beatles,
   cd:      '|let it be|,
},

{
   artist:  '|peter gabriel|,
   cd:      'so,
},

{
   artist:  '|peter gabriel|,
   cd:      'us,
},

{
   artist:  '|brian eno|,
   cd:      '|wrong way up|,
},

{
   artist:  '|john cale|,
   cd:      '|wrong way up|,
},
```

The fifth and last type of entries is CD-*song* relations. They would need an index on both the `song` and `theCD` slots:

```
{
   song:    '|lay my love|,
   theCD:   '|abbey road|,
},

{
   song:    '|come together|,
   theCD:   '|abbey road|,
},

...
```

```
{
   song:     '|crime in the desert|,
   theCD:    '|wrong way up|,
},
{
   song:     '|the river|,
   theCD:    '|wrong way up|,
},
```

With this brand new relational design, let us see how our application requirements would be met:

Display All CD Titles Alphabetically

We would use the `cdTitle` index as the `indexPath` of a `Query` on the CD-*info* entries.

What Songs Are on a Particular CD?

We could do a `Query` using the `cdTitle` index in the CD-*info* entries to find the corresponding `cdKey`. We would then use that key as the `startKey` of a `Query` that relies on the `theCD` index on the CD-*song* entries. Finally, for each entry, we would do a lookup in the *song-info* entries to find the title.

For example, to find what songs are on the cd "Wrong Way Up", we would make the following `Query` in the soup containing the CD-*info* entries:

```
Query(soup, {type: 'index,
   indexPath: 'cdTitle,
   startKey: "Wrong Way Up"})
```

The `Query` would yield the following entry:

```
{
   cdKey:    '|wrong way up|,
   cdTitle: "Wrong Way Up",
},
```

Then, we'd make another `Query` in the CD-*song* entries:

```
Query(soup, {type: 'index,
   indexPath: 'theCD,
   startKey: '|wrong way up|,
   endTest: func(e) e.cdKey <> '|wrong way up|})
```

The `Query` would yield the following entries (among others):

```
{
   song:     '|crime in the desert|,
   theCD:    '|wrong way up|,
},
{
   song:     '|the river|,
   theCD:    '|wrong way up|,
},
```

For each of these entries, we would use the `song` slot to do a `GotoKey` in the *song-info* entries:

```
cursor := Query(soup, {type: 'index,
   indexPath: 'songKey});
cursor:GotoKey('|crime in the desert|);
cursor:GotoKey('|the river|);
```

The `GotoKey` would go to each of the following entries:

```
{
   songKey:    '|crime in the desert|,
   songTitle:  "Crime in the Desert",
},
{
   songKey:    '|the river|,
   songTitle:  "The River",
},
```

Who the Artist Was for a Particular CD?

We would do a `Query` using the `cdTitle` index on the CD-*info* entries. This would allow us to find the corresponding `cdKey`. We would then use that key as the `startKey` of a `Query` based on the `cd` index on the CD-*artist* entries. Finally, we would do a number of `GotoKeys` in the artist-info entries to obtain the name of the artist(s).

For example, to find the artists for "Wrong Way Up," we'd do a `Query` with the startKey `"Wrong Way Up"` in the CD-*info* entries, yielding:

```
{
   cdKey:   '|wrong way up|,
   cdTitle: "Wrong Way Up",
},
```

Then, we'd use the key `'|wrong way up|` as the `startKey` of a `Query` in the CD-*artist* entries, yielding the following entries:

```
{
   artist:   '|brian eno|,
   cd:       '|wrong way up|,
},
{
   artist:   '|brian eno|,
   cd:       'neroli,
},
{
   artist:   '|john cale|,
   cd:       '|wrong way up|,
},
```

For each of these entries, we would use the `artist` slot to do a `GotoKey` in the *artist-info* entries, yielding:

```
{
   artistKey:   '|brian eno|,
   artistTitle:"Brian Eno",
},
{
   artistKey:   '|john cale|,
   artistTitle:"John Cale",
},
```

On Which CD Was a Particular Song?

We would do a `Query` using the `songTitle` index on the *song-info* entries to find the corresponding `songKey`. Next, we would use that key as the `startKey` of a `Query` using the `song` index on the CD-*song* entries. Finally, we would use the CD-*info* entries to find the title of the CD.

For example, to find the cd that contains the song "Come Together," we'll do a Query using a `startKey` of `"Come Together"` in the song-info entries, yielding:

```
{
   songKey:     '|come together|,
   songTitle:   "Come Together",
},
```

Then, we would use the `songKey` as the `startKey` of a `Query` in the CD-*song* entries, yielding:

```
{
   song:    '|come together|,
   theCD:   '|abbey road|,
},
```

Finally, we would use `theCD` as the `startKey` of a `Query` in the CD-*info* entries, yielding:

```
{
   cdKey:   '|abbey road|,
   cdTitle: "Abbey Road",
},
```

What CDs has an Artist Done?

We would do a `Query` using the `artistTitle` index on the *artist-info* entries to find the corresponding `artistKey`. We would use that as the `startKey` of a `Query` using the `artist` index on the CD-*artist* entries. Finally, we would do a number of `GotoKey`s in the CD-*info* entries to obtain the CD title(s).

For example, to find what CDs Brian Eno has done, we would do a `Query` in the *artist-info* entries with a `startKey` of `"Brian Eno"`, yielding:

```
{
   artistKey:   '|brian eno|,
   artistTitle:"Brian Eno",
},
```

Then, we would do a `Query` in the CD-*artist* entries using a `startKey` of `'|brian eno|`, yielding:

```
{
   artist:    '|brian eno|,
   cd:        '|wrong way up|,
},
```

For each of these entries, we would use the `cd` slot to do a `GotoKey` in the CD-*info* entries, yielding:

```
{
   cdKey:   '|wrong way up|,
   cdTitle: "Wrong Way Up",
},
```

Notice that in addition to nicely handling the current requirements, this design is very extensible. If we need to keep additional information about a CD (*e.g.*, date released, publisher), an artist (*e.g.*, birthdate or nationality) or about a song (*e.g.*, author) we can put it in corresponding info entries. If we need to keep additional relations (*e.g.*, what bands has an artist been in, or what songs has an artist played on), we can do so by creating additional sets of entries to represent that relation.

Summary

You should now understand most of the aspects of good soup design. The first important point we covered was how indexes work and the best ways to create them. Next, we looked at how the search and display requirements of an application impact the structure of its indexes. This discussion included such issues as how to create synthetic slots for special types of indexes and how to identify entries uniquely. Last of all, we took up the more general issue of soup design; we saw that there are a gamut of approaches you can take to structuring your soup entries; you can be anything from naive to sophisticated.

Above all, what should be clear by now is that a good application can either be ruined or strengthened by the soup design you give it. You goal as a designer is straightforward:

- *Figure out how users will want to search and display data and then make sure that your soup design allows these things to be accomplished quickly and easily.*

As you saw when we discussed Performance Tuning (see Chapter 9), there is a lot that you can do to speed up crucial areas in your application, but all of this will be for naught if your data handling is doltish.

Appendix A

Important Messages

View/Proto Messages
Store Messages
Soup Messages
Cursor Messages
Endpoint Messages

This appendix discusses many of the important messages that you will use in your applications. It is divided into several sections corresponding to the types of messages they are. Within each section, the messages are arranged alphabetically.

As the names of the sections indicate, you send messages to view, proto, and data objects.

View/Proto Messages

view:ChildViewFrames()

This returns the child views of `view` in an array. If `view` has no children, this function returns `nil`.

view:Close()

This closes *`view`* and all of *`view`*'s descendants. The `viewQuitScript` of the view and each child is called though the order of the calls for the descendants is undefined. The `viewCObject` for each view is destroyed and the views themselves are no longer referenced by the view system (and thus available for garbage collection). If a view is still referenced elsewhere (for instance, declared to some open ancestor), the view is not destroyed.

view:CopyBits(*picture, x, y, mode*)

Draws the bitmap object *`picture`* with its top left at the coordinates (*x*, *y*). The drawing mode is one of the following transfer modes: `modeCopy` (or equivalently, `nil`), `modeOr`, `modeXor`, `modeBic`, `modeNotCopy`, `modeNotOr`, `modeNotXor`, `modeNotBic`.

view:Delete(*deleteMethodSymbol, parameterArray*)

This method gives you the visual effect of crumpling *`view`* and throwing it into a trash can. Underneath the crumpling view, the contents of a new view show simultaneously.

This method works by allocating a new offscreen bitmap. The current screen contents are saved in this new bitmap. Then, `Delete` locks the screen and sends the *`deleteMethodSymbol`* message to *`view`*, passing as parameters each of the elements of *`parameterArray`*. *`deleteMethodSymbol`* should either directly draw, or cause drawing to occur (for example, by calling `Dirty` or `SetValue`). At this point, `Delete` has a bitmap containing the old contents of *`view`*, and another containing the new contents of *`view`*. Animation occurs using those two bitmaps. At the conclusion, you are provided with the screen displaying the new contents of *`view`*.

rootView:Dial(*numberString*, *where*)

This method dials the number specified in *numberString* (which can contain numbers, the letters 'A'-'D', '*', '#', ',', or '-'). The parameter, where, is either 'speaker or 'modem.

view:Dirty()

This method marks *view* as dirty and in need of redrawing. It is redrawn at the next idle time (when there are no other pending events).

view:DoDrawing(*drawMethodSymbol*, *parameterArray*)

You will call this message when you want to draw somewhere other than within your viewDrawScript. This routine sets up clipping and then allows you to do your drawing by sending the *drawMethodSymbol* message to *view*. It uses the parameters specified in *parameterArray*. Afterward, it restores the clipping.

view:Drag(unit, bounds)

This method drags view as the user moves the stylus on the screen. It is called from within a viewClickScript. The *unit* parameter is the same one passed to viewClickScript. The *bounds* is a bounding box (in global coordinates) limiting the dragging of *view*. If *bounds* is nil, *view* can be dragged anywhere on screen.

view:DrawShape(*shape*, *styleFrame*)

This message draws in view. The *shape* parameter is either a shape (line, rectangle, rounded-rectangle, oval, arc, polygon, region, picture, or text) or an array of shapes and style frames. The *styleFrame* parameter is a frame specifying a style to use for drawing *shape*. The slots in *styleFrame* are transferMode, penSize, penPattern, fillPattern, font, and justification.

view:Effect(*effect*, *reveal*, *sound*, *messageSymbol*, *parameterArray*)

This method does a visual animation effect on `view` as specified by the *`effect`* parameter. (These are the same effects as in `viewEffect`.) It also plays the sound specified by *`sound`* if it is not `nil`.

This method works by allocating a new offscreen bitmap. The current screen contents are saved in this new bitmap. Then, `Effect` locks the screen and sends the *`messageSymbol`* message to *`view`*, passing as parameters each of the elements of *`parameterArray`*. *`messageSymbol`* should directly draw or cause drawing to occur (for example, by calling `Dirty` or `SetValue`). At this point, `Effect` has a bitmap containing the old contents of *`view`* and another containing the new contents of *`view`*. Animation occurs using those two bitmaps and you end up with the screen displaying the new contents of *`view`*. If *`reveal`* is non-`nil`, then `Effect` uses a reveal line at the edge between the old contents and the new contents.

protoPicker:GetItemMark(*itemIndex*)

This returns the mark associated with the item in the picker at *`itemIndex`*. It returns `nil` if there is no mark associated with that item.

view:GlobalBox()

This method returns the bounds of *`view`* in global screen coordinates. It returns a frame with `left`, `right`, `top`, and `bottom` slots.

view:Hide()

This method hides *`view`*, so that it no longer displays on screen. The `viewCObject` still exists, however. It also sends *`view`* the `viewHideScript` message.

view:Hilite(*hiliteOn*)

If *`hiliteOn`* is non-`nil`, this method highlights *`view`* (according to the highlight specified in `viewFormat`). If *`hiliteOn`* is `nil`, this method unhighlights *`view`*. You can use this as a toggling message.

`view:HiliteUnique(hiliteOn)`

If *`hiliteOn`* is `nil`, this method unhighlights *`view`*. If *`hiliteOn`* is non-`nil`, this method highlights *`view`* (according to the highlight specified in `viewFormat`). It also ensures that all siblings of *`view`* are unhighlighted.

`view:LayoutColumn(childTemplateArray, childIndex)`

This method is used to calculate a subset of child templates from *`childTemplateArray`* and display them in a column in *`view`*. The *`childIndex`* parameter specifies the index of the first child to display from the *`childTemplateArray`*.

`LayoutColumn` returns an array of children from *`childTemplateArray`* that are within the bounds of *`view`*. You can use this array as a `stepChildren` array.

`view:LayoutTable(tableDefinition, columnIndex, rowIndex)`

This method creates a matrix of child templates from a larger matrix of templates. The *`tableDefinition`* frame describes the matrix of templates that will be shown in *`view`*; the *`columnIndex`* parameter specifies which column appears at the left of view; and the *`rowIndex`* parameter specifies which row appears at the top of the view.

The *`tableDefinition`* frame contains:

`tabAcross`	The number of columns in the matrix.
`tabDown`	The number of rows in the matrix.
`tabWidths`	The integer width of the columns or an array of integer widths.
`tabHeights`	The integer height of the rows or an array of integer heights.
`tabProtos`	The template to be used for each matrix element or an array of templates.
`tabValueSlot`	A symbol which stores the names of each matrix element view.

`tabValues`	The value to be stored in each matrix element, or an array of those values.
`tabSetup`	A method (`func (childTemplate, columnNumber, rowNumber)`) that is called before each matrix element view is created.

For any slot that can take an array, the elements of the array are mapped down the first column of the matrix, then down the second column, and so on. If there aren't enough elements in the array, the mapping starts over with the array again.

view:LocalBox()

This method returns the bounds of *view* in coordinates local to the view. It returns a frame with `left`, `right`, `top`, and `bottom` slots. The `left` and `top` slots will always be `0`. The `right` and `bottom` slots contain the width and height of *view*.

view:LockScreen(*lock*)

All drawing is done to an offscreen bitmap and then copied from off screen to on screen. Before a view (or hierarchy of views) draws, the screen is locked so that the copy from off screen to on screen does not take place until the drawing is completed. When all the drawing is completed, the screen is unlocked and updated. The visual effect is of a single update, rather than a succession of views with drawing in bits and pieces.

If you draw within your `viewDrawScript`, the screen has already been locked. If you draw from within any other method, you should call `:LockScreen(true)` to lock the screen before drawing, and then call `:LockScreen(nil)` after drawing to unlock it.

rootView:Notify(*level*, *headerStr*, *messageStr*)

This method notifies the user of an event. The *headerStr* is a header at the top of the notification slip. The *messageStr* is the actual message (which should be `EnsureInternaled` in case the user scrolls up to a message from an application that is no longer present). The *level* parameter specifies the urgency of the message. It is one of the following:

`kNotifyLog`	The user is not formally alerted, but the error is put in the notification log. Users can see the message if another notification occurs and they scroll back through the notification slip.
`kNotifyMessage`	An icon blinks on-screen until the user taps it. The notification slip is displayed then.
`kNotifyQAlert`	The notification slip is displayed immediately.
`kNotifyAlert`	The notification slip is displayed immediately, and the system beep is played.

`view``:Open()`

This method opens *`view`* and all its descendants (unless they are not `vVisible`). It sends the `viewSetupFormScript`, `viewSetupChildrenScript`, `viewSetupDoneScript` messages to *`view`* and all its descendants. It then sends the `viewShowScript` message to *`view`*.

`rootView:NetChooser:OpenNetChooser(`*`zone,entityType,entity, notifyView, buttonText, singularType, pluralType`*`)`

This method opens a slip to allow the user to select an entity on the network. When the user closes the slip (with either the button, or the close box), the `NetworkChooserDone` message is sent to *`notifyView`*.

The *`zone`* parameter is a string specifying the initial zone; use `nil` to specify the local zone. The *`entityType`* parameter specifies what to look for (as an NBP entity string). The *`entity`* parameter specifies what entity should initially be selected. The *`buttonText`* string is used as the text of the button. The *`singularType`* and *`pluralType`* parameters are strings specifying what is being searched for.

`view``:Parent()`

This method returns the parent of *`view`*. Use this method rather than directly accessing the `_parent` slot of *`view`*.

view:RedoChildren()

This method destroys all the children of *view*, sends *view* the viewSetupChildrenScript message, and then recreates them. Consider using SyncChildren instead, because it does not destroy and recreate child views that remain inside the bounds of *view*.

view:RevealEffect(*numPixels*, *bounds*, *sound*, *methodSymbol*, *parameterArray*)

This method does a reveal animation effect on the portion of view specified by *bounds*. It also plays the sound specified by *sound* (unless it is nil). It moves the specified on-screen portion of *view* by *numPixels* pixels (a positive number moves the bits up, a negative number moves them down). This method doesn't affect *view* itself, but only the bits on screen.

This method works by allocating a new offscreen bitmap. The current screen contents are saved in this new bitmap. Then, RevealEffect locks the screen and sends the *messageSymbol* message to *view*, passing as parameters each of the elements of *parameterArray*. *messageSymbol* should either directly draw or cause drawing to occur (for example, by calling Dirty or SetValue). At this point, RevealEffect has a bitmap containing the old contents of *view* and another containing the new contents of *view*. It then does animation using those two bitmaps, ending up with the screen displaying the new contents of *view*.

protoRadioCluster:SetClusterValue(*whichButton*)

This method changes the selected radio button to the one whose buttonValue slot is equal to *whichButton*. It updates the radio buttons on screen, sends itself the ClusterChanged message, and is undoable (that is, the user can undo the change in radio button).

rootView:extrasDrawer:SetExtrasInfo(*appSymbol*, *newInfo*)

This method changes the extras drawer information of the application specified by *appSymbol* to *newInfo*. The *newInfo* parameter is a frame with the following slots:

icon	A bitmap or pict that displays in the extras drawer.

`text`	The text which displays below the picture in the extras drawer.

`clParagraphView:SetHilite(start, end)`

This method highlights the characters in *`clParagraphView`* in the range `start` to *`end`*. Before calling this method, the view must have been set as a key view using `SetKeyView`.

`protoPicker:SetItemMark(itemIndex, markCharacter)`

This method sets the mark for the item at *`itemIndex`* to *`markCharacter`*.

`protoLabelInputLine:SetLabelCommands(labelArray)`

This updates the `labelCommands` slot to the array of strings in *`labelArray`*.

`protoLabelInputLine:SetLabelText(label)`

This updates the label to the string in *`label`*.

`view:SetOrigin(originX, originY)`

This method changes the origin of *`view`* to (*`originX`*, *`originY`*). It dirties *`view`* so that it is redrawn. Likewise, *`view`*'s children are redrawn relative to the new origin. This method is one way to implement scrolling.

`view:SetPopup()`

This method causes *`view`* to be a popup view. A popup view is closed on the next pen tap.

`view:SetupIdle(delay)`

This method sets up idling for *`view`*. The `viewIdleScript` message will be sent to *`view`* after the number of milliseconds specified in *`delay`*.

view:Show()

This message shows *view* on screen and sends it the `viewShowScript` message.

view:SlideEffect(*amtToScroll*, *amtToMove*, *sound*, *messageSymbol*, *parameterArray*)

This method does a scrolling animation effect on *view*. It also plays the sound specified by *sound* (unless it is `nil`). The sound parameter is commonly one of `scrollUpSound` or `scrollDownSound`. It scrolls the bits on screen within the bounds of *view* by *amtToScroll* pixels (a positive number moves the bits down, a negative number moves them up). It moves the entire view on-screen by *amtToMove* pixels (a positive number moves the bits up, a negative number moves them down). This method does not affect *view* itself, but only the bits on screen.

This method works by allocating a new offscreen bitmap. The current screen contents are saved in this new bitmap. Then, `SlideEffect` locks the screen and sends the *messageSymbol* message to *view*, passing as parameters each of the elements of *parameterArray*. *messageSymbol* should either directly draw or cause drawing to occur (for example, by calling `Dirty` or `SetValue`). At this point, `SlideEffect` has a bitmap containing the old contents of *view* and another containing the new contents of *view*. It then does animation using those two bitmaps, ending up with the screen displaying the new contents of *view*.

view:SyncChildren()

This method redraws all the children of *view*, creating children that are now within the bounds and deleting any child views that are no longer within the bounds of *view*.

First, it sends *view* the `viewSetupChildrenScript` message and then creates and deletes necessary children. If any children have moved, they are sent the `SyncView` message.

view:SyncScroll(*childViewArray*, *childIndex*, *upOrDown*)

This method scrolls the children in *childViewArray* up the height of the top child (if *upOrDown* is `-1`), or down the height of the bottom child (if *upOrDown* is 1). The *childIndex* parameter is the index within *childViewArray* of the top child within *view*.

The method plays the scrollUpSound or scrollDownSound as appropriate and returns the array of child views that are visible within *view* after scrolling. SyncScroll looks for the optional slots, allCollapsed and collapsedHeight, in view. It looks for the optional slot collapsed and the required slot height in each child view.

The optional slots are used for views that support children and that have both expanded and collapsed modes. The slots values are:

allCollapsed	Contains true if all children are collapsed; nil otherwise.
collapsedHeight	Contains the height of child views in the collapsed state.
collapsed	Contains true if this child view is collapsed; nil otherwise.
height	Contains the height in pixels of this child view in the expanded state.

view:SyncView()

This method updates the viewCObject to reflect changes in *view* slot values (for example, the viewBounds). It sends the viewSetupFormScript message to *view* before updating the viewCObject.

If you use the SetValue function to set viewBounds, viewFlags, viewFormat, viewJustify, or viewFont, you need not explicitly call SyncView. If you change the values of one of these slots directly, you must call SyncView before that view reflects the changes.

rootView:SysBeep()

This method plays the system beep sound (or flashes the screen if the volume is off).

view:Toggle()

If *view* is open, this method sends it the `Close` message. If view is closed, the reverse happens and this method sends the `Open` message.

protoCheckbox:ToggleCheck()

This method toggles the state of its checkmark. The `viewValue` slot holds the current state of the checkmark: `true` if on, `nil` if off.

view:TrackButton(*unit*)

This method is called from within a `viewClickScript`. The *unit* parameter is the same one passed to `viewClickScript`. The method makes a click sound and then starts tracking the pen. While the pen is in the bounds of view, *view* is highlighted; if the pen goes outside the bounds, *view* is unhighlighted. While the pen is in the bounds, `TrackButton` repeatedly sends view the `button-PressedScript` message (if it exists). If the pen is released while within the bounds of *view*, this method sends view the `buttonClickScript` message. In any case, `TrackButton` unhighlights *view* before returning.

If an error occurs while tracking (if the *buttonPressedScript* method causes an error), this method will unhighlight *view*.

view:TrackHilite(*unit*)

This method is called from within a `viewClickScript`. The *unit* parameter is the same as that passed to `viewClickScript`. The method makes a click sound and then starts tracking the pen. While the pen is in the bounds of view, *view* is highlighted; if the pen goes outside the bounds, *view* is unhighlighted. While the pen is in the bounds, `TrackHilite` repeatedly sends *view* the `button-PressedScript` message (if it exists). `TrackHilite` returns `true` if the pen is released while within the bounds of *view* (and leaves view highlighted); it returns `nil` otherwise.

protoFilingButton:Update()

This method updates the filing button to display a triangle if the current target is on an external store. It erases the triangle if the target is `nil` or if the target is not on an external store.

protoLabelPicker:UpdateText(*newText*)

This method changes the currently selected item to *`newText`*.

protoLabelInputLine:UpdateText(*newText*)

This method changes the text in the input line to `newText`. It is undoable (the user can undo the change).

protoFolderTab:UpdateFilter(*oldFolder, newFolder*)

This method updates the show bar to display a new folder. You need to call this if you change the folder programmatically (rather than allowing the user to change it). The *`oldFolder`* and *`newFolder`* parameters are symbols describing folders.

Store Messages

store:CreateSoup(*soupName, arrayOfIndexes*)

This store method creates a soup named *`soupName`* on `store`. The parameter, *`arrayOfIndexes,`* is an array of frames. Each frame in that array specifies a soup index and has the slots `structure` (with the value `'slot`), `path` (containing a symbol or path expression), and `type` (one of `'int`, `'string`, `'char`, `'real`, or `'symbol`).

store:Erase()

This completely erases `store`.

`store:GetKind()`

This method returns the media type of *`store`*. The possibilities (for Newtons localized for the U.S.) are `"Internal"`, `"Flash storage card"`, `"Storage card"` and `"Application card"`.

`store:GetName()`

This method returns the name of *`store`*.

`store:GetSignature()`

This method returns the integer signature of *`store`*.

`store:GetSoup(soupName)`

This method gets the soup named *`soupName`* from *`store`*. If no such soup exists, the method returns `nil`.

`store:GetSoupNames()`

This method returns an array containing the names of all the soups on a specified *`store`*.

`store:HasSoup(soupName)`

This method returns `true` if *`store`* contains a soup named *`soupName`*. It returns `nil` otherwise.

`store:IsReadOnly()`

This returns `true` if *`store`* is in a read-only state. This returns `nil` if *`store`* can be written.

`store:SetName(newName)`

This sets the name of *`store`* to *`newName`* (if the store is writable).

`store:SetSignature(newSignature)`

This method sets the integer signature of `store`.

`store:TotalSize()`

This returns the total size of `store` in bytes.

`store:UsedSize()`

This returns the number of bytes currently used in `store`.

Soup Messages

`unionSoup:AddToDefaultStore(frame)`

This method takes `frame` and adds it to `unionSoup`. The entry frame is placed on the default store. The user specifies the default store in the Card slip. An error will be generated if the default store does not contain that soup.

`soup:CopyEntries(destSoup)`

This method copies all the entries from `soup` to `destSoup`. Neither soup nor `destSoup` may be a union soup.

`soup:GetIndexes()`

This method returns an array of frames, one for each index in `soup`. This method is not supported for union soups.

`soup:GetNextUid()`

This returns the value of the `_uniqueID` that will be assigned to the next entry added to `soup`. This method is not supported for union soups.

soup:GetSignature()

This method returns the signature of *soup* and is not supported for union soups.

soup:RemoveIndex(*indexPath*)

You use this to remove a *soup* index on the path *indexPath*. This method works on both union and nonunion soups.

soup:SetSignature(*signature*)

When a soup is created, it is assigned a random integer as a signature. With this signature, you can tell whether a soup has been removed and a new soup added with the same name. This method sets the signature of *soup* to *signature* and is not supported for union soups.

Cursor Messages

cursor:Clone()

The `Clone` method returns a new cursor that shares the same query as *cursor*. The new cursor has an independent location and, as such, provides you with a way to have multiple iterators over the same entries.

cursor:Entry()

This returns the current entry from *cursor*. If the entry gets deleted while the cursor is on it, `Entry` returns `'deleted` and no longer rests on a soup entry.

cursor:Goto(*entry*)

This function moves *cursor* to *entry*. The *entry* must be an entry in the soup and have already been returned by a call to `Entry`, `Next`, or `Prev`.

cursor:GotoKey(*keyValue*)

This moves *cursor* to the first entry with an indexed value equal to *keyValue*. If no such entry is present, the *cursor* is put on the next indexed entry. This method should only be called for queries that have specified an `indexPath`.

cursor:Move(*numEntries*)

This moves *cursor* forward or backward by *numEntries*. `Move(1)` is equivalent to `Next()` and `Move(-1)` is equivalent to `Prev()`. This function is faster than repeatedly calling `Next` or `Prev`.

cursor:Next()

This moves *cursor* to the next entry satisfying the query and then returns it. If there is no next entry, `Next` returns `nil`.

cursor:Prev()

This moves *cursor* backward to the previous entry that satisfies the query and returns it. If you are at the beginning and there is no previous entry, `Prev` returns `nil`.

cursor:Reset()

This moves *cursor* back to the first entry that satisfies the query.

Endpoint Messages

endPoint:Abort()

If a `Connect` or `Listen` hasn't yet completed, this method will abort it. If they have completed, `Abort` will abort any input or output in progress. To abort input, you should use the following lines of code:

```
endPoint.nextInputSpec := nil;// no next input spec
ep:SetInputSpec(nil);// replace present input spec
ep:Abort();           // abort present input spec
```

To abort input, output, or an in-progress `Connect` or `Listen`, use the `Abort` method.

`Abort` will cause your exception handler to execute. This exception can usually be ignored.

endPoint:Connect(*address*, *options*)

This method opens a connection. If *address* is `nil`, the address from *endPoint*.`configOptions` is used (if this service requires an address). Otherwise, *address* is an address frame which specifies the connecting address. The *options* parameter is not used and should be `nil`.

This method may be time-consuming (for instance, it may require dialing the modem). For this reason, it should normally be executed from within a deferred action so that your user interface continues to operate and thus allows the user to abort, if desired.

endPoint:Disconnect()

This method closes the connection (the opposite of `Connect` or `Listen`).

endPoint:FlushOutput()

This routine sends buffered data, returning when more data can be sent.

endPoint:FlushPartial()

This method removes all the buffered input through the end of the last call to the `partialScript`.

endPoint:Input()

This method returns the data in the input buffer, formatted according to the current input spec. It terminates the current input spec.

Note that the `inputScript` method of the current input spec is not called.

endPoint:Instantiate(*endPoint, options*)

This method creates data structures and does initialization for the specified service. If *options* is nil, options are used from *endPoint*.configOptions. Otherwise, the options in *options* are used, while the options found in the *endPoint*.configOptions are ignored. Thus, this parameter is a useful way of overriding default settings.

Only one endpoint can be instantiated at a time. If one is already instantiated when this call is made, then an error is returned.

Instantiate will throw an exception on any error.

endPoint:Listen(*options*)

This method waits for a connection to be made to this endpoint. The *options* parameter is unused and should be nil.

This method may be time consuming (for instance, it may require waiting for a phone call, or for an IR transmission to arrive). For this reason, it should normally be executed from within a deferred action. This way your user interface continues to operate and the user can abort, if necessary.

endPoint:Output(*data, flags*)

This method sends *data*; the type of *data* will determine how it is sent. The *data* parameter is one of the following four possibilities:

integer	Sends the low 8 bits of the integer as a byte.
character	Sends that character. Each character is converted from Unicode to ASCII using the Macintosh character encodings (for example '…' is converted from Unicode \u2026 to ASCII 0x7e). Characters that don't have an equivalent in the Macintosh character encoding are converted to 0xff.
string	Sends the characters of the string (each character is converted to ASCII).

`array`	Sends each element of the array (each element must be an integer). The low 8 bits of each integer is sent.

The *`flags`* parameter is normally `nil` and is ignored by most endpoints. The IR endpoint uses this parameter, but the details of this parameter aren't covered in this book.

Note that data is buffered; the ADSP endpoint, in particular, allows at most three `Output` calls before you need to call `FlushOutput` to send the buffered data.

`endPoint``:OutputFrame(`*`frame, flags`*`)`

This method sends *`frame`* as a flattened sequence of bytes. The format of this sequence is undocumented.

`endPoint``:Partial()`

This method returns the data since the last call to the `partialScript`. `Partial` doesn't remove data from the input buffer.

`endPoint``:Release()`

This method closes the connection (the opposite of `Connect` or `Listen`) after waiting for all pending input and output to complete.

`endPoint``:SetInputSpec(`*`inputSpec`*`)`

This method makes *`inputSpec`* the active input spec. If *`inputSpec`* is `nil`, this aborts the current input spec.

Appendix B

Important Methods

This appendix provides a brief description of the various methods you may need to provide.

`buttonClickScript()`

This is called when a `protoTextButton`, `protoPictureButton`, `protoCloseBox`, `protoLargeCloseBox`, or `protoPictRadioButton` is clicked on. It will also be called for any view that calls `TrackButton`.

For a `protoCloseBox` or a `protoLargeCloseBox`, make sure to call `inherited:buttonClickScript()` so that the standard action takes place.

`buttonPressedScript()`

This is called repeatedly while a `protoTextButton`, `protoPictureButton`, `protoCloseBox`, `protoLargeCloseBox`, or `protoPictRadioButton` is pressed. It will also be called for a view that calls `TrackButton`.

`changedSlider()`

This is a required method for `protoSliders`. It is called whenever the user moves the gauge and lifts the pen. The current gauge setting is in the `viewValue` slot.

`clusterChanged(`*`whichRadioButton`*`)`

This is for a `protoRadioCluster`. This routine is called when a different radio button is turned on. The `buttonValue` slot of the newly turned-on radio button is passed as the *`whichRadioButton`* parameter.

`DateFind(`*`findTime, filter, results, scope, statusContext`*`)`

This is a method of the application's base view. It is called when the user does a "dates before" or "dates after" find. If an application doesn't have a `DateFind` method, a local Find will generate a "Date find not supported" slip. This method may be called when your application is closed.

The parameters are:

`findTime`	Time to look for in number of minutes since midnight January 1, 1904.
`filter`	Either `'dateBefore` or `'dateAfter`.
`results`	Array of frames, one for each successful find. Append a frame to this frame if your application finds at least one item.
`scope`	Either `'localFind` or `'globalFind`.
`statusContext`	The Find view to which you can send status messages with the `SetStatus(`*`statusString`*`)` message.

`ExceptionHandler(`*`error`*`)`

This method of an endpoint gets called whenever a communications error occurs. The *`error`* parameter is a frame with a `name` slot that contains the exception type, along with a `data` slot containing an integer error code.

`FilingChanged()`

This is called when the user files an item in a different folder. By the time this routine executes, the `labels` slot of the item has been modified.

`FilterChanged()`

This is called when the user selects a new folder from the folder tab. By the time this routine executes, the `labelsFilter` slot has already been changed.

`Find(`*`what, results, scope, statusContext`*`)`

This is a method of the application's base view. It is called when the user selects a "Find". If an application doesn't have a `Find` method, a local Find will generate a "Find not supported" slip.

The parameters are:

`what`	The string to Find.
`results`	Array of frames, one for each successful find. Append a frame to this frame if your application finds at least one item.
`scope`	Either `'localFind` or `'globalFind`.
`statusContext`	The Find view to which you can send status messages with the `SetStatus(`*`statusString`*`)` message.

This method may be called when your application is closed.

`FindSoupExcerpt(entry, finder)`

This is called to return the string displayed in the Find results slip for `entry`.

`FlushEdits()`

This routine is called for a `protoExpandoShell` when an expanded edit field is closed, and that edit field had been modified.

`FolderChanged(soupName, oldLabel, newLabel)`

This application base view method is called when the user deletes or renames a folder. It is sent to all applications in the `soupNotify` global array. The routine should update all entries in the `soupName` soup—the `labels` slot value needs to be changed from `oldLabel` to `newLabel`. It should only do this if the application is responsible for that soup.

This method may be called when your application is closed.

`inputScript(endpoint, data)`

This input spec method is called whenever the termination condition of the input spec occurs. The `inputScript` has the following parameters:

endPoint	The endpoint that had input.
data	The data that was input. The format of the data is dependent on the value of `inputForm` (the slot in the input spec).

`keyPressScript(keyResult)`

This routine is called for a `clKeyboardView` when the user presses a key. The default (if this method isn't present) converts *keyResult* into a sequence of characters that are posted as key events to the key receiver view.

`labelActionScript(`*`itemIndex`*`)`

This routine is called for a `protoLabelInputLine` or `protoLabelPicker` when the user selects an item from the picker. *`itemIndex`* is the index of the chosen item.

For a `protoLabelInputLine`, if this method returns `true`, it signifies that it has handled it. If it returns `nil`, the default action (setting the input field to the chosen item) is taken.

For a `protoLabelPicker`, the return result is ignored.

`labelClick(`*`unit`*`)`

This routine is called for a `protoLabelInputLine` when the user clicks on the label. The *`unit`* parameter is the same as is passed to `viewClickScript`. If this method returns `true`, it signifies that it has handled the click. If it returns `nil`, the default action (displaying a picker and then calling `labelActionScript`) is taken.

`monthChangedScript()`

This method is called for a `clMonthView` when the user changes the day.

`NetworkChooserDone(`*`currentSelection, currentZone`*`)`

This message is sent to the view specified as a parameter to the `NetworkChooser` method when the user closes the Chooser slip. If the user closes with the close box, *`currentSelection`* is the empty string. Otherwise, *`currentSelection`* contains the name of the currently selected entity; *`currentZone`* contains the zone in which it is located (an empty string specifies the current zone).

`partialScript(`*`endpoint, data`*`)`

This input spec method is called based on the frequency specified in the `partialFrequency` slot of the input spec.

The `partialScript` has the following parameters:

endPoint	The endpoint that had input.
data	The data that was input. The format of the data is dependent on the value of `inputForm` (the slot in the input spec).

`pickActionScript(itemNumber)`

This is used for a `clPickView`. The `pickActionScript` should be a method in the view located in the `callbackContext` slot of the pick view. *itemIndex* is the index of the picked item in the `pickItems` array.

`pickerSetup()`

This method is called for a `protoLabelPicker` when the user taps on the label. If it returns `true`, it signifies it has handled the tap. If it returns `nil`, the default action (displaying a picker, setting the text field, and then calling the `labelActionScript`) is taken.

`powerOffScript(what)`

This method is called for a view that has registered itself as a power-off handler with `AddPowerOffHandler`. If *what* is `'okToPowerOff`, the method should return `true` to signify that it is okay to power off, or `nil` to abort the power off. If *what* is `'powerOff`, the Newton is about to power off.

`printNextPageScript()`

This method of a `protoPrintFormat` is called at the end of each page while printing or faxing to prepare the next page. It should return `nil` if there are no more pages to print or fax.

`PutAway(item)`

This is a method of the application's base view. It is a message that is called to handle a frame that has been beamed from another Newton, store it, and redraw it if necessary. *item*.body contains the frame to store. This method may be called when your application is closed.

SetupRoutingSlip(*fields*)

This is called to prepare a title and embed data for faxing or printing. Set *fields*.text to the title, and *fields*.body to the data you'll need for faxing or printing (for beaming or mailing, *fields*.body will be set to target after this method executes).

ShowFoundItem(*entry, finder*)

This is called when the user taps on an item in the Find results slip (or if the user does a find that finds exactly one item). This routine should display the found item.

By the time this routine executes, the application is already open. If the result found in the results array (see discussion of DateFind or Find) protos from ROM_compatibleFinder, the second parameter is not present.

soupChanged(*soupName*)

When a soup changes, the BroadcastSoupChange global function should be called with the name of the changed soup. This function sends the soupChanged message to each application that registered with that soup name in the soupNotify global array (an application registers with two entries: its application symbol and a soup name). Other applications may call BroadcastSoupChange, or the system will call it when a card is inserted or removed.

textChanged()

This routine is called for a protoLabelPicker or protoLabelInputLine when the value of the text slot is changed.

textSetup()

This routine is called for a protoLabelInputLine or a protoLabelPicker when it is created. It returns a string which is the initial value of the input field.

`trackSlider()`

This routine is called for a `protoGauge` whenever the `viewValue` slot changes. In particular, it is called repeatedly while the user moves the gauge knob.

`valueChanged()`

This routine is called when the value of a `protoCheckbox` changes.

`viewAddChildScript(`*`newChildView`*`)`

This method is called for `clEditViews` when *`newChildView`* is about to be added as a child. The method can modify *`newChildView`* before it is added. If the routine returns `true`, it signifies it handled adding the child. If it returns `nil`, the default action (adding the view) is taken.

 *Caution:* At the time this book was written, if this method returns `true` it must: add *`newChildView`* as a child, or it must add a new view which protos from *`newChildView`*.

`viewChangedScript(`*`slotSymbol, view`*`)`

When `SetValue` is called to change a view's slot value, it first changes the value and then sends the view the `viewChangedScript` message. This message contains the symbol of the slot, along with the view that was changed.

Override this message to be notified when slots in a view are changed (as long as they are changed by `SetValue`). The *`view`* parameter specifies the view that changed. It will usually, but need not always, equal `self`.

`viewClickScript(`*`unit`*`)`

This is called when a view is tapped on. To receive this message, the `vClickable` view flag must be set. *`unit`* contains information about the stroke. To turn off ink, call `InkOff(`*`unit`*`)`.

This method should return `true` if it handled the tap, and then no further processing of the tap will be done (strokes, gestures, words). If it returns `nil`, the closest ancestor that is `vClickable` will be sent the same message. If no such ancestor is found, the default action is taken, which may include calls to `viewStrokeScript`, `viewGestureScript`, or `viewWordScript`.

`viewDrawScript()`

This is called just after a view is drawn. Augment the standard view drawing by overriding this message.

`viewDropChildScript(`*`childView`*`)`

This method is called for `clEditViews` when *`childView`* is about to be deleted. If the routine returns `true`, it signifies that it removed the child from the `viewChildren` or `stepChildren` array. If it returns `nil`, the default action (removing the child from the `viewChildren` or `stepChildren` array) is taken.

`viewGestureScript(`*`unit, gestureCode`*`)`

If gesture recognition is turned on, the `viewGestureScript` will be called for gestures that the primitive view class does not handle. For example, a `clEditView` handles double tap, scrub, hilite, caret, and line gestures. On the other hand, a `clView` handles no gestures. Therefore, a scrub gesture in a `clEditView` will not cause its `viewGestureScript` to be called, whereas the same gesture in a `clView` does cause its `viewGestureScript` to be called.

The *`unit`* parameter is the same stroke unit passed to `viewClickScript`. The *`gestureCode`* parameter is 49 (tap), 50 (double-tap), 13 (scrub), 47 (hilite), 15(caret), 16 (line).

`viewHideScript()`

The view system sends this message to a view when that view is closed or hidden. Subviews of a view that are closed or hidden do not receive the `viewHideScript` message.

`viewIdleScript()`

This message is sent for views that have set up idling. To set up idling for a view, call *`view`*`:SetupIdle(`*`milliseconds`*`)` where *`milliseconds`* is the number of milliseconds before the next idle call. This routine can do whatever processing it desires, but should return the number of milliseconds before the next call to `viewIdleScript`. Returning `nil` stops idling.

`viewOverviewScript()`

This message is sent when the user taps the overview button. The message is sent to the frontmost view whose `viewFlags` has the `vApplication` bit set (normally your application's base view).

`viewQuitScript()`

The view system sends this message to a view when that view, or any ancestor view, is closed. When you send the `Quit` message to a view, that view will receive the `viewQuitScript` message before any descendants. However, the order in which descendants receive `viewQuitScript` messages is undefined.

`viewScrollDownScript()`

This message is sent when the user taps the down arrow. The message is sent to the frontmost view whose `viewFlags` has the `vApplication` bit set (normally your application's base view).

`viewScrollUpScript()`

This message is sent when the user taps the up arrow. The message is sent to the frontmost view whose `viewFlags` has the `vApplication` bit set (normally your application's base view).

`viewSetupChildrenScript()`

This message is sent just before any children of this view are created. Override this message to modify the `stepChildren` array when you want to create child views dynamically.

`viewSetupDoneScript()`

This message is sent after any children of this view are created, but before the view is drawn.

`viewSetupFormScript()`

This message is sent before the graphical representation of the view (`viewCObject`) is created. Override this message to set the size of the view dynamically.

`viewShowScript()`

The view system sends this message to a view when that view is opened or shown. Subviews of that view do not receive the `viewShowScript` message.

`viewStrokeScript(`*`unit`*`)`

If `vStrokesAllowed` is set, this method gets called when the user lifts the pen after a pen tap (for instance, when the user lifts the pen after writing a word in cursive). It is not called, however, if the `viewClickScript` returns `true`, since that implies the `viewClickScript` handled all pen interaction. In addition, it is only called for the first stroke of a multistroke letter or word.

This method should return `true` if it handled the tap; no further processing of the tap will be done (gestures, words). If it returns `nil`, the closest ancestor that has `vStrokesAllowed` will be sent the same message. If no such ancestor can be found, the default action is taken, which may include calls to `viewGestureScript` or `viewWordScript`.

The *`unit`* parameter is the same as in the `viewClickScript` method.

`viewWordScript(`*`unit`*`)`

If word recognition is on, the `viewWordScript` will be called in cases where the primitive view class does not already handle words. For example, a `clEditView` handles words by entering them into a child `clParagraphView`. On the other hand, a `clView` does nothing with words. Therefore, entering a word in a `clEditView` will not cause its `viewWordScript` to be called, but entering a word in a `clView` will cause its `viewWordScript` to be called.

This method should return `true` if it handled the word. If it returns `nil`, the closest ancestor that has word recognition enabled will be sent the same message. If there is no such ancestor, the default action is taken, which may include calls to `viewGestureScript` or `viewWordScript`.

The *`unit`* parameter is the same stroke unit passed to `viewClickScript`. `GetWordArray(unit)` returns an array of words. The first entry is the most likely word. Subsequent entries are less probable possibilities for the word.

Appendix C

Global Functions

This appendix covers the documented global functions (each of which is a slot in `GetGlobals().functions`).

`Abs(number)`

This returns the absolute value of *number*.

`Acos(number)`

This returns the arccosine of *number*.

`Acosh(number)`

This returns the arc-hyperbolic-cosine of *number*.

`AddArraySlot(array, value)`

This adds `value` as a new element in `array`. The new element is at the end of the array.

`AddDeferredAction(function, parameterArray)`

This calls `function` with the parameters in `parameterArray` the next time through the event loop. This provides a way to defer execution of code for a short period.

If `function` uses the value of `self` (directly, or does inheritance lookup), then (due to a bug in `AddDeferredAction`) you must call `AddDeferredAction` as:

```
AddDeferredAction(
   func() Apply(function, parameterArray),
   '[]);
```

`AddDelayedAction(function, parameterArray, delay)`

This queues `function` to be called with the parameters in `parameterArray` after `delay` milliseconds.

If `function` uses the value of `self` (directly, or does inheritance lookup), then (due to a bug in `AddDelayedAction`) you must call `AddDelayedAction` as:

```
AddDelayedAction(
   func() Apply(function, parameterArray),
   '[],
   delay);
```

`AddPowerOffHandler(view)`

This registers `view` to be notified when the Newton is going to sleep. When the Newton is about to go to sleep, `view` will be sent the `powerOffScript` message.

`AddStepView(view, childTemplate)`

This function creates a view from `childTemplate`, and adds it as a child view to `view`.

`AddToUserDictionary(`*`wordString`*`)`

This function adds *`wordString`* to the user's personal dictionary. It returns `true` if it succeeded, `nil` otherwise.

`AddUndoAction(`*`methodSymbol, parameterArray`*`)`

This adds an entry to the undo stack. If the user chooses undo, the message *`methodSymbol`* will be sent to the value of self at the time `AddUndoAction` is called with the parameters *`parameterArray`*.

`AddWordToDictionary(`*`dictionary, wordString`*`)`

This function adds *`wordString`* to *`dictionary`*. It returns `true` if it succeeded, `nil` otherwise.

`Annuity(`*`ratePerPeriod, numberPeriods`*`)`

Calculates the present value factor of an annuity at an interest rate of *`ratePerPeriod`* over *`numPeriods`*.

`Apply(`*`function, argumentArray`*`)`

Executes the function object *`function`* with the arguments in *`argumentArray`*. The number of parameters to *`function`* and the number of elements in *`argumentArray`* must match. The return result is the return result of *`function`*.

`Array(`*`numberElements, initialValue`*`)`

This function returns a new array with *`numberElements`* elements. Each element has the value *`initialValue`*.

`ArrayMunger(`*`dstArray, dstStart, dstCount, srcArray, srcStart, srcCount`*`)`

This function destructively modifies *`dstArray`*, replacing *`dstCount`* elements starting at *`dstStart`* (`0` specifies the first elements) with *`srcCount`* elements from *srcArray*, starting at *srcStart* (0 specifies the first elements). If *dst-Count* (or *srcCount*) is nil, all elements to the end of *dstArray* (or *srcArray*) are specified. If *srcArray* is nil, elements are deleted from *dstArray*. The function returns the munged array.

ArrayPos(*array, value, startIndex, testFunction*)

This function searches for *value* in *array*, starting at *startIndex*. *testFunction* is a function that does equality testing on its two parameters. If *testFunction* is nil, = is used for equality testing.

ArrayRemoveCount(*array, startIndex, count*)

This removes *count* elements from *array*, starting at *startIndex*.

ArrayToPoints(*pointsArray*)

This converts the points in *pointsArray* to a binary object of class polygonShape. The array should be of the form returned by PointsToArray.

Asin(*number*)

This returns the arcsine of *number*.

Asinh(*number*)

This returns the arc-hyperbolic-sine of *number*.

Atan(*number*)

This returns the arctangent of *number*.

Atan2(*numberX, numberY*)

This returns the arctangent of the Cartesian coordinates *numberX* and *numberY* as a real. It is equivalent to Atan(*numberX*/*numberY*) except that *numberX* can be 0.

`Atanh(`*`number`*`)`

This returns the arc-hyperbolic-tangent of *`number`*.

`BAnd(`*`integerA, integerB`*`)`

This returns the bit-wise AND of *`integerA`* and *`integerB`*.

`BeginsWith(`*`string, subString`*`)`

Returns `true` if *`string`* begins with *`subString`* (ignoring case and diacritical marks).

`BNot(`*`integer`*`)`

This returns the bit-wise complement of *`integer`*.

`BOr(`*`integerA, integerB`*`)`

This returns the bit-wise OR of *`integerA`* and *`integerB`*.

`BreakLoop()`

This enters a break loop.

`BroadcastSoupChange(`*`soupName`*`)`

This notifies applications (those that are interested) of the change to *`soupName`*.

`BuildContext(`*`template`*`)`

This function creates a view that protos from *`template`* and makes it a child of the root view. The return result is the view.

`BXor(`*`integerA, integerB`*`)`

This returns the bit-wise EXCLUSIVE-OR of *`integerA`* and *`integerB`*.

`Capitalize(`*`string`*`)`

This function destructively modifies *`string`* by capitalizing its first character. The modified string is returned.

`CapitalizeWords(`*`string`*`)`

This function destructively modifies *`string`* by capitalizing the first character of each word. The modified string is returned.

`Ceiling(`*`real`*`)`

This returns the smallest integer greater than or equal to *`real`*.

`CheckThatFolderExists(`*`frame`*`)`

This function checks the value of the `labels` slot in *`frame`*. If it is present and the value is not a valid folder, it sets the `labels` slot to `nil`.

`Chr(`*`integer`*`)`

Converts *`integer`* into its Unicode equivalent.

`ClassOf(`*`value`*`)`

Returns the class (type) of *`value`*. The common classes are: `'int`, `'char`, `'boolean`, `'string`, `'array`, `'frame`, `'function`, `'symbol`, `'pattern`, `'pathExpr`.

`Clone(`*`value`*`)`

This makes a duplicate of *`value`*, one-level deep. The frame map (if *`value`* is a frame) is not copied.

`CloseAppletalk()`

This closes AppleTalk. You would call this routine after you are finished calling all other AppleTalk global functions.

`Compile(string)`

This compiles *`string`* and returns a function object that will evaluate it.

`Compound(`*`ratePerPeriod`*`, `*`numPeriods`*`)`

This returns the compound interest factor for an interest rate of *`ratePerPeriod`* over *`numPeriods`*.

`CopySign(`*`numberX`*`, `*`numberY`*`)`

This returns a real whose sign is that of *`numberY`* and whose magnitude is that of *`numberX`*.

`Cos(`*`number`*`)`

This returns the cosine of *`number`*.

`Cosh(`*`number`*`)`

This returns the hyperbolic cosine of *`number`*.

`Date(`*`time`*`)`

This returns the given time (in minutes since midnight, January 1, 1904) as a date frame with the following slots:

`time`	The integer number of minutes since midnight, January 1, 1904.
`year`	The year (for example, 1993).
`month`	The month (from 1 to 12).
`date`	The day of the month.
`dayOfWeek`	The day of the week (0 is Sunday).
`hour`	The hour (from 0 to 23).

`minute`	The minute (from 0 to 59).
`second`	The second (from 0 to 59).

`DateNTime(`*`time`*`)`

This converts time (number of minutes since midnight, January 1, 1904) to a string displaying both the date and time.

`Debug(`*`value`*`)`

This prints *`value`* to the Inspector, followed by a newline. Strings are surrounded with `"  "`, and characters are preceded by `$`.

`DeepClone(`*`value`*`)`

This returns a recursive clone of *`value`*. Frame maps are not copied.

`DeleteWordFromDictionary(`*`dictionary, wordString`*`)`

This function deletes *`wordString`* from *`dictionary`*. It returns `true` if it succeeded, `nil` otherwise. If *`dictionary`* is `kUserDictionary`, make sure to call `SaveUserDictionary` so that the dictionary is saved in a soup.

`Display(`*`value`*`)`

This prints *`value`* to the Inspector. Strings are surrounded with `"  "`, and characters are preceded by `$`.

`DisposeDictionary(`*`dictionary`*`)`

This function disposes of the RAM-based *`dictionary`* created previously with a call to `NewDictionary`.

`DoPopup(`*`list, insetH, insetV, notifyView`*`)`

This creates a popup view filled with elements from the array *`list`*. A bounds rectangle is created by insetting the bounds of the current view (`self`) by *`insetH`* and `insetV`. The popup is then aligned to one of the four corners of that bounds rectangle. The `notifyView` is sent the `pickActionScript` message when an item is selected from the popup (it is passed the picked element number), and the `pickCancelledScript` message when the picker is cancelled (by tapping outside of it).

An entry in the list can be a :

string	Specifies the string to be drawn.
bitmap	Specifies a bitmap to draw.
`'pickSeparator`	A non-pickable dividing line is drawn.
frame	Containing: `item` (string or bitmap), `pickable` (a boolean specifying whether the item can be picked), and `mark` (a character used to display a checkmark).

The function returns the template of the view. To access the view itself, use `GetView`.

`Downcase(string)`

This function destructively modifies *`string`* so that each uppercase character is replaced with its lowercase equivalent; non-uppercase letters are not modified. The modified string is returned.

`DrawXBitmap(bounds, bitmap, index, drawingMode)`

Draws a vertical slice of *`bitmap`*. *`bounds`* specifies the location to which the bitmap will be drawn and *`index`* specifies which slice (`0` is the leftmost slice). Each slice is the width of *`bounds`*. *`drawingMode`* specifies the mode in which the bitmap will be drawn. `self` should be the view to which the bitmap will be drawn.

`DV(view)`

This displays a one-line summary of *`view`* and each of its descendants in the Inspector. It also briefly flashes *`view`* on the Newton.

`EndsWith(string, subString)`

Returns `true` if *`string`* ends with *`subString`* (ignoring case and diacritical marks).

`EnsureInternal(value)`

If *`value`*, and everything within *`value`* (recursively), is in either internal memory or on the system ROM, then this just returns *`value`*. Otherwise, it `TotalClones` as necessary to return a copy of *`value`*. It ensures that the copy, and everything within it, are either in internal memory or on the system ROM.

`EntryChange(entry)`

This writes *`entry`* back into its soup and updates its modification time.

`EntryChangeWithModTime(entry)`

Saves *`entry`* in its soup, keeping the `_modTime` slot as it was in the soup.

`EntryCopy(entry, newSoup)`

This copies *`entry`* from its current soup to *`newSoup`*.

`EntryModTime(entry)`

This returns the last modification time of *`entry`*. It returns `0` if *`entry`* has not been modified.

`EntryMove(entry, newSoup)`

This moves *`entry`* from its current soup to *`newSoup`*.

`EntryRemoveFromSoup(entry)`

This removes *`entry`* from its soup.

`EntryReplace(`*`entry, newEntry`*`)`

This replaces *entry* with the contents of *newEntry*, while retaining the original unique ID.

`EntryReplaceWithModTime(`*`entry, newEntry`*`)`

Replaces *entry* with *newEntry* in the soup, keeping the `_modTime` slot as it was in *entry*.

`EntrySize(`*`entry`*`)`

This returns the number of bytes *entry* occupies in its soup.

`EntrySoup(`*`entry`*`)`

This returns the soup to which *entry* belongs.

`EntryStore(`*`entry`*`)`

This returns the store of the soup to which *entry* belongs.

`EntryTextSize(`*`entry`*`)`

This returns the number of bytes in a soup that the string slots of *entry* occupy.

`EntryUndoChanges(`*`entry`*`)`

This reads *entry* from its soup, wiping out any changes that have been made to *entry* since it was last read or written.

`EntryUniqueID(`*`entry`*`)`

This returns the unique ID of *entry*.

`Erf(`*`number`*`)`

This returns the error function of *number*.

`Erfc(number)`

This returns the complementary error function of `number`.

`EvalStringer(frame, array)`

This concatenates all the elements of `array` into a string. Numbers are converted to strings, while frames, arrays, and booleans are ignored. Symbols are looked up as slot values within `frame`, and then the slot value is converted to a string.

`ExitBreakLoop()`

This exits the current break loop.

`Exp(number)`

This returns e^{number}.

`Expm1(number)`

This function returns e^{number}-1.

`ExtractByte(data, offset)`

This returns the byte at `offset` from the binary object `data`.

`ExtractBytes(data, offset, length, class)`

This returns a binary object composed of `length` bytes starting at `offset` from the binary object `data`. The returned binary object is of class `class`.

`ExtractChar(data, offset)`

This returns the byte at `offset` from the binary object `data` converted to a Unicode character.

`ExtractCString(data, offset)`

This returns a Unicode string created from a C string starting at `offset` bytes within the binary object `data`. A C string contains characters terminated by a zero byte.

`ExtractLong(data, offset)`

This returns an integer composed of four bytes starting at `offset` bytes within the binary object `data`. The high two bits are ignored.

`ExtractPString(data, offset)`

This returns a unicode string created from a pascal string starting at `offset` bytes within the binary object `data`. A pascal string contains a length byte followed by that many bytes of string.

`ExtractUniChar(data, offset)`

This returns a character composed of two bytes starting at `offset` bytes within the binary object `data`.

`ExtractWord(data, offset)`

This returns an integer composed of two bytes starting at `offset` bytes within the binary object `data`.

`ExtractXLong(data, offset)`

This returns an integer composed of four bytes starting at `offset` bytes within the binary object `data`. The low two bits are ignored.

`Fabs(number)`

This returns the absolute value of `number` as a real.

`Fdim(numberX, numberY)`

This returns:

numberX - *numberY*	If *numberX* - *numberY* ≥0.
0	If *numberX* - *numberY* <0.
NaN	If *numberX* or *numberY* is NaN (not a number).

`FindStringInArray(array, string)`

This returns the index of *string* within *array*, or `nil` if *string* isn't present in *array*. This function is case sensitive and diacritical sensitive.

`FindStringInFrame(frame, stringArray, reportPaths)`

This recursively searches *frame* for the strings in *stringArray*. If all the strings are not found, it returns `nil`. Otherwise, if *reportPaths* is `nil`, the return result is `true`; if *reportPaths* is `true`, the return result is an array of match triplets. Each triplet contains:

- The value of the slot containing string.
- The path expression of the slot.
- The character offset within the string at which the match was found.

This function ignores case and diacritical marks, and only matches beginnings of words.

`Floor(real)`

This returns the greatest integer less than or equal to *real*.

`Fmax(numberX, numberY)`

This returns the maximum of *numberX* and *numberY* as a real.

`Fmin(numberX, numberY)`

This returns the minimum of *numberX* and *numberY* as a real.

`Fmod(numberX, numberY)`

This returns the floating point modulus of *numberX* and *numberY*.

`FontAscent(fontSpec)`

This returns the ascent (the distance from the baseline to the ascent line) of *fontSpec* in pixels.

fontSpec can be:

a font frame	Similar to `{family: 'geneva, face: 0, size: 9}`.
one of the ROM fonts	Like `ROM_fontsystem9` or `ROM_fontsystem14underline`.
a packed integer	This specifies a font like `tsSimple + tsSize(18) + tsBold`.

`FontDescent(fontSpec)`

This returns the descent (the distance from the baseline to the descent line) of *fontSpec* in pixels.

fontSpec can be:

a font frame	Like `{family: 'geneva, face: 0,size: 9}`.
one of the ROM fonts	Like `ROM_fontsystem9` or `ROM_fontsystem14underline`.
or a packed integer	This specifies a font like `tsSimple + tsSize(18) + tsBold`.

`FontHeight(fontSpec)`

This returns the height (the ascent plus descent plus leading) of *`fontSpec`* in pixels.

`fontSpec` can be:

a font frame	Like `{family: 'geneva, face: 0,size: 9}`.
one of the ROM fonts	Like `ROM_fontsystem9` or `ROM_fontsystem14underline`.
or a packed integer	This specifies a font like `tsSimple + tsSize(18) + tsBold`.

`FontLeading(fontSpec)`

This returns the leading (the distance from the descent line to the ascent line of the next line) of *`fontSpec`* in pixels.

`fontSpec` can be:

a font frame	Like `{family: 'geneva, face: 0,size: 9}`.
one of the ROM fonts	Like `ROM_fontsystem9` or `ROM_fontsystem14underline`.
or a packed integer	This specifies a font like `tsSimple + tsSize(18) + tsBold`.

`FormattedNumberStr(number, formatSpecString)`

This converts *`number`* to a string as specified by *`formatSpecString`*, which is used like the format string of `printf` in C. For example,

```
FormattedNumberStr(3.5, "%0.2f")
```

yields:

```
3.50
```

`FrameDirty(`*`frame`*`)`

This returns `true` if *`frame`* has been modified since it was last saved in a soup and returns `nil` otherwise.

`Gamma(`*`numberX`*`)`

This returns the gamma function of *`numberX`*.

`GC()`

This causes an immediate garbage collection. Other than taking a little time, this function should have no effect on your application. The only objects that will be garbage collected are those that can no longer be reached. You will rarely, if ever, need to call this routine.

`GetAppParams()`

This returns a frame with information about the Newton. Currently, the information includes the size of the screen and the location of the icon bar. More information may be provided in this frame in the future.

`GetCaretBox()`

Returns the bounding frame (in global coordinates) of the caret (receiving key input). If the caret isn't displayed (if, for instance no keyboard is open), it returns the bounding frame of where written input would be inserted. If no view is receiving key input, the function returns `nil`.

`GetDictionaryData(`*`dictionary`*`)`

This function returns a binary object that contains all the words in *`dictionary`*. The words can be retrieved from the binary object using `SetDictionaryData`.

`GetGlobals()`

This returns the globals frame that contains one slot for each global variable. The frame includes a `functions` slot that contains an array of all global functions.

`GetLocalFromStack(symbol, stackLevel)`

While within a break loop, this returns the value of the local variable or parameter whose symbol is `symbol`. A `stackLevel` of 2 is the current function.

`GetMyZone()`

This function returns the name of the current AppleTalk zone (or `"*"` if the zone has no name).

`GetNames(entityOrEntities)`

This function extracts the entity name from the NBP address `entityOrEntities`. If `entityOrEntities` is an array of NBP address entities, this function returns an array of extracted entity names.

For example, `GetNames("Ralph:Pager@*")` returns `"Ralph"`. `GetNames(["Ralph:Pager@*", "Bob:Pager@Fastzone"])` returns the array `["Ralph", "Bob"]`.

`GetPackages()`

This returns an array of package frames describing each package installed in the system.

`GetPoint(selector, unit)`

This returns a point from the stroke. `selector` can be:

`unitFirstX`	Returns the X coordinate of the pen down.
`unitFirstY`	Returns the Y coordinate of the pen down.
`unitFinalX`	Returns the last X coordinate in the stroke (so far, if the stroke isn't completed).
`unitFinalY`	Returns the last Y coordinate in the stroke (so far, if the stroke isn't completed).

`unit` is the same parameter used by `viewClickScript`, `viewStrokeScript`, or `viewWordScript`.

`GetPointsArray(unit)`

This returns an array of points describing the stroke. Each point is specified in two successive elements of the array. The first is a Y coordinate and the second an X coordinate. They are in global coordinates. *`unit`* is the same parameter used by `viewClickScript`, `viewStrokeScript`, or `viewWordScript`.

`GetRandomWord(minLength, maxLength)`

This returns a random word chosen from the common word dictionary whose length is between *`minLength`* and *`maxLength`* (but in any case no longer than 20 characters).

`GetRoot()`

This returns the root view. Within this view, there is a slot for each application. The slot symbol is the application symbol; the value is the application base view.

`GetScoreArray(unit)`

This returns an array of confidence score for each of the words returned by `GetWordArray(unit)`. The confidence score is a number between 1 and 1000; smaller numbers represent better scores. *`unit`* is the same parameter used by `viewClickScript`, `viewStrokeScript`, or `viewWordScript`.

`GetSelfFromStack(stackLevel)`

While in a break loop, this returns the value of `self` associated with the specified *`stackLevel`*. A *`stackLevel`* of 2 is the current value of `self`.

`GetSlot(frame, slotSymbol)`

This returns the value of the slot whose symbol is *`slotSymbol`* in *`frame`*. It returns `nil` if no such slot is found. This function does not use proto or parent inheritance.

`GetStores()`

This returns an array of stores. The first entry in the array is the internal store.

`GetUnionSoup(soupName)`

This returns a union soup composed of all soups with the name `soupName`.

`GetUnitEndTime(unit)`

This returns the time in ticks that the stroke completed. `unit` is the same parameter used by `viewClickScript`, `viewStrokeScript`, or `viewWordScript`. Don't call this routine until `StrokeDone(unit)` returns true.

`GetUnitStartTime(unit)`

This returns the time in ticks that the stroke was started. `unit` is the same parameter used by `viewClickScript`, `viewStrokeScript`, or `viewWordScript`.

`GetVariable(frame, slotSymbol)`

This returns the value of the slot whose symbol is `slotSymbol` in `frame`. It returns `nil` if no such slot is found. This function uses both proto and parent inheritance.

`GetView(template)`

This returns the view associated with `template`. Returns `nil` if no such view is open.

The following constants can also be passed:

`'viewFrontMost`	Returns the frontmost view with the `vApplication` bit set.
`'viewFrontMostApp`	Returns the frontmost view with the `vApplication` bit set which doesn't have the `vFloating` bit set.
`'viewFrontKey`	Returns the view accepting keystrokes.

`GetViewFlags(template)`

Returns the value of the `viewFlags` slot in the view corresponding to `template` (if it exists) or in `template` itself (if the view doesn't exist).

`GetVolume()`

This returns the current volume represented as an integer from 0 (volume off) to 4 (maximum volume).

`GetWordArray(`*`unit`*`)`

This returns an array of strings which are the recognition choices for the stroke. The strings are ordered by confidence; earlier words in the array are more likely candidates than later words (to obtain the actual confidence values, use `GetScoreArray(`*`unit`*`)`). *`unit`* is the same parameter used by `viewClick-Script`, `viewStrokeScript`, or `viewWordScript`.

`GetZoneFromName(`*`entity`*`)`

This function extracts the zone name from the NBP address entity.

For example, `GetZoneFromName("Ralph:Pager@*")` returns `"*"`. `Get-ZoneFromName("Bob:Pager@Fastzone")` returns `"Fastzone"`.

`GetZoneList()`

This function returns an array containing the names of all the available zones. If there are no named zones, then it returns an array containing the string `"*"`. If this zone list exceeds the memory available, this function may run out of memory.

`HasSlot(`*`frame, slotSymbol`*`)`

This returns true if *`frame`* contains a slot whose symbol is *`slotSymbol`*. It returns `nil` if no such slot is found. This function does not use proto or parent inheritance.

`HasVariable(`*`frame, slotSymbol`*`)`

This returns true if *`frame`* contains a slot whose symbol is *`slotSymbol`*. It returns `nil` if no such slot is found. This function uses both proto and parent inheritance.

`HaveZones()`

This function returns `true` if there are zones on the network. It returns `nil` if there are none to be found.

`HiliteOwner()`

This returns the view containing hilited data (or the lowest common parent if multiple views contain hilited data). If no view contains hilited data, this function returns `nil`.

`HitShape(`*`shape, x, y`*`)`

This function tests whether the point described by *x* and *y* (in local coordinates) is in the shape (or shape array) described by *`shape`*. It returns `true` if the point is in the shape, or `nil` otherwise.

`HourMinute(`*`time`*`)`

This returns the *`time`* (in minutes since midnight, January 1, 1904) as a string specifying the time, for example `"12:45"`.

`Hypot(`*`numberX, numberY`*`)`

This returns $\sqrt{numberX^2 + numberY^2}$.

`InkOff(`*`unit`*`)`

This turns off the electronic ink that is drawn as the user moves the stylus. *`unit`* is the same parameter used by `viewClickScript`, `viewStrokeScript`, or `viewWordScript`.

`InkOn(`*`unit`*`)`

This turns on the display of electronic ink which has been turned off by a previous call to `InkOff(`*`unit`*`)`. *`unit`* is the same parameter used by `viewClickScript`, `viewStrokeScript`, or `viewWordScript`.

`Intern(`*`string`*`)`

This returns the symbol that has the name *`string`*.

`IsA(`*`template, typeSymbol`*`)`

This returns `true` if template is of the type specified by *`typeSymbol`* (that is, *`template`* has an `isa` slot which directly or indirectly contains the value *`typeSymbol`*).

`IsFinite(`*`number`*`)`

This returns true if *`number`* is finite.

`IsNan(`*`number`*`)`

This returns `true` if *`number`* is NaN (not a number).

`IsNormal(`*`number`*`)`

This returns `true` if *`number`* is normalized.

`IsSoupEntry(`*`frame`*`)`

This returns `true` if *`frame`* is an entry in a soup; it returns `nil` otherwise.

`Length(`*`array`*`)`

This returns the number of elements in *`array`*. Do not use this for strings; use `StrLen` instead.

`LessEqualOrGreater(`*`numberX, numberY`*`)`

This returns `true` if neither *`numberX`* nor *`numberY`* are NaN (not a number).

`LessOrGreater(numberX, numberY)`

This returns `true` if neither *`numberX`* nor *`numberY`* are NaN (not a number).

`LGamma(number)`

This returns the logarithm of the absolute value of the gamma function of *`number`*.

`Log(number)`

This returns $\log_e$*`number`*.

`Log10(number)`

This returns $\log_{10}$*`number`*.

`Log1p(number)`

This returns $\log_e$(1+*`number`*).

`Logb(number)`

This returns the integral portion of $\log_2$*`number`*.

`LongDateStr(time, dateStrSpec)`

This returns a string representing *`time`* formatted according to *`dateStrSpec`*. *`dateStrSpec`* can be `kIncludeAllElements` (which prints the year, month, day, and day of week), an entry from the frame `ROM_DateTimeStrSpecs`, or the result of a call to the compile-time function `GetDateStringSpec`.

`LookupWordInDictionary(dictionary, wordString)`

This function returns `true` if *`wordString`* is found in *`dictionary`*, `nil` otherwise.

`MakeLine(`*`x1, y1, x2, y2`*`)`

This makes a line shape starting at (*`x1`*, *`y1`*) and ending at (*`x2`*, *`y2`*).

`MakeOval(`*`left, top, right, bottom`*`)`

This makes an oval shape with the specified *`left`*, *`top`*, *`right`*, and *`bottom`* bounds.

`MakePict(`*`shapeArray, styleFrame`*`)`

This makes a picture shape which, when drawn, draws just as `:DrawShape(`*`shapeArray, styleFrame`*`)` would have.

`MakePolygon(`*`pointsArray`*`)`

This makes a polygon shape from the points specified in *`pointsArray`*. Each point is represented by two integers: an `X` coordinate and a `Y` coordinate. Thus, the number of elements in *`pointsArray`* is twice the number of points.

`MakeRect(`*`left, top, right, bottom`*`)`

This makes a rectangle shape with the specified *`left`*, *`top`*, *`right`*, and *`bottom`* bounds.

`MakeRegion(`*`shapeArray`*`)`

This makes a region shape based on a series of other line and fill shapes as specified in *`shapeArray`*.

`MakeRoundRect(`*`left, top, right, bottom, diameter`*`)`

This makes a rounded rectangle whose bounds are *`left`*, *`top`*, *`right`*, and *`bottom`*, and whose corners are 90-degree arcs from a circle of *`diameter`* pixels.

`MakeShape(`*`object`*`)`

This makes a shape based on *`object`*. The kind of shape created depends on *`object`*:

bounds frame	Returns a rectangle shape.
polygonShape binary object	Returns a polygon shape.
bitmap frame	Returns a bitmap shape.
picture	Returns a picture shape.

`MakeText(`*`string, left, top, right, bottom`*`)`

This returns a text shape whose bounds are *`left`*, *`top`*, *`right`*, and *`bottom`*, and whose text is *`string`*.

`MakeWedge(left, top, right, bottom, startAngle, arcAngle)`

This returns the wedge portion of an oval (whose bounds are *`top`*, *`left`*, *`right`*, and *`bottom`*) as specified by the *`startAngle`* and *`arcAngle`*. A *`startAngle`* of `0` specifies straight up, positive values are clockwise, negative are counter clockwise. The *`arcAngle`* specifies the angle subtended by the arc.

`MapCursor(cursor, applyFunction)`

This applies *`applyFunction`* to each entry in *`cursor`*. It returns an array with the (non-`nil`) return results of *`applyFunction`*.

`Max(`*`numberA, numberB`*`)`

This returns the maximum of *`numberA`* and *`numberB`*.

`Min(`*`integerA, numberB`*`)`

This returns the minimum of *`numberA`* or *`numberB`*.

`NBPGetLookupNames()`

This function returns an array of the entity names that have been found so far by the current NBP lookup. Since the NBP lookup is dynamic (some entities take longer than others to appear, and entities may appear and disappear from the network); this function may return different values on successive calls.

`NBPLookupCount()`

This function returns the number of entities that have been found so far by the current NBP lookup. Since the NBP lookup is dynamic (some entities take longer than others to appear, and entities may appear and disappear from the network), this function may return different values on successive calls.

`NBPStartLookup(`*`entity`*`)`

This function begins a lookup for entities as specified by *`entity`*. The entity parameter can be a complete entity (specifying a name, type, and zone), or a wildcard '=' as the name. Such a wildcard would specify all entities of the given type found in the given zone. Since you cannot also specify a wildcard for the zones, this call only finds entities within a given zone.

For example, `NBPStartLookup("Ralph:Pager@*")` looks for the entity named "Ralph" of the type "Pager" in the current zone. On the other hand, `NBPStartLookup("=:Pager@*")` looks for all entities of the type "Pager" in the current zone.

`NBPStopLookup()`

This function stops a lookup started by `NBPStartLookup`.

`NearbyInt(`*`number`*`)`

This rounds *`number`* to the nearest integer and returns it as a real. It doesn't raise the inexact exception (as does `RInt`).

`NewDictionary(dictionaryKind)`

This function creates a new RAM-based dictionary. *`dictionaryKind`* should be `'custom`.

`NextAfterD(numberX, numberY)`

This returns the next representable number after *`numberX`* that is closer to *`numberY`*.

`NumberStr(number)`

This converts *`number`* to a string.

`OffsetShape(shape, deltaH, deltaV)`

This translates the *`shape`* (or shapes if *`shape`* is an array) by *`deltaH`* pixels horizontally and *`deltaV`* pixels vertically.

`OpenAppletalk()`

This opens AppleTalk. You would call this routine before any other AppleTalk global functions.

`Ord(character)`

This function converts *`character`* from Unicode to its integer equivalent.

`ParamStr(baseString, paramStrArray)`

This destructively modifies *`baseString`*, replacing all occurrences of the characters `^0`, `^1`, ..., `^9` within `baseString` with a corresponding string from `paramStrArray`. `^0` is replaced with `paramStrArray[0]`, `^1` with `paramStrArray[1]`, and so on.

`ParseUtter(`*`string`*`)`

This calls Intelligent Assistance with *`string`* as if the user had written it in the Assist slip and tapped Do. It returns the result frame.

`Perform(`*`frame, messageSymbol, parameterArray`*`)`

This sends the *`messageSymbol`* message to *`frame`* with the parameters in *`parameterArray`*. The return result is the return result of the method that executes.

`PlaySound(`*`soundFrame`*`)`

This plays the sound specified in *`soundFrame`*. It returns before the sound has finished playing.

`PlaySoundSync(`*`soundFrame`*`)`

This plays the sound specified in *`soundFrame`*. It returns after the sound has finished playing.

`PointsToArray(`*`polygonShape`*`)`

This extracts the data from *`polygonShape`* (a binary object of class `polygonShape`) and returns the data in an array. The first element of the array contains an integer describing the type of the shape. The second element is an integer specifying the number of points. The remaining elements are pairs of coordinates: the first is an x coordinate; the second a Y coordinate.

`PostKeyString(`*`view, keystrokeString`*`)`

Sends the characters in *`keystrokeString`* to *`view`* as if they had been entered at the keyboard.

`Pow(numberX, numberY)`

This function returns $numberX^{numberY}$ as a real.

`PrimClassOf(value)`

This returns the primitive class (type) of *`value`*. Possible results are `'immediate`, `'binary`, `'array`, or `'frame`.

`PtInPicture(x, y, bitmap)`

This function attempts to determine whether the point *`x`*, *`y`* is within *`bitmap`*. It doesn't work correctly, however. Use the `PtInBitmap` platform function instead.

`Query(soup, querySpec)`

This creates a cursor for *`soup`* based on the *`querySpec`*. Types of queries are index, words, and text.

`Random(lowerBoundInteger, upperBoundInteger)`

This returns a random integer in the range [*`lowerBoundInteger`*, *`upperBoundInteger`*].

`RandomX(number)`

This returns an array where the first element is a pseudo random number based on the seed *`number`*. The second element is the new seed.

`Real(integer)`

This function returns *`integer`* converted to a real number.

`RefreshViews()`

This function redraws any dirty views immediately.

`RegTaskTemplate(`*`taskTemplate`*`)`

This registers the Intelligent Assistance template specified by *`taskTemplate`*. It returns a value that can be used to unregister using `UnRegTaskTemplate`

`RelBounds(`*`left`*`, `*`top`*`, `*`width`*`, `*`height`*`)`

This returns a frame `{left:` *`left`*`, right:` *`left`* `+` *`width`*`, top:` *`top`*`, bottom:` *`top`* `+` *`height`*`}`.

`Remainder(`*`numberX`*`, `*`numberY`*`)`

This returns the remainder of *`numberX`* divided by *`numberY`*.

`RemovePackage(`*`packageFrame`*`)`

Removes the package specified by *`packageFrame`* (as returned by `GetPackages`).

`RemovePowerOffHandler(`*`view`*`)`

This unregisters *`view`* so that it will no longer be notified when the Newton is going to sleep. *`view`* should have previously been registered with `AddPowerOffHandler`.

`RemoveSlot(`*`arrayOrFrame`*`, `*`arrayIndexOrSlotSymbol`*`)`

This function removes a specified slot from a frame or element from an array.

`RemoveStepView(`*`parentView`*`, `*`childView`*`)`

This removes *`childView`* from *`parentView`* and closes it.

`RemQuo(numberX, numberY)`

This function returns in an array both the remainder and the quotient of `numberX` divided by `numberY`. Only the seven low-order bits of the quotient are retained (plus the sign bit) to yield a value between -127 and 127).

`ReplaceObject(originalBinaryObject, targetBinaryObject)`

This function causes all references to `originalBinaryObject` (a nonimmediate) to be redirected to `targetBinaryObject` (a nonimmediate).

`RInt(number)`

This rounds `number` to the nearest integer and returns it as a real.

`RIntToL(number)`

This rounds `realNumber` to its nearest integral value. The return result is an integer.

`Round(realNumber)`

This rounds `realNumber` to its nearest integral value by adding 0.5 and truncating. The return result is a real number with the fraction part equal to `0`.

`SaveUserDictionary()`

This function saves any changes that have been made to the user's personal dictionary to the system soup.

`Scalb(numberX, numberY)`

This returns $numberX \cdot 2^{numberY}$ as a real.

`ScaleShape(`*`shape, srcRect, dstRect`*`)`

This modifies the location and size of *`shape`* (or, if *`shape`* is an array, all the shapes in the array). The location is modified by the difference from the top left of *`dstRect`* from the top left of *`srcRect`*. The size is scaled by the ratio in widths (and heights) of *`dstRect`* to *`srcRect`*.

`SetAdd(`*`array, value, uniqueFlag`*`)`

If *`uniqueFlag`* is non-`nil`, this function adds *`value`* to *`array`* when it is not already present. If *`uniqueFlag`* is `nil`, the function adds *`value`* to the end of *`array`*.

`SetBounds(`*`left, top, right, bottom`*`)`

This function returns a bounds frame: `{left:` *`left`*`, top:` *`top`*`, right:` *`right`*`, bottom:` *`bottom`*`}`.

`SetClass(`*`arrayFrameOrBinaryObject, classSymbol`*`)`

This functions sets the class of *`arrayFrameOrBinaryObject`* to *`classSymbol`*. The modified object is returned. You cannot set the class of an immediate.

`SetContains(`*`array, value`*`)`

This returns `true` if *`value`* is present in *`array`*; it returns `nil` otherwise.

`SetDictionaryData(`*`dictionary, binaryObject`*`)`

This function retrieves the words from *`binaryObject`* (which should have been created by a previous call to `GetDictionaryData`). It stores the word in *`dictionary`*, an existing dictionary.

`SetDifference(arrayA, arrayB)`

This returns an array containing those elements that are in `arrayA` but are not in `arrayB`.

`SetInkerPenSize(inkThickness)`

This function sets the thickness of the electronic ink drawn by the stylus to `inkThickness` (an integer from 1 to 4).

`SetKeyView(view, characterOffset)`

This sets `view` to accept keystrokes. It sets the caret to be located `characterOffset` characters within any text in the view. `view` must be a `clEditView` or `clParagraphView`.

`SetLength(array, newLength)`

This sets the length of `array` to `newLength`. If the array has to grow, the new elements are `nil`.

`SetOverlaps(arrayA, arrayB)`

This function returns the index of the first element of `arrayA` that is also present in `arrayB`. If no such element is found, it returns `nil`.

`SetRandomSeed(number)`

This function sets the starting seed of the pseudorandom number generator to `number`.

`SetRemove(array, value)`

If `value` is present in `array`, this function removes it.

`SetTime(time)`

This function sets the system clock to `time` (specified in the number of seconds since midnight, January 1, 1904).

`SetUnion(arrayA, arrayB, uniqueFlag)`

This function returns an array with all the elements from ***arrayA*** and ***arrayB***. If ***uniqueFlag*** is non-`nil`, then duplicates are not added.

`SetValue(view, slotSymbol, value)`

This function does the following: sets the ***slotSymbol*** slot of ***view*** to ***value***; sends the `viewChangedScript` message to the view; and redraws the view if an important slot has changed (such as `viewBounds`, `viewFlags`, `text`, and so on).

`SetVolume(volume)`

This sets the current volume to ***volume***. Acceptable values are integers from 0 (volume off) to 4 (maximum volume).

`ShapeBounds(shape)`

This returns the bounding box enclosing shape. If shape is an array, this returns the bounding box enclosing all the shapes.

`ShortDate(time)`

This function converts time (an integer specifying the number of minutes since midnight, January 1, 1904) into a string. The format of the string depends on the locale; in the U.S. it is of the form `"Mon 1/2"`.

`ShortDateStr(time, dateStrSpec)`

This returns a string representing ***time*** formatted according to ***dateStrSpec***. ***dateStrSpec*** can be `kIncludeAllElements` (which prints the numeric year, month, and day), an entry from the frame `ROM_DateTimeStrSpecs`, or the result of a call to the compile time function `GetDateStringSpec`.

`SignBit(number)`

This returns zero if ***number*** is negative, nonzero otherwise.

`Signum(number)`

This function returns 0 if *number* is zero, -1 if *number* is negative, and 1 if *number* is positive. The return result is an integer if *number* is an integer and a real if *number* is a real.

`Sin(number)`

This returns the sine of *number*.

`Sinh(number)`

This returns the hyperbolic sine of *number*.

`SmartStart(length)`

This routine allocates `length` bytes for a smart string (a string which has extra bytes at the end so that adding to it doesn't necessarily require a memory allocation) and returns it.

`SmartStop(string, length)`

This routine truncates the smart string *string* at `length` characters.

`SmartConcat(string, length, newText)`

This routine appends *newText* to the smart string *string* `length` characters within the string. If necessary, it resizes *string* to reserve enough room for *newText*.

`Sort(array, test, key)`

This sorts *array* and returns it based on the supplied `test` and *key*. The *key* specifies a path expression, which is the key within each array element; `nil` specifies the entire array element. `test` can be a function taking two array elements and returning 0 if the two are equal, a positive integer if the first sorts after the second, and a negative number if the first sorts before the second. `test` can also be one of:

`'|<|`	Does a numeric ascending sort.
`'|>|`	Does a numeric descending sort.
`'|str<|`	Does a string ascending sort.
`'|str>|`	Does a string descending sort.

`SplitString(`*`string`*`)`

This breaks up *`string`* into separate words, returning an array of words.

Caution: `SplitString` is fragile. It doesn't split on tabs or newlines, handles multiple spaces badly, and splits Japanese text into individual characters rather than words.

`SPrintObject(`*`value`*`)`

This converts *`value`* to a string. Characters, numbers, and symbols are converted to their natural values; other values (like frames, booleans, and arrays) are converted to empty strings.

`Sqrt(`*`number`*`)`

This returns $\sqrt{number}$.

`Stats()`

This function returns the number of bytes of frame heap that are available. It also prints to the Inspector that number and the largest number of bytes available. Call `GC` before calling `Stats` to obtain an accurate number.

`StrCompare(stringA, stringB)`

This returns `true` if the characters in *`stringA`* are equal to the characters in *`stringB`*. It returns a negative number if *`stringA`* sorts before *`stringB`*, and a positive number if *`stringA`* sorts after *`stringB`*. This function ignores case and diacritical marks.

`StrConcat(stringA, stringB)`

This returns a new string composed of the characters in *`stringA`* followed by the characters in *`stringB`*.

`StrEqual(stringA, stringB)`

This returns `true` if *`stringA`* contains the same characters as *`stringB`*, and `nil` otherwise. This function ignores case.

`StrFontWidth(string, fontSpec)`

This returns the width of *`string`* in pixels if it were drawn with the font specified by *`fontSpec`*.

`fontSpec` can be:

a font frame	Like `{family: 'geneva, face: 0: size: 9}`.
one of the ROM fonts	Like `ROM_fontsystem9` or `ROM_fontsystem14underline`.
or a packed integer	This specifies a font like `tsSimple + tsSize(18) + tsBold`.

`Stringer(array)`

Returns a string composed of each of the elements of *`array`* concatenated together. Numbers and symbols are converted to strings before concatenation.

`StringToDate(string)`

This converts *`string`* to a time (number of minutes since midnight, January 1, 1904). The function returns *`nil`* if *`string`* doesn't represent a date.

If *`string`* contains a time, but no date, the current day is used. In a fit of foolish consistency, if string contains a date, but no time, the current time is used.

`StringToNumber(string)`

This converts *`string`* to a real number. The function returns `nil` if *`string`* doesn't represent a number.

`StringToTime(timeString)`

This function converts the time information specified in *`timeString`* along with today's date to an integer representing the number of minutes since midnight, January 1, 1904.

`StrLen(string)`

This returns the number of characters in *`string`*.

`StrMunger(dstString, dstStart, dstCount, srcString, srcStart, srcCount)`

This function destructively modifies *`dstString`*, replacing *`dstCount`* characters starting at *`dstStart`* (`0` specifies the first character) with *`srcCount`* characters from *`srcString`*, starting at *`srcStart`* (`0` specifies the first character). If *`dstCount`* (or *`srcCount`*) is `nil`, all characters to the end of *`dstString`* (or *`srcString`*) are specified. If *`srcString`* is `nil`, characters are deleted from *`dstString`*. The function returns the munged string.

`StrokeBounds(`*`unit`*`)`

This returns the bounds of the stroke in global coordinates. *`unit`* is the same parameter used by `viewClickScript`, `viewStrokeScript`, or `viewWordScript`.

`StrokeDone(`*`unit`*`)`

This returns `true` if the stroke has been completed. *`unit`* is the same parameter used by `viewClickScript`, `viewStrokeScript`, or `viewWordScript`.

`StrPos(`*`string, subString, startPosition`*`)`

This function searches for *`subString`* within *`string`* starting at *`startPosition`* (`0` specifies the first character). It returns the position within *`string`* at which *`subString`* is found (`0` specifies the first character), or `nil` if *`subString`* isn't found. This search ignores case and diacritical marks.

`StrReplace(`*`string, subString, replacementString, count`*`)`

This function destructively modifies *`string`* by replacing *`count`* occurrences of *`subString`* with *`replacementString`*. If count is `nil`, all occurrences are replaced. Case and diacritical marks are ignored. This function returns the number of replacements made.

`StrTruncate(`*`string, widthInPixels`*`)`

This function returns a new string composed of characters from *`string`*, truncated so that its width is no more than *`widthInPixels`*. If characters must be truncated, an ellipsis (…) is added to the end of the string. This function needs a current font to determine width. It gets the current font by looking for a slot named `viewFont`, using inheritance. If no such slot is found, an exception is thrown.

`StrWidth(`*`string`*`)`

This returns the width of *`string`* (in pixels). This function needs a current font to determine width. It gets the current font by looking for a slot named `view-Font`, using inheritance. If no such slot is found, an exception is thrown.

`StuffByte(`*`data, offset, integerContainingByte`*`)`

This sets the byte at *`offset`* in the binary object *`data`* to the low-order byte of *`integerContainingByte`*.

`StuffChar(`*`data, offset, intOrChar`*`)`

This sets the byte at *`offset`* in the binary object *`data`* to the low-order byte of *`intOrChar`*.

`StuffLong(`*`data, offset, theLong`*`)`

This sets the four bytes starting at *`offset`* bytes within the binary object *`data`* to *`theLong`*. The two high-order bits of *`theLong`* are sign-extended.

`StuffUniChar(`*`data, offset, intOrChar`*`)`

This sets the two bytes starting at *`offset`* bytes within the binary object *`data`* to the two low-order bytes of *`intOrChar`*.

`StuffWord(`*`data, offset, integer`*`)`

This sets the two bytes starting at *`offset`* bytes within the binary object *`data`* to the two low-order bytes of *`integer`*.

`SubStr(`*`string, startPosition, numberCharacters`*`)`

This returns a new string containing *`numberCharacters`* characters starting at position *`startPosition`* within *`string`*. A *`startPosition`* of `0` specifies starting with the first character.

`Tan(`*`number`*`)`

This returns the tangent of `number`.

`Tanh(`*`number`*`)`

This returns the hyperbolic tangent of `number`.

`Ticks()`

This returns numbers in sixtieths of a second, which can be used to measure durations.

`TieViews(`*`mainView, dependentView, methodSymbol`*`)`

This function sets up a dependency between `mainView` and `dependentView`. Whenever `mainView` changes (*ala* `viewChangedScript`), the `methodSymbol` message is sent to `dependentView` with two parameters: the `mainView` and the symbol of the slot within `mainView` that changed.

If the function succeeds in setting up a dependency, it returns `true`, otherwise it returns `nil`.

`Time()`

This returns the number of minutes since midnight, January 1, 1904, as an integer.

`TimeInSeconds()`

This returns the number of seconds since midnight January 1, 1993, as an integer.

`TimeStr(`*`time, timeStrSpec`*`)`

This returns a string representing `time` formatted according to `timeStrSpec`. `timeStrSpec` can be `kIncludeAllElements` (which prints the hour, minutes, seconds, and AM/PM), an entry from the frame `ROM_DateTimeStrSpecs`, or the result of a call to the compile time function `GetDateStringSpec`.

`TotalClone(value)`

This returns a recursive clone of `value`. Frame maps are copied.

`TotalMinutes(dateFrame)`

This returns the time in minutes since midnight, January 1, 1904, when passed `dateFrame`. The slots in `dateFrame` should be as described in `Date()`, although the `time` and `dayOfWeek` slots are not used.

`TrimString(string)`

This destructively strips any white space (spaces, tabs, or newlines) from the beginning and ending of `string`. The function returns the trimmed `string` as a result.

`Trunc(number)`

This returns number with its fractional part truncated (toward zero). The return result is a real. For example, `Trunc(3.6)` is `3.0`, `Trunc(-3.6)` is `-3.0`.

`Unordered(numberX, numberY)`

This returns `true` if *numberX* or *numberY* is a NaN (not a number).

`UnorderedGreaterOrEqual(numberX, numberY)`

This returns `true` if *numberX* is greater than or equal to *numberY* or if *numberX* or *numberY* is a NaN (not a number).

`UnorderedLessOrEqual(numberX, numberY)`

This returns `true` if *numberX* is less than or equal to *numberY* or if *numberX* or *numberY* is a NaN (not a number).

`UnorderedOrEqual(`*`numberX, numberY`*`)`

This returns `true` if *`numberX`* is equal to *`numberY`* or if *`numberX`* or *`numberY`* is a NaN (not a number).

`UnorderedOrGreater(`*`numberX, numberY`*`)`

This returns `true` if *`numberX`* is greater than *`numberY`* or if *`numberX`* or *`numberY`* is a NaN (not a number).

`UnorderedOrLess(`*`numberX, numberY`*`)`

This returns `true` if *`numberX`* is less than *`numberY`* or if *`numberX`* or *`numberY`* is a NaN (not a number).

`UnRegTaskTemplate(`*`templateID`*`)`

This unregisters a registered Intelligent Assistance task template. The parameter, *`templateID`*, is the return result of a previous call to `RegTaskTemplate`.

`Upcase(`*`string`*`)`

This function destructively modifies *`string`* so that each lowercase character is replaced with its uppercase equivalent; non-lowercase letters are not modified. The modified string is returned.

`Visible(`*`view`*`)`

This returns `true` if *`view`* is visible on screen (open and not hidden).

`Write(`*`value`*`)`

This prints *`value`* to the Inspector.

Appendix D

Important Global Variables

This appendix discusses most of the important global variables (each of which is a slot in `GetGlobals()`).

`findApps`

An array of application symbols that support global find. Don't directly modify this variable; use the platform file functions `RegFindApps` and `UnregFindApps` instead.

`functions`

This frame contains one entry for each global function. The name of each slot is the name of a function and its value is a code block.

`international`

This frame contains slots used for internationalization. One of the slots is `locales`, which is an array containing all the locales available on this Newton. Don't modify or access this array directly; use the platform file functions `AddLocale` and `FindLocale` instead.

`motorola`

This frame (present only on the Marco) containing a number of methods and variables.

`printDepth`

An integer containing the depth that print statements use. The default value is 1.

`routing`

A frame containing one slot for each application that supports routing. The slot name is the symbol of the application, while its value is a routing frame.

`soupNotify`

This is an array to which an application add entries if it wants to be notified about changes to a particular soup. The application registers in `soupNotify` by adding two entries to the array: the first is a particular soup name (a string) and the second is an application symbol. When `BroadcastSoupChange` is called with a soup name, it sends the `soupChanged` message to each application that registered with that soup name.

`trace`

If `trace` is equal to `'functions`, all function calls are printed. If it is equal to `true`, all function calls and variable assignments are printed. The default value of `trace` is `nil`.

`userConfiguration`

This frame contains information entered by the user in the Preferences application, such as name, phone number, etc. Don't directly modify or access this frame; use the platform file functions `SetUserConfig` and `GetUserConfig` instead.

Appendix E

Platform File Functions

The Platforms folder which comes with NTK (and Demo NTK) contains a file named "MessagePad" (normally called the "Platform" file). This file contains a number of functions that you might find useful in your applications.

How to Use Functions in the Platform File

Functions in the Platform file are not global functions; instead, they are predefined constants (each function starts with `k` and ends with `func`). If you call one, it will be included as part of your package. Only the platform file functions you call are included in your package.

You'll normally call a Platform file function using the standard NewtonScript `call` syntax. Here is an example of calling the `RegisterCardSoup` Platforms file function:

```
call kRegisterCardSoupFunc with
   (soupName, soupIndexes, appSymbol, appObject);
```

(Many of you may notice that this function looks quite familiar. It is after all how we create and register soups in an application.)

Platform File Functions

Here are the available functions in the Platform file (in alphabetical order). Along with the function name and its argument list is a brief description of what the function does.

Caution: Remember that each of these Platform file functions actually has a `k` at the beginning and a `func` at the end. So, for instance, the Platform file function `Send` is actually named `kSendFunc`.

`AddAlarm(key,time,notifyArgs,callBackFn,callBackArgs)`

This function registers an alarm starting at *`time`* (in minutes since midnight, January 1, 1904). When the alarm goes off, the following happens:

- If *`notifyArgs`* is non-`nil`, then *`notifyArgs`* is used as an array of arguments to `Notify`.
- If *`callBackFn`* is non-`nil`, then it is called with the array of arguments in *`callBackArgs`*.

The key is a string that identifies the alarm. It should use your unique signature as a suffix to guarantee uniqueness.

Caution: Make sure that *`callBackFn`* has an empty lexical environment. Otherwise, when the function is copied to internal memory (in order to save it to a soup), the entire lexical environment will be copied.

`AddLocale(theLocaleBundle)`

This function adds *`theLocaleBundle`* to the `locales` array in the `international` global frame.

`ClearSelectionHilites(theView)`

This removes from *`theView`* any highlighting due to a selection. *`theView`* must be a `clEditView` or a `clParagraphView`.

`CloseRemoteControl(irConnectionID)`

This closes the remote control channel *`irConnectionID`* that had been opened by `OpenRemoteControl`.

`EnableIRModule(enable)`

This function allows the Newton infrared module to operate as a half-duplex serial port. It will enable or disable the infrared module based on the value of *`enable`*.

`FindLocale(titleString)`

This finds the frame in the `locales` array in the `international` global frame whose title matches *`titleString`*.

`FlushUserConfig()`

The method saves any changes made to the global `userConfiguration` frame into the System soup. You should call this function after you modify `userConfiguration` with `SetUserConfig`.

`GetAlarm(key)`

This returns a frame with information about the alarm whose key value is *`key`*. The alarm must have already been added with `AddAlarm`. The documented slots in the frame are `key`, `time`, `notifyArgs`, `callBackFn`, `callBackParams`.

The values of these slots should be the same as those of the arguments for the function `AddAlarm`.

`GetAppAlarmKeys(keySuffix)`

This method returns an array of all alarm key strings having the suffix *`keySuffix`*. This provides a way to obtain all alarms associated with your application. The alarms key strings are stored by the order in which the alarms will execute, with the alarm for the first entry to execute stored first.

`GetDefaultStore()`

This returns the store frame which is the current default store.

`GetLanguageEnvironment()`

This returns the language of the ROM of the particular Newton. Current, possible return values are:

0	English
1	French
3	German
14	Japanese

`GetPrinterName(printerFrame)`

This function returns the name of the printer associated with the printer frame object *`printerFrame`*. You can obtain a printer frame object with:

```
call kGetUserConfigFunc with ('currentPrinter)
```

`GetSCCSideB()`

This function is used to allow the Newton infrared module to operate as a half-duplex serial port. This function returns a value which is used as the data for a special endpoint option.

This endpoint will be a plain half-duplex serial endpoint, running over the infrared module instead of the serial port.

Here is an example of how to create such an endpoint:

```
foo := {
   _proto: protoEndpoint,
   configOptions: [
      {
         label: "aser",
         type: 'service,
         opCode: opSetRequired
      },
      {
         label: "scc ",
         type: 'option,
         opCode: opSetRequired,
         data: call kGetSCCSideBFunc with ()
      }
   ]
};
```

`GetUserConfig(`*`configSym`*`)`

This returns the value of a slot (specified by *`configSym`*) in the `userConfiguration` frame. You should use this function rather than directly reading from the `userConfiguration` global frame.

`OpenRemoteControl()`

This function initializes the remote control functionality. It returns an object, *`irConnectionID`*, that must be passed to subsequent remote control calls. It returns `nil` if the function failed.

`PtInBitMap(`*`x`*`,`*`y`*`,`*`bitMap`*`)`

This function tests whether the point (*x*, *y*) within *`bitmap`* has been set. It returns `true` if the point is on, and `nil` if it is off. Note that if *`bitmap`* has a mask, the mask is tested instead.

RegFindApps(*appSymbol*)

This registers `appSymbol` for global Find. The Find and DateFind messages will be sent to `GetRoot().(kAppSymbol)` when the user chooses Find All or Date Find All. You should use this function instead of directly modifying the global `findApps` array.

RegisterCardSoup(*soupName,indexArray,appSymbol,appObject*)

This function does all the preparation necessary to use a union soup. The parameters are:

soupName	A string that represents the name of your soup.
soupIndexes	An array of frames. Each describes a soup index that needs to be created.
appSymbol	The unique application symbol.
appObject	An array of two strings. The first is the singular version of the items in the soup. The second is the plural version (for example: `["Name", "Names"]` or `["Goose", "Geese"]`).

RemoveAlarm(*key*)

This function removes the alarm whose key is *key*.

RemoveAppAlarms(*keySuffix*)

This function removes all alarms having key strings that end with the string *keySuffix*. This provides a way to remove all alarms for a given application.

Send(*transportSym, itemFrame*)

This function posts `itemFrame` to the Out Box. It uses the transport specified by `transportSym`.

`SendRemoteControlCode(`*`irConnectionID, command, count`*`)`

This function sends a remote control command through the infrared channel specified by *`irConnectionID`*. The command is sent *`count`* times. The *`command`* parameter is a binary object which is broken up in the following order:

name	This four bytes is a command code. Set this to whatever you want.
timeBase	This four bytes is the time base in microseconds.
leadIn	This four bytes is the duration (in timeBase units) of the lead bit cell.
repeat	This four bytes is the duration (in timeBase units) of the last bit cell for loop commands. This value is used only when the *`count`* parameter is greater than 1.
leadOut	This four bytes is the duration (in timeBase units) of the last bit cell for nonloop commands.
transitionCount	This four bytes specifies the number of following (four-byte) transitions.
transitions	This is the transitionCount four-byte duration values (in timeBase units).

The actual values to use for a particular manufacturer's device should be obtained from that manufacturer.

`SetDefaultStore(`*`newDefaultStore`*`)`

This function makes *`newDefaultStore`* the default store.

`SetUserConfig(`*`configSym, theValue`*`)`

This sets the value of the slot in the system `userConfiguration` frame whose symbol is *`configSym`* to *`theValue`*. You should use this function instead of directly modifying the global `userConfiguration` frame.

Make sure to call `FlushUserConfig` to save the changes to the System soup.

`SimpleTextHeight(`*`text,width,fontSpec`*`)`

This function calculates the height (in pixels) that the string, **`text`**, will occupy if it were drawn into a `clParagraphView` of width *`width`* with the font set of *`fontSpec`*, no `styles` slot, and no `viewLineSpacing` slot.

`UnionSoupIsNull(`*`unionSoup`*`)`

This function returns `true` if *`unionSoup`* contains no constituent soups. This situation should not normally occur with a normal application, but could happen during development. Since a union soup with no soups on any stores can't be used, this function is provided so that you can detect this case. Most applications will have no need to use this function.

`UnRegFindApps(`*`appSymbol`*`)`

This function unregisters *`appSymbol`* from global finds. You should use this function instead of directly modifying the global `findApps` array.

`UnRegisterCardSoup(`*`soupName`*`)`

This function unregisters the soup whose name is *`soupName`*. You should use this function after you are finished with a union soup.

`ViewIsOpen(`*`view`*`)`

This function returns a boolean specifying whether *`view`* is open. You should use this function instead of accessing the `viewCObject` slot of a view.

Appendix F

Using Newton Toolkit

Newton Toolkit:
Development Environment of Champions

—*The Newton Toolkit Team*

This appendix is about Newton Toolkit (NTK), the development environment for Newton. NTK manages the entire life cycle of an application: You create projects, layout templates, build your application, download to the Newton, and debug all within NTK.

System Requirements

To run NTK, you need:

- System 7 (or later).
- A Macintosh with a 68030, 68040, or Power PC processor.
- 32-bit addressing. For older Macintoshes (the Macintosh II, IIcx, IIci, and Macintosh SE/30) whose ROMs only support 24-bit addressing, you'll need the "Mode32 7.5" extension, which allows 32-bit addressing on those machines. For Macintoshes that support both 24-bit and 32-bit addressing, you can select 32-bit addressing from the Memory control panel (see FIGURE F.1). Many newer Macintoshes (like the Power Macintosh) run only in 32-bit mode and do not provide the 32-bit addressing option in the Memory control panel.

Note: The "Mode32 7.5" extension is available online in the Connectix forum on America Online, eWorld, Compuserve, and Applelink. It can also be ordered from Connectix at (800) 950-5880 or (415) 571-5100.

Caution: If you are using System 7.5, do not use the outdated "Mode32" control panel; it will corrupt your system folder. Use the newer "Mode32 7.5" extension instead, because it will run on System 7.5 (and earlier systems).

- A serial cable (use a Macintosh/Imagewriter II cable, available from an Apple dealer as model M0197—part number 590-0552-A).

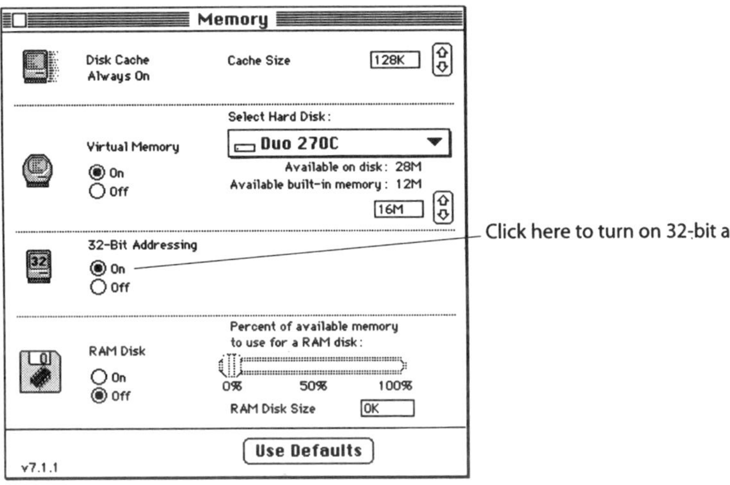

FIGURE F.1 Memory control panel.

Installing Demo NTK

Two files need to be installed in your system folder for demo NTK to work properly:

Apple Modem Tool	This Comm Toolbox communications tool is used by NTK to communicate with your Newton using a serial port. Install it in the Extensions folder.
Newton Toolkit Font	This font is used to preview within NTK how things will look on the Newton. Install it in the Fonts folder (if using System 7.1 or later), or in the System file (if using System 7.0).

These are the remaining files in your Newton Toolkit Demo folder:

Newton Toolkit Demo	The Development Environment of Champions.
NTK Demo Release Notes	Late-breaking information about Demo NTK.

EditorCommands	File defining commands used for editing text within NTK.
NTK Definitions	A text file containing a huge number of constants that you can use in your code.
Toolkit App	A package for the Newton that allows you to connect the Inspector and download NTK applications.
Platforms Folder	A folder that contains files for each of the Newton platforms.
MessagePad	The platform file for the Newton Message Pad, first Newton in a family of products.

Differences between Demo NTK and NTK

There are a number of limitations in Demo NTK as compared to NTK:

- You can only have a maximum of 10 files per project.
- The Project Settings dialog cannot be edited (clicking OK from the Project Settings dialog yields the alert shown in FIGURE F.2). This means that the application name and application symbol cannot be changed. It follows from this that you can have at most one package built using Demo NTK existing on a Newton at any single time.

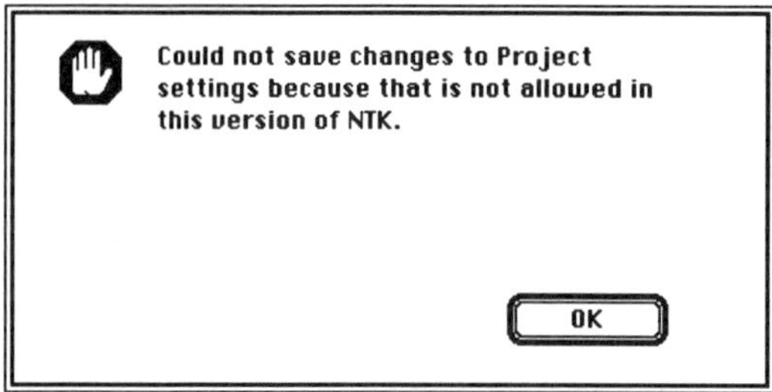

FIGURE F.2 Alert shown when the Project Settings dialog is OKed.

- Each time a package built with Demo NTK is installed on the Newton, the slip shown in FIGURE F.3 is displayed.

FIGURE F.3 Slip shown when a package built with Demo NTK is installed.

- The license for Demo NTK restricts its use to "doing the exercises contained in the book with which it is distributed and does not include the right to use the Apple software for purposes of building applications for any Newton-based products." NTK can be used for creating Newton-based products.
- Demo NTK contains no AppleTalk ADSP Tool.
- Demo NTK doesn't contain the Book Maker application (used to create digital books).
- NTK ships with a set of four manuals: The NewtonScript Programming Language, Newton Programmer's Guide, Newton Book Maker User's Guide, and Newton Toolkit User's Guide.

Please read the Apple Computer, Inc. Software License (found at the back of this book) in its entirety.

NTK can be purchased from APDA at (800) 282-2732 or (716) 871-6555.

Using NTK

Now that you have installed NTK, it is time to learn how to use it. Notice that the information we cover is identical for both demo NTK and NTK. Because it makes no difference in how you use the product we have dispensed with any reference to a demo version; we will simply refer to the environment as NTK.

Launching NTK

The first time you launch NTK you'll see the dialog shown in FIGURE F.4. If you already have an existing project, you can select it from this dialog. New projects are handled differently—for now, simply click the Cancel button.

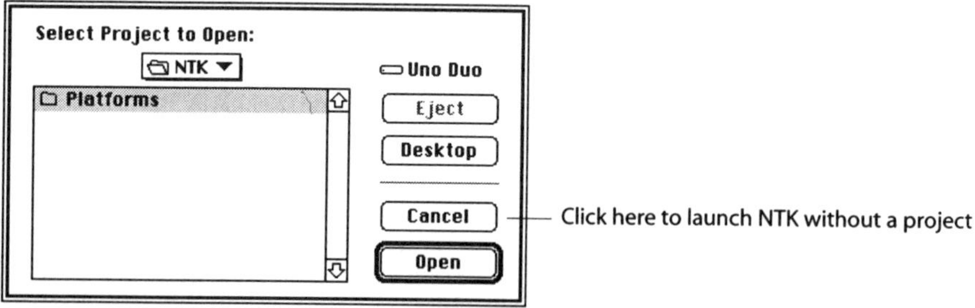

FIGURE F.4 Initial NTK dialog.

Setting Your Connection Type

The Newton does not need to be connected to the Macintosh while you write and build your applications. But if you want to download or debug the application, you need to connect the two machines. You usually use a serial connection to connect the Macintosh to the Newton (although it is possible to download applications using LocalTalk if you have the "AppleTalk ADSP Tool" installed in your Extensions folder). You specify your connection choice using the Toolkit Preferences dialog, which is found in NTK's Edit menu (see FIGURE F.5).

Choosing a Serial Connection

If you are using a serial connection, you'll need to specify which serial port you are using. For most Macintoshes, the available choices are the printer or modem port (see FIGURE F.6).

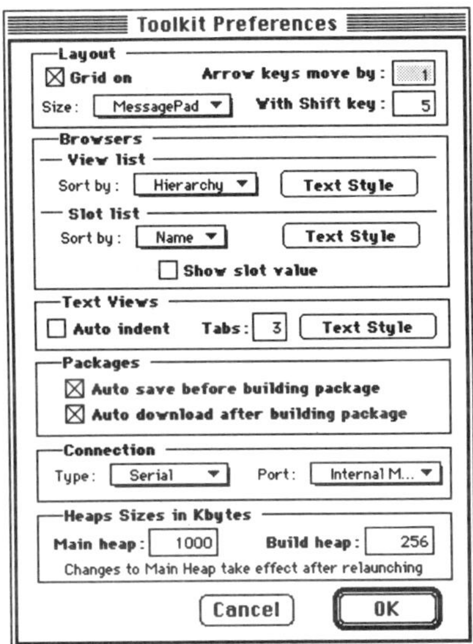

FIGURE F.5 Toolkit Preferences dialog.

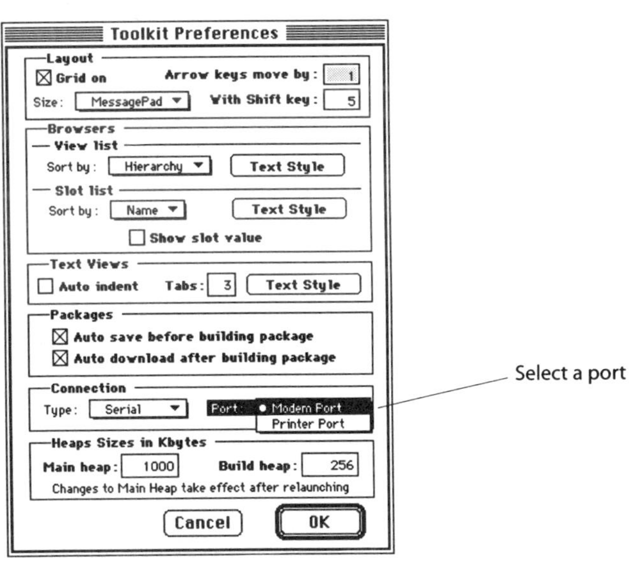

FIGURE F.6 Choice of ports in NTK Preferences dialog.

Note: If you are using the built-in serial port of a PowerBook Duo, select the Printer-Modem port and make sure AppleTalk is off. If you have a PowerBook internal modem, make sure to use the Express Modem control panel to set the Serial Port Setting to "Use External Modem", and not "Use Express Modem" (see FIGURE F.7).

Hook up a serial cable from your Newton to the port you specified on your Macintosh. Then, choose Install Toolkit App from the Project menu in NTK. On the Newton, open the Extras drawer and tap the Connection icon (see FIGURE F.8).

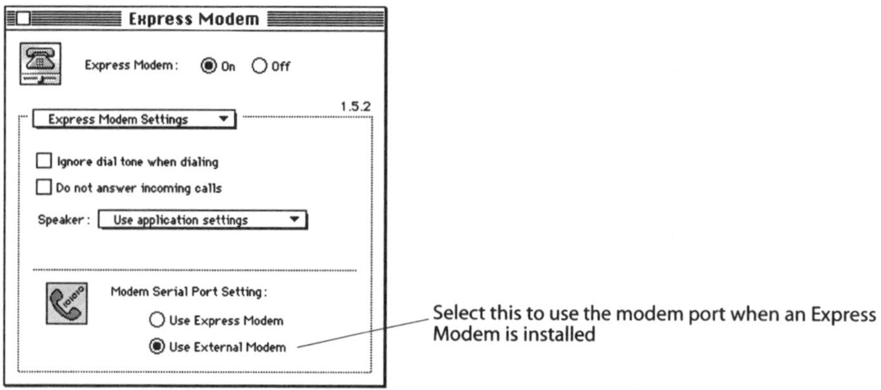

FIGURE F.7 Choice of serial port settings in the Express Modem control panel.

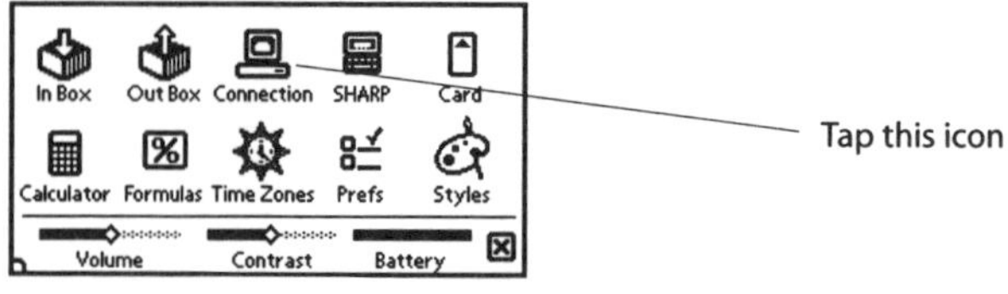

FIGURE F.8 Connection application in the Extras drawer.

After choosing the "Macintosh serial" radio button, tap Connect within the slip (see FIGURE F.9). In a few seconds, the NTK Toolkit App should appear in the Extras drawer.

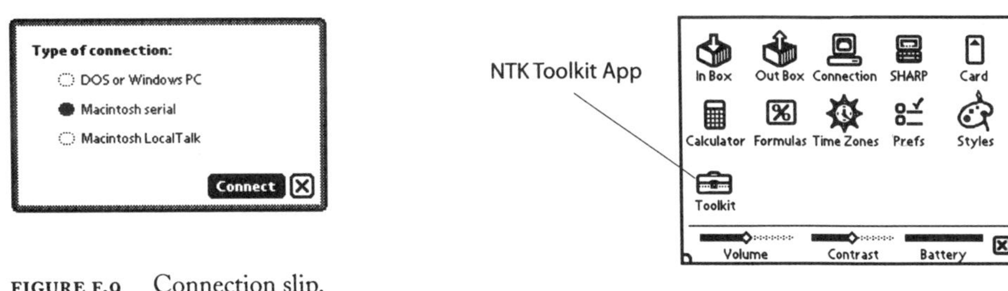

FIGURE F.9 Connection slip.

NTK Menus

Here is a brief look at the menu options available in NTK. Many of them are standard to the Macintosh and require no additional information. This section is just an overview of the NTK-specific items so that you can get the flavor of the whole environment. Specific items are covered later in this appendix.

The File Menu

In the File menu, you select what type of layout file you wish to create (see FIGURE F.10). You will most commonly choose New Layout; this is where you draw out your application's view templates. You select New Proto Template when you want to make a user proto.

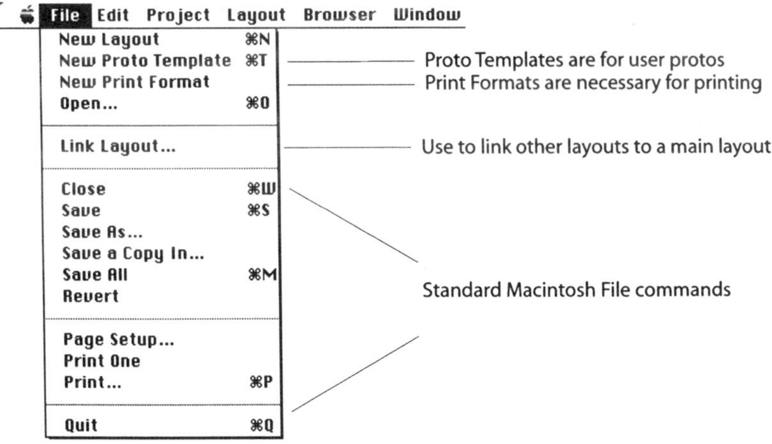

FIGURE F.10 NTK's File menu.

The Edit Menu

The Edit menu contains standard copying and editing commands as well as some selection and find tools (see FIGURE F.11). The Toolkit Preferences dialog is also accessed from this menu. As shown in FIGURE F.5, Preferences is where you specify connection settings and other project settings such as auto saving and layout screen size.

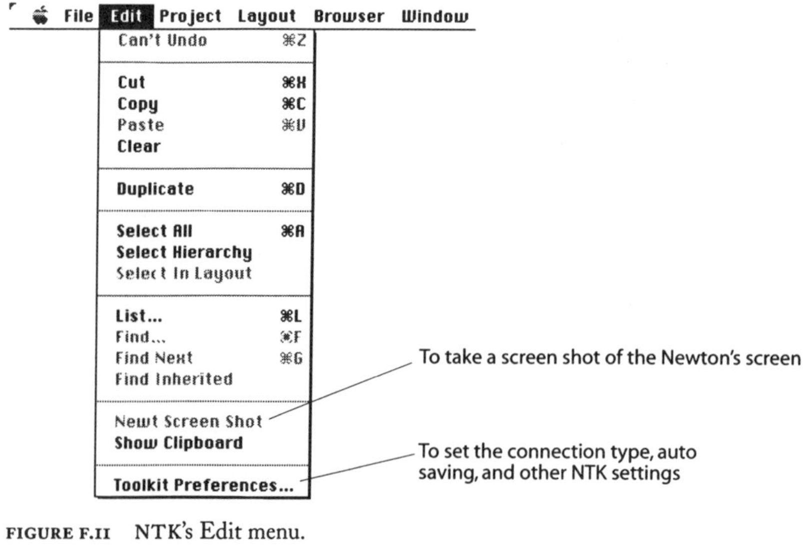

FIGURE F.11 NTK's Edit menu.

The Project Menu

The Project menu contains NTK's project management tools (see FIGURE F.12). You need to have the project window frontmost to be able to access many of these tools, and you must add the files to your project before you can either build or download the application.

Further, if a file is not added to a project it will not be part of a build. You can tell which files are part of a project by the list in the project window (see FIGURE F.20 later in this appendix).

Four important items in this menu include Mark as Main Layout, Project Data, Settings, and Export Package to Text. You should only have one file marked as a main layout, and this should be your main application layout. The Project Data is the file where you keep various settings like symbol and constant definitions and scripts that get run prior to installation and removal of your application.

In the Settings dialog, you specify various aspects of the application, like the name, symbol, and icon. Export Package to Text creates a text version of your project (this conversion is one way; NTK cannot read text versions of a project).

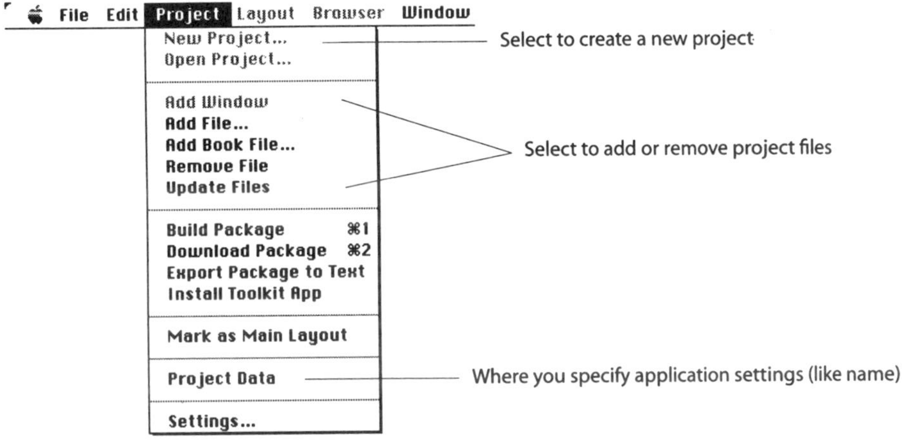

FIGURE F.12 NTK's Project menu.

The Layout Menu

The Layout menu contains items that control layout and template graphical arrangement (see FIGURE F.13). Preview is a particularly nice feature. It gives you a rough idea of what your application views will look like on the Newton without actually having to download. ⌘-Y toggles Preview on and off.

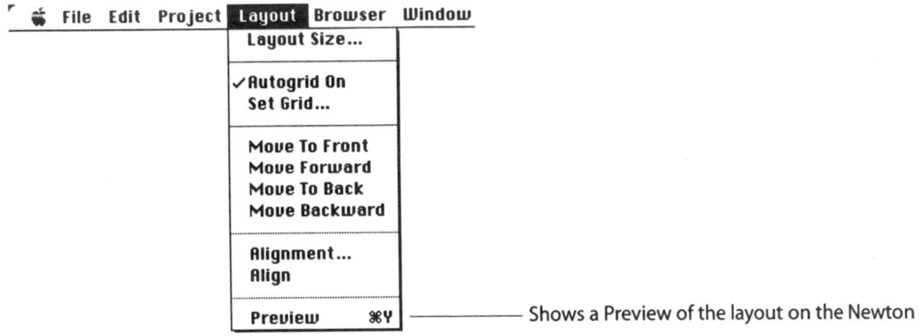

FIGURE F.13 NTK's Layout menu.

The Browser Menu

Items in the Browser menu consist of a slot creation dialog and various display options for templates and slots (see FIGURE F.14). The Template Info dialog is used to name and declare templates.

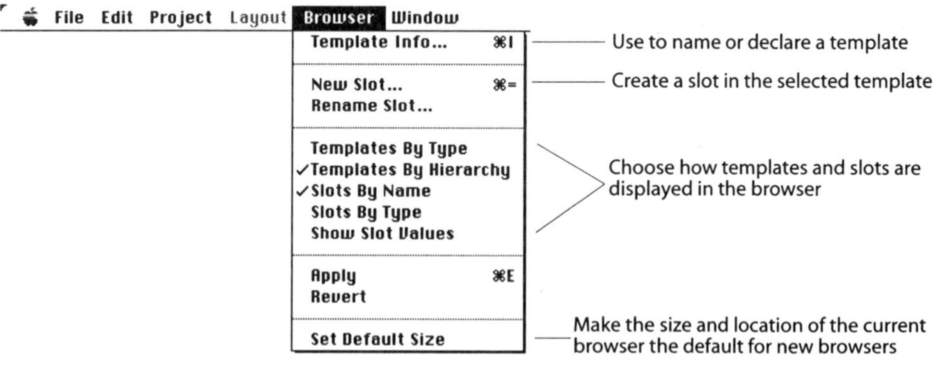

FIGURE F.14 NTK's Browser menu.

The Window Menu

Within the Window menu, the most important items are connecting to the Inspector and creating a New Browser (see FIGURE F.15). You select Connect Inspector (not Open Inspector) to establish your connection to a tethered Newton. Once you have created a layout window with drawn-out templates, you access the templates and their slots in Browsers.

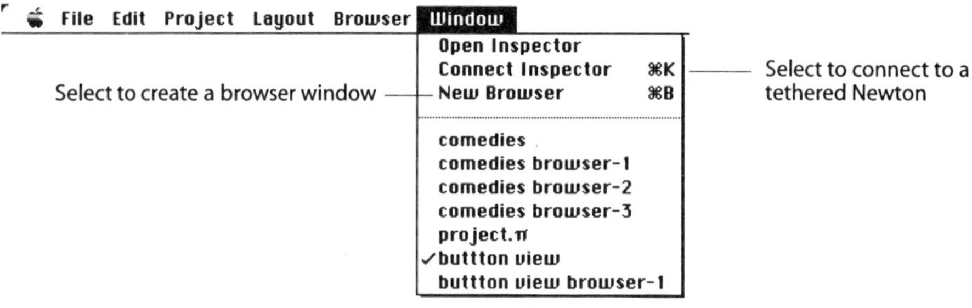

FIGURE F.15 NTK's Window menu.

Creating a Project

To create a project, use New Project from the Project menu (see FIGURE F.12). You are prompted with a dialog in which you must specify a name and folder for the project (see FIGURE F.16).

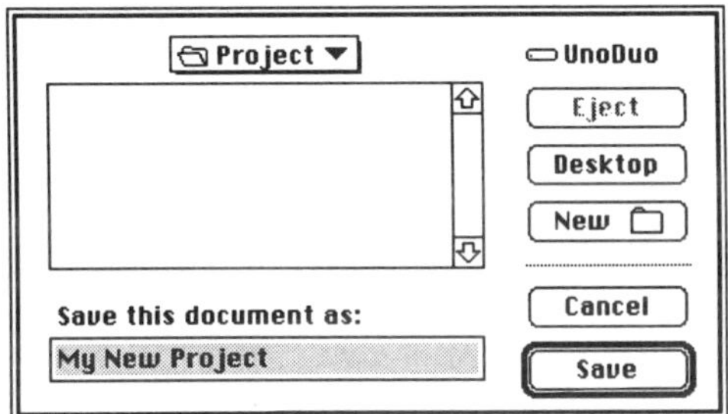

FIGURE F.16 New Project dialog.

Because there are no files in the project, the project window is empty (see FIGURE F.17).

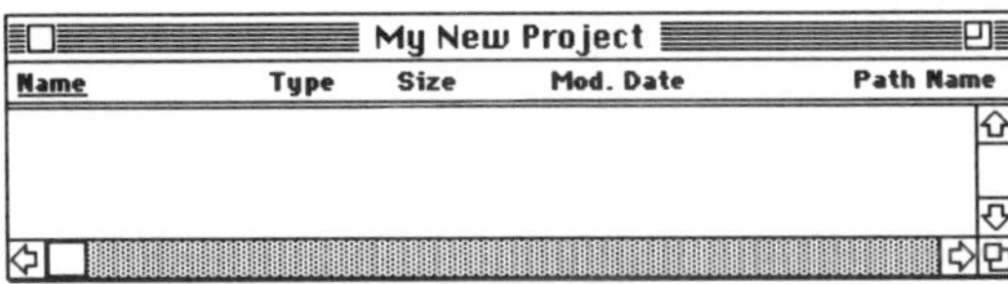

FIGURE F.17 Empty project window.

NTK Project Files

There are five different types of files that you might have in your NTK projects: Layout files, Proto files, Resource files, Book files, and Print Format files (see TABLE F.1). Later in this appendix, we will discuss the three types of project files that you create within NTK—Layout, Proto, and Print Format files.

Adding Files to a Project

You can add files to the project in one of two ways: using Add File or Add Window.

File type	*Contents*
Layout files	Contain a collection of views.
Resource files	Contain sounds or pictures.
Proto files	Contain a collection of views that can be used as a basis for other views.
Book files	A Newton Book Maker file that is added to the project. NTK builds a Newton Book using this file.
Print Format files	Contains a collection of views that are compiled before layout files.

TABLE F.1 Files in an NTK project.

Add File

Add File is the most general-purpose way to add a file to a project and is what you usually use. To use Add File to add a layout window to a project, do the following:

1. Save your layout by choosing Save from the File menu. Name the file.
2. Save the layout in the same folder as your project (see FIGURE F.18).
3. Choose Add File from the Project menu.
4. Select the file you just saved in the Add File dialog (see FIGURE F.19).

Your file is now part of the project, as you can see in the project window in FIGURE F.20.

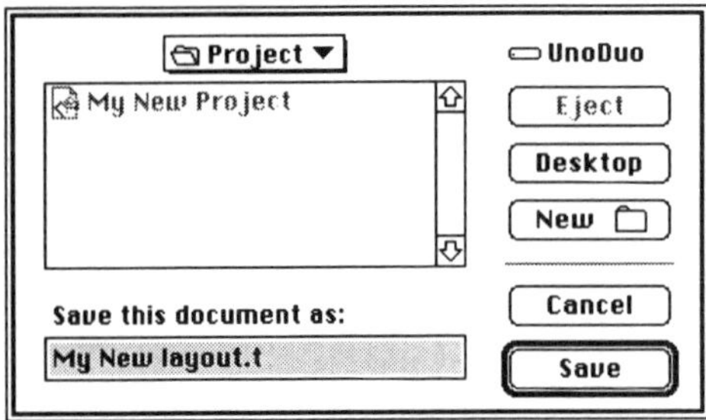

FIGURE F.18 Saving a layout.

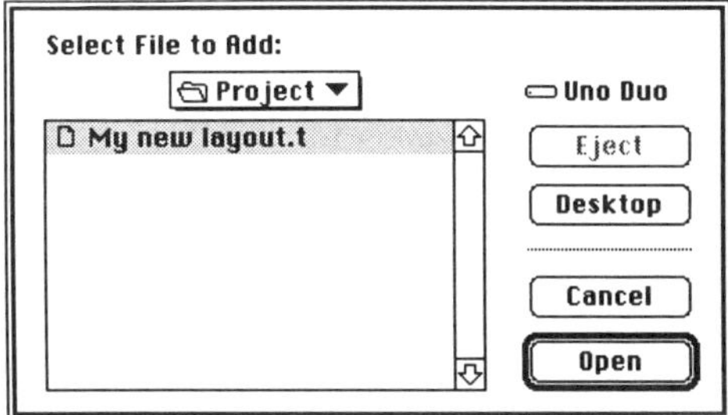

FIGURE F.19 Adding a file to a project.

My New Project

Name	Type	Size	Mod. Date	Path Name
•My New layout.t	Layout	1057	12/30/94-11:24 AM	UnoDuo:Pr

FIGURE F.20 Project window with a new file.

 *Note:* By convention, layout files end with ".t". One way to remember this is that the word layout ends in "t".

Add Window

Add Window is a shortcut for adding a new layout window to the project. To use it, simply make sure that the new layout window is the frontmost window on your screen. Then select Add Window from the Project menu. Add Window is dimmed if the frontmost window is not a layout or if it is already part of the project.

The Add Window command will prompt you with a Save dialog if the file has never been saved.

Creating a Layout

You create the templates that correspond to Newton views in NTK layout windows.

1. To create a layout window, select New Layout from the File menu.

2. NTK then displays an empty layout window, along with a Tool Palette (see FIGURE F.21). You create templates inside this layout by graphically dragging them out on the layout screen.

3. To begin, select protoApp in the palette by clicking on its icon or selecting it from the popup menu (see FIGURE F.21). Notice that in the popup menu the proto prefix is lacking from the words.

4. Once the protoApp is selected, go to the window and drag out a rectangular shape (see FIGURE F.22).

This template shape will correspond to the view displayed on the Newton; it will have the same dimensions and screen location.

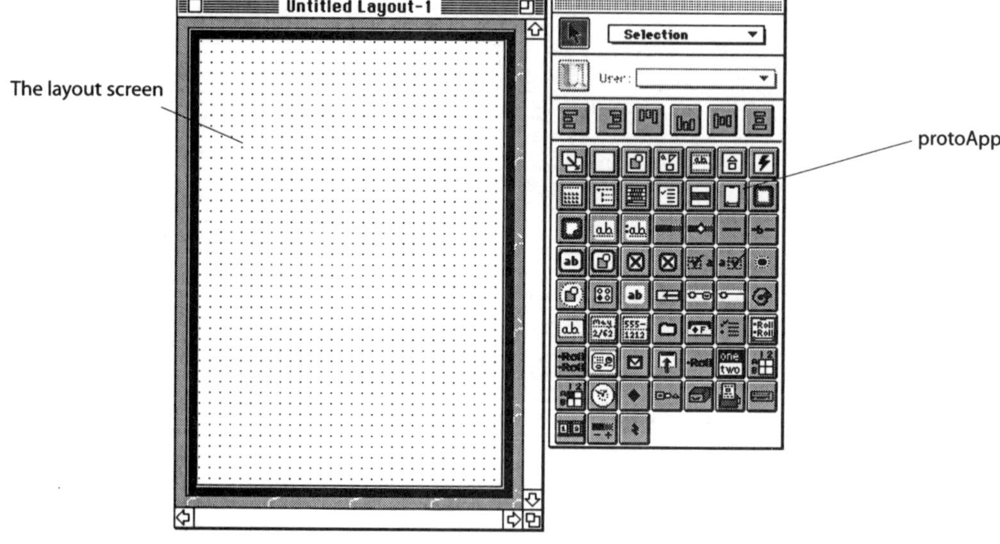

FIGURE F.21 Empty layout window with tool palette.

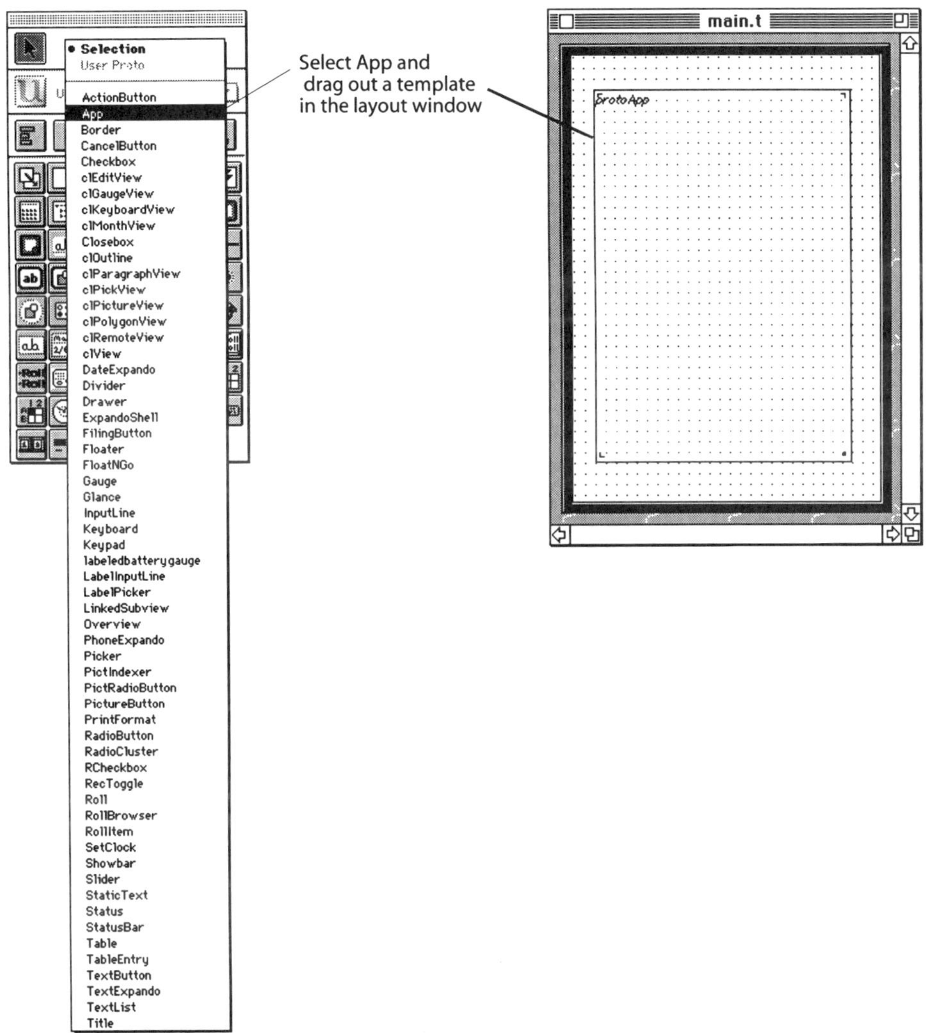

FIGURE F.22 Selecting protoApp in the tool palette and creating a template.

All of the views that you have in your application are usually created in NTK as templates in these layout windows. For each template in your application, you go through the same twofold process:

- First, you select a template type.
- Second, you drag out its shape in the layout window.

Linking Layouts

When you have too many templates to fit comfortably in one layout, or you have templates that will be visible in the same screen location, it is usually easier to create them in different layout files and then link the layouts to each other. As a rule of thumb, you will have a main layout and your other layouts will be linked to it. To link layouts, follow these steps:

1. Make sure that the main layout file and any sub-layout files are added to the project (see FIGURE F.23).
2. While in the main view, select the Linked Subview tool in the Tool palette (see FIGURE F.23).
3. Drag out a small shape in the main layout (see FIGURE F.24). The size of this linking template does not affect the view size—location and size are determined by the templates in the other layout file.
4. Select this new template and select Link Layout from the File menu (see FIGURE F.24).
5. In the dialog, select the name of the layout file you wish to link (see FIGURE F.24).

The linked subview template in the main layout should now be linked to the layout (see FIGURE F.25).

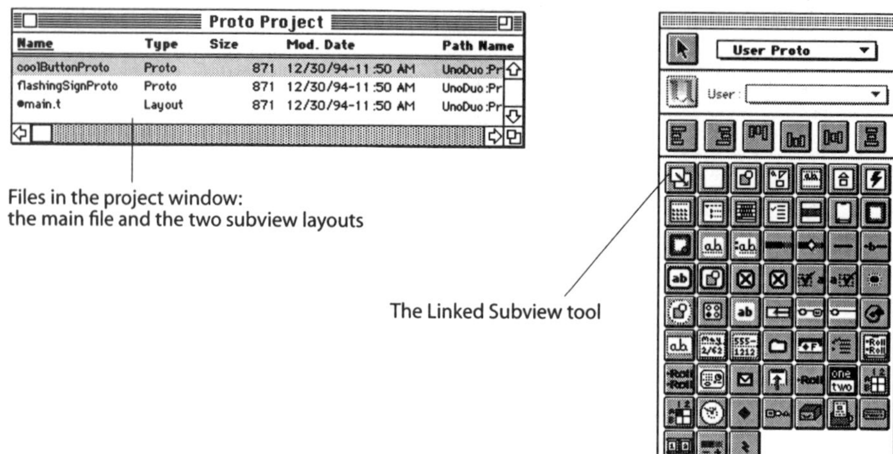

FIGURE F.23 The project window and the Tool Palette.

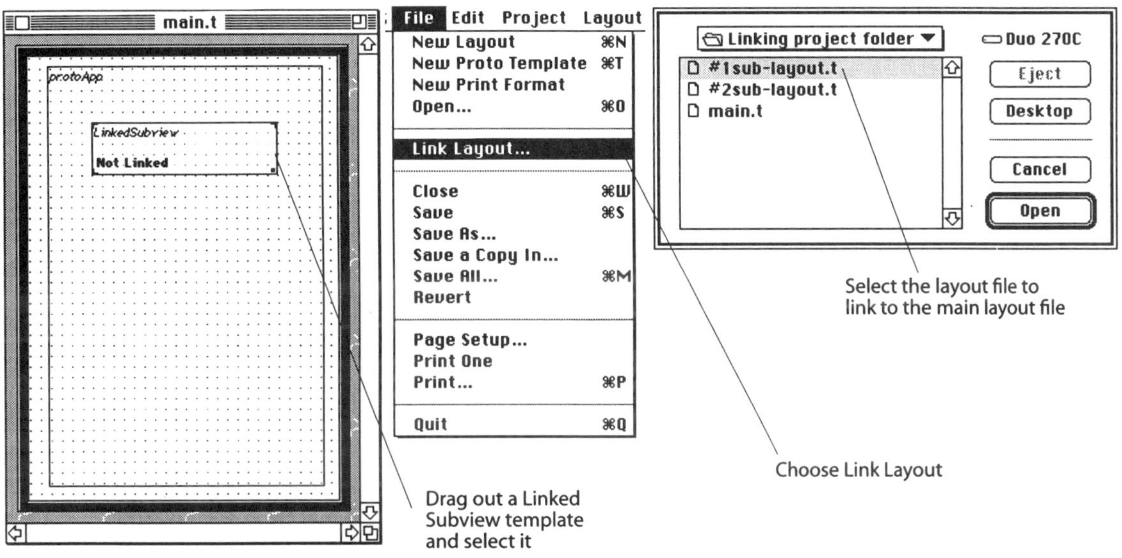

FIGURE F.24 Linking "#1sub-layout.t" to the `protoApp` template in "main.t".

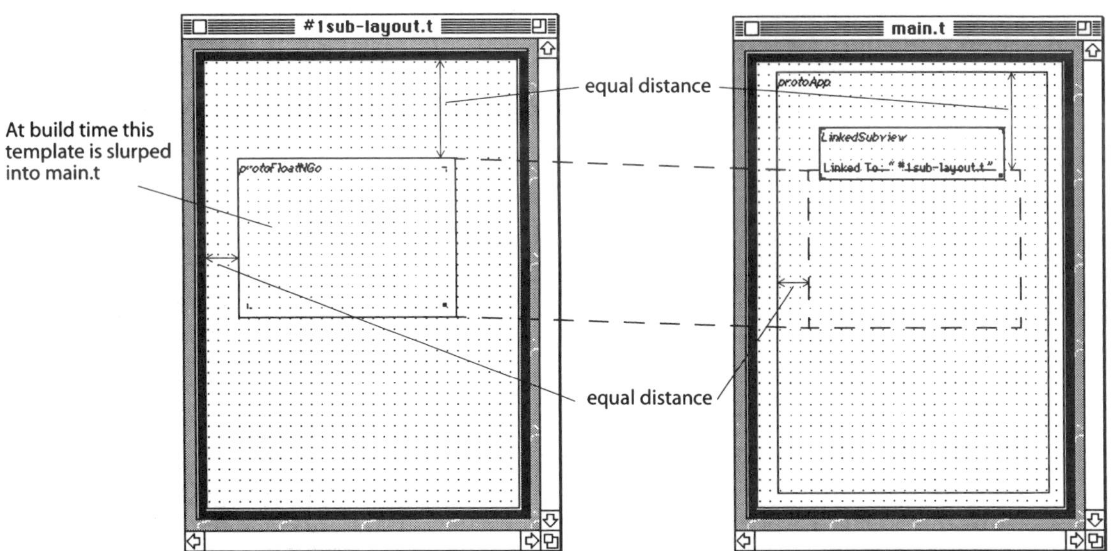

FIGURE F.25 Main layout, "main.t", and linked layout, "#1sub-layout.t".

NTK and Linked Subviews

Linking was added to NTK as a convenience for Newton programmers and to aid in application design. This should be fairly comprehensible when you understand what NTK does with your layouts. At build time, NTK reads all the templates for the main layout and for any linked layouts. NTK continues to read those layout files (recursively continuing through each linked layout and its linked layouts) and then creates a large tree of templates, which it sends to the Newton (see FIGURE F.25).

Creating a Print Format

To create a layout that can be referenced at compile-time from other layouts, you use a print format. Print formats are compiled before other layout files; each print format file defines a constant whose name is `printFormat_fileName` that can be used within other layouts. This facility is used most often when supporting printing where you create a frame that looks like:

```
{
   title:       "Title of print format",
   mainFormat: printFormat_thePrintFormatFile,
   ...
}
```

To create a print format, select new Print Format from the File menu. FIGURE F.26 shows the resulting window. Edit the print format just like you would edit any layout.

Note: Print format file names should contain no special characters or spaces. Unlike layout files, you will use the name of the print format in your code. If you have special characters in the name, then you must reference the print format with vertical bars. It is easier to avoid special characters and spaces in the first place.

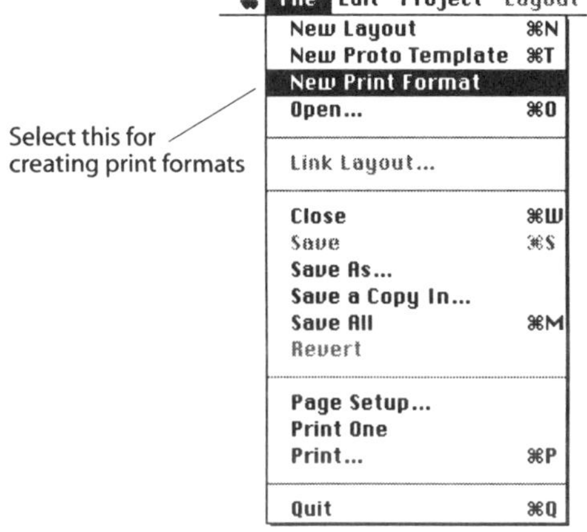

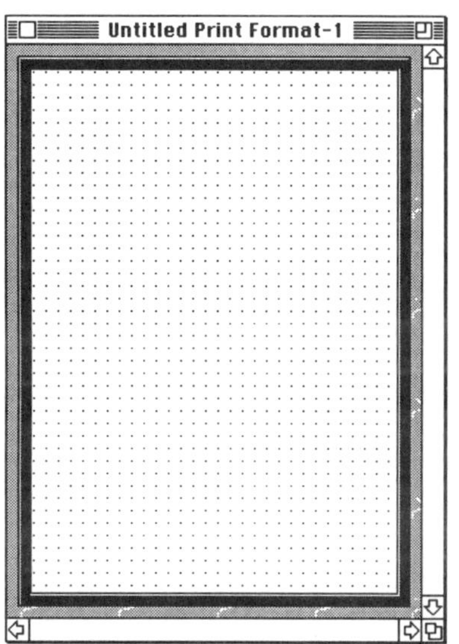

FIGURE F.26 Creating a new print format.

Creating a User Proto

The way to reuse templates and build code libraries of application components is to use protos. Proto creation is remarkably similar to regular layout creation; the only difference is the type of layout file you select in the File menu.

To create a proto, select New Proto Template from the File menu and you will get the different-looking layout screen you see in FIGURE F.27.

You have the same tool palette available to you in creating your proto. Further, the actual dragging out of templates in the proto is identical to template creation in a regular layout window. Once you have completed your proto you must add it to your project using Add Window or Add File from the Project menu.

Once the new proto is added to the project, it is available for use in the tool palette with the other tools (see FIGURE F.28). User protos are displayed and picked from their own list in the tool palette, however (see FIGURE F.28).

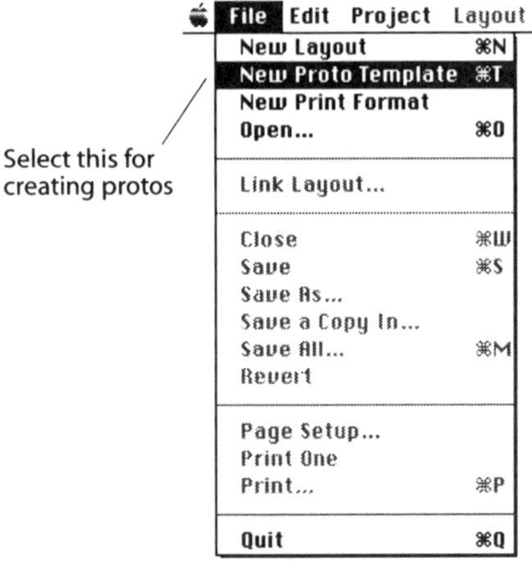

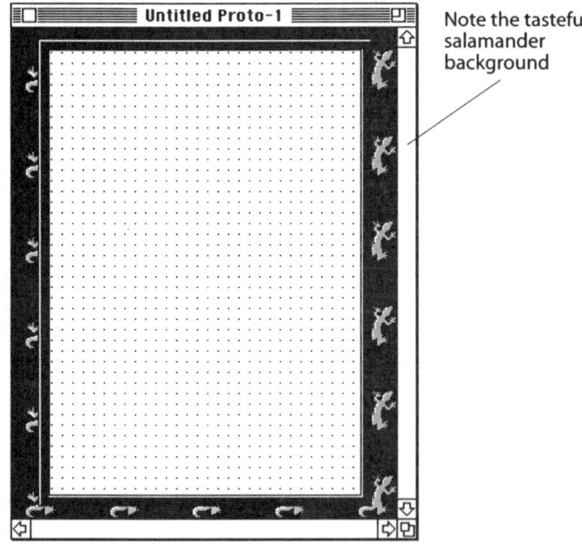

FIGURE F.27 Creating a new proto template.

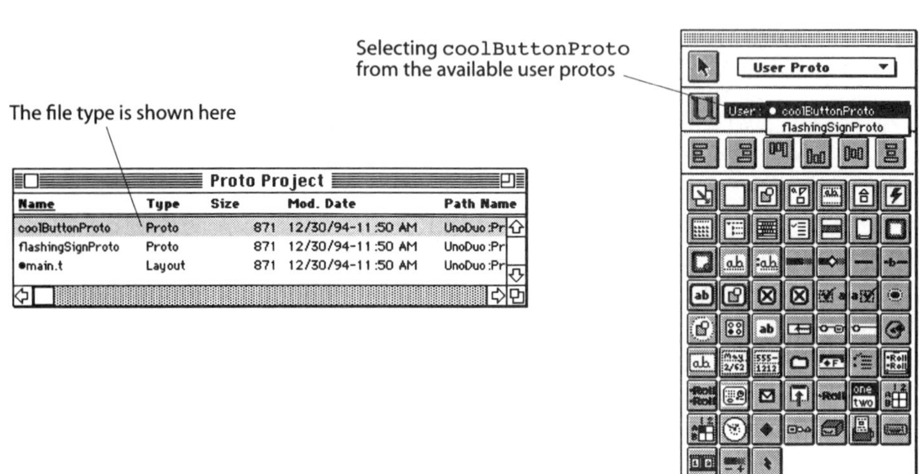

FIGURE F.28 A project and tool palette with user protos.

Note: Proto file names should contain no special characters or spaces. Unlike layout files, you might use the name of the proto in your code. If you have special characters in the name, then you must reference the proto with vertical bars. It is easier to avoid special characters and spaces in the first place.

Adding a User Proto to a Layout

It is very simple to add a user proto to a layout:

1. Select the user proto in the list of user protos (for example, `cool-Buttonproto`).
2. Select the user proto icon [U] and drag out a template in the layout window (see FIGURE F.29). You must reselect the user proto icon each time you want to drag out another proto template.

Notice that it does not matter what size you attempt to make the proto template. It will draw out with the screen location and size specified in the original proto (see FIGURE F.29). The proto template is created with only one slot, the `_proto` slot. If you want to change the size of the template, add a `viewBounds` slot to the newly created template.

Note: If you have no `viewBounds` slot in your proto, then NTK will create your template with both a `_proto` slot and a `viewBounds` slot.

Creating and Modifying Templates

Once you have your layout window created you need to populate it with some templates. As we said before, template creation is a two-step process that you repeat for each template you want to create:

1. Select a user proto, view class, or proto from the tool palette that corresponds to the type of template you want to create.

2. Drag out a shape in the layout window that you want the template to have. Move the template to the correct screen location if it is not already in the right place.

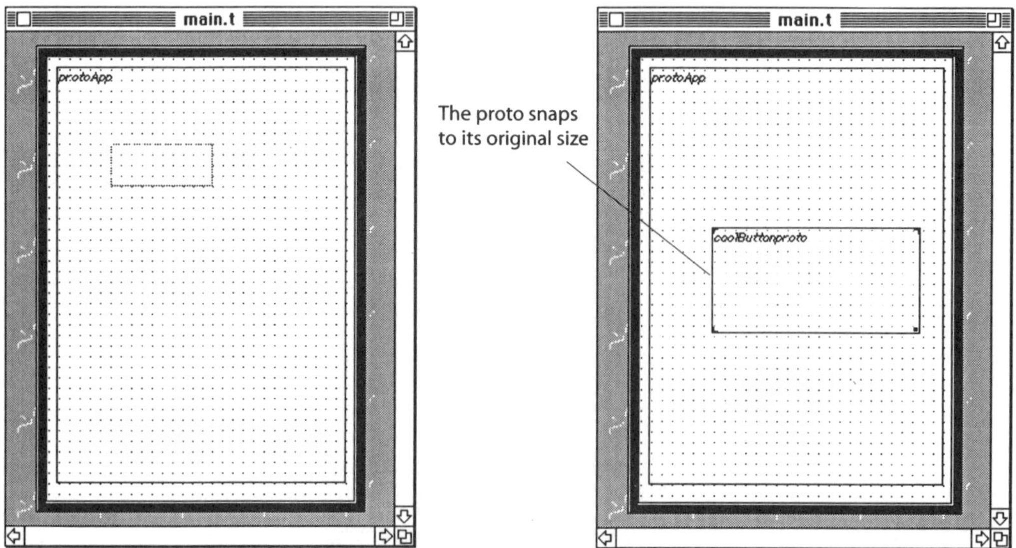

FIGURE F.29 Dragging out a `coolButtonproto` template.

Putting Templates in a Parent–Child Hierarchy

The first template that you create in a layout window serves as the main parent template. There can only be one parent of this type in a layout. If you try to create another template that isn't a child of the first template, you will get the NTK alert shown in FIGURE F.30.

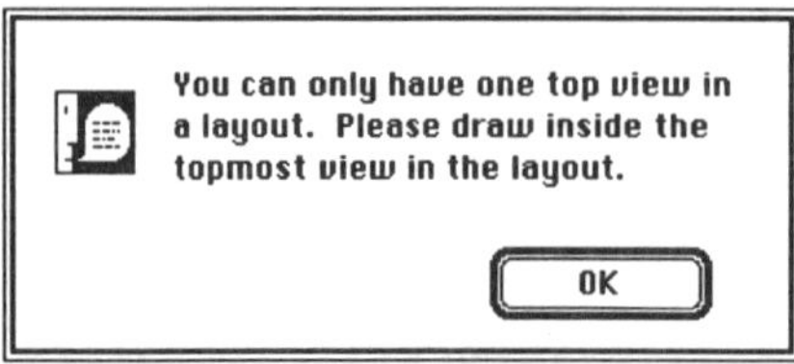

FIGURE F.30 NTK alert warning of a parent clash.

Just remember that every layout file must have a base parent template—NTK insists on the single-head-of-household filing status. Inside of that parent, however, you can have as many children as you wish.

To make sure that a template is created as a child of another template, you simply drag out the child template shape inside the parent as in FIGURE F.31. You can also change the template hierarchy if you don't get it positioned correctly on the first try (see "The Template Area of the Browser" on page 423).

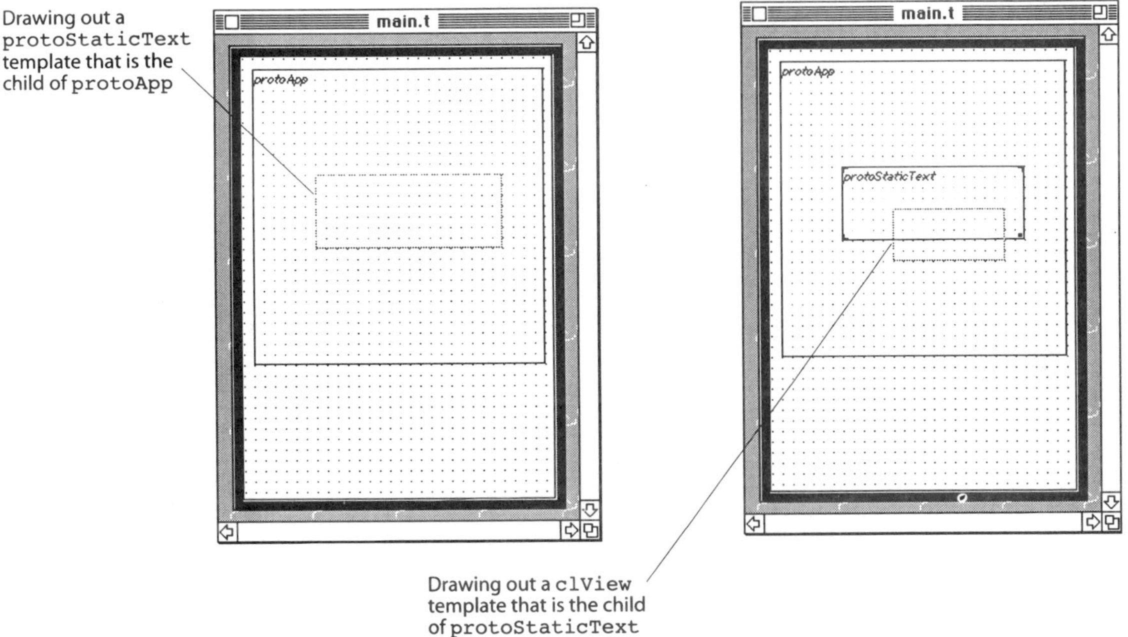

FIGURE F.31 Dragging out child templates.

Notice in FIGURE F.31 that the child template of the `protoStaticText` template is drawn longer than its parent. This does not change its place in the hierarchy, which is set by where you first start to draw a template. If a child view is longer or wider than its parent, it will draw outside its parent on the Newton, unless the parent had its `vClipping` bit set in `viewFlags`. In the latter case, the child view will be clipped off at the edge of the parent. NTK always displays children clipped to their parent, regardless of the `vClipping` bit setting.

Moving and Changing Template Size Graphically

To move a template, simply click on it, and its four corner edit squares will become active. If you want to move the template, click, hold the mouse down, and drag the template to the new position. You can tell that you are moving and not resizing the template because the cursor icon turns into a little moving hand.

If you want to resize the template, drag the bottom right edit square (see FIGURE F.32) to be smaller or larger. You can tell that you are resizing and not moving because the cursor turns into a diagonal line with arrow heads.

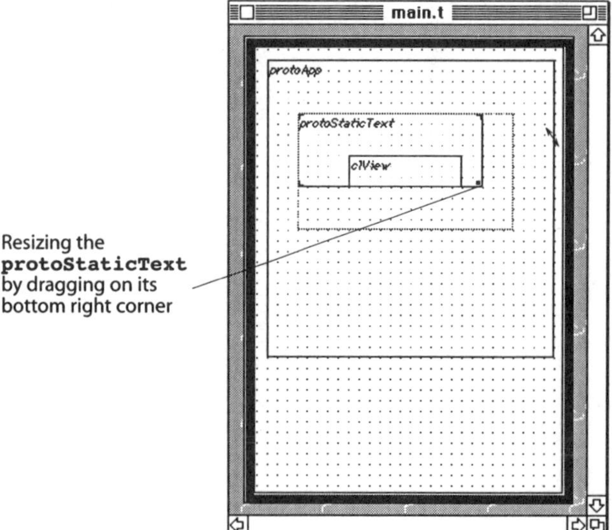

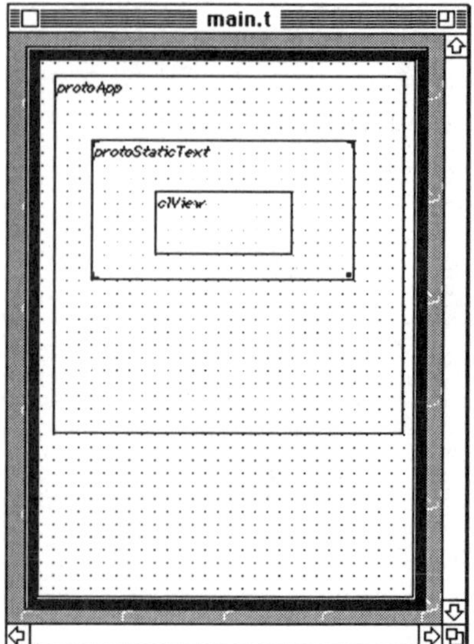

FIGURE F.32 Resizing the `protoStaticText` template.

Naming Templates

To name a template, select it in the layout window and choose Template Info from the Browser menu. Type the name of the template in the dialog (see FIGURE F.33). You can see the name of the template in the layout window if the template is large enough.

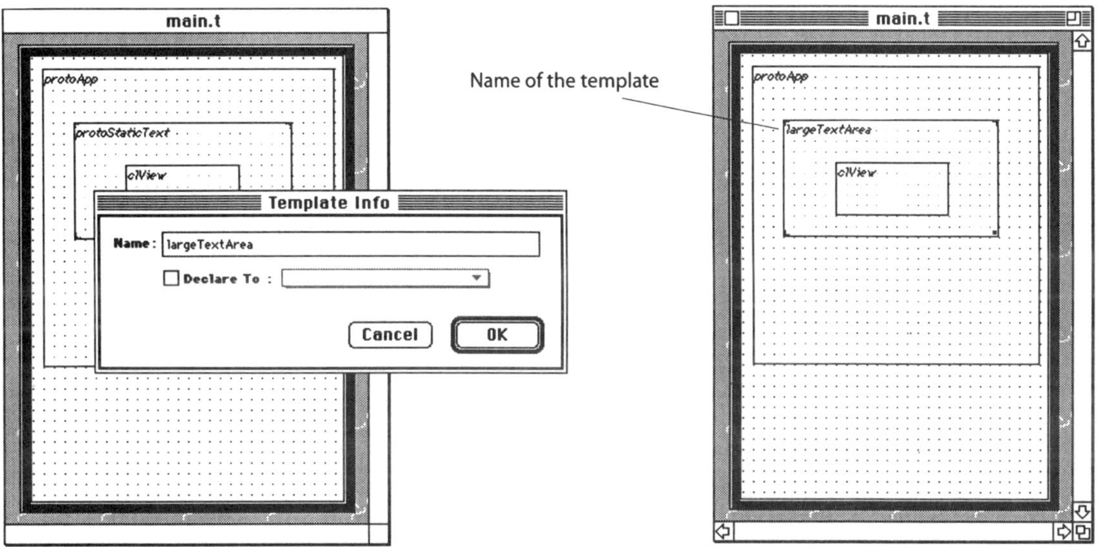

FIGURE F.33 Naming a template.

Using Declare To on a Template

You can also declare templates in the Template Info dialog. You declare a child template to a parent or other ancestor template if it needs access to the child. Here is how you declare a template:

1. Make sure the parent is named. Select the template and name it in Template Info (see FIGURE F.34).

2. Open Template Info on the child template, name it, and check Declare To. Select the parent's name in the popup (see FIGURE F.34).

If the child template is large enough, it will indicate its declared status in the layout window as well (see FIGURE F.35). You can declare a child to any ancestor template in its layout—not just its immediate parent. *If you forget to name the child, the Declare To checkbox remains dim. If you forget to name the parent, its name won't appear in the popup.*

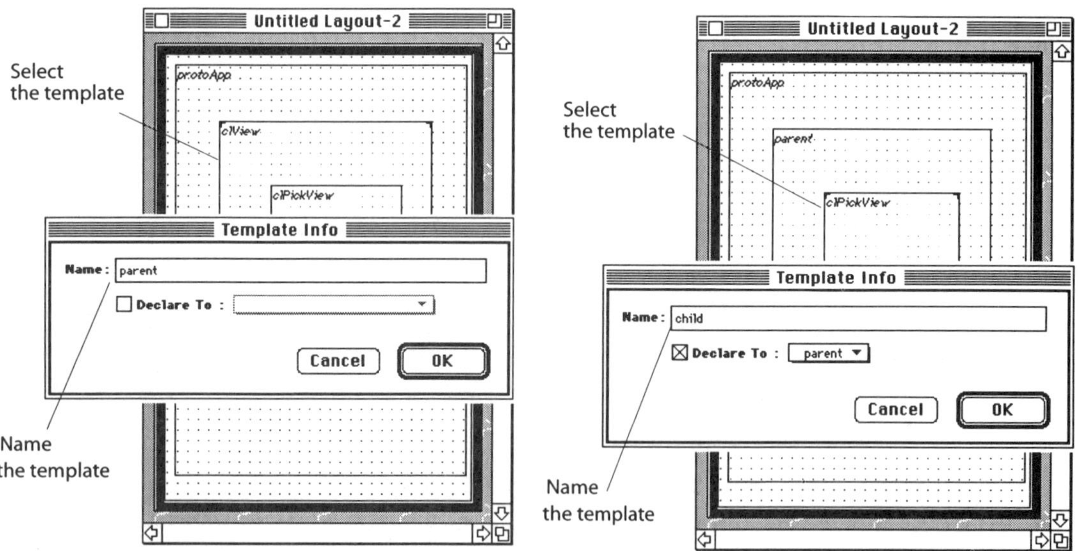

FIGURE F.34 Declaring a child to a parent.

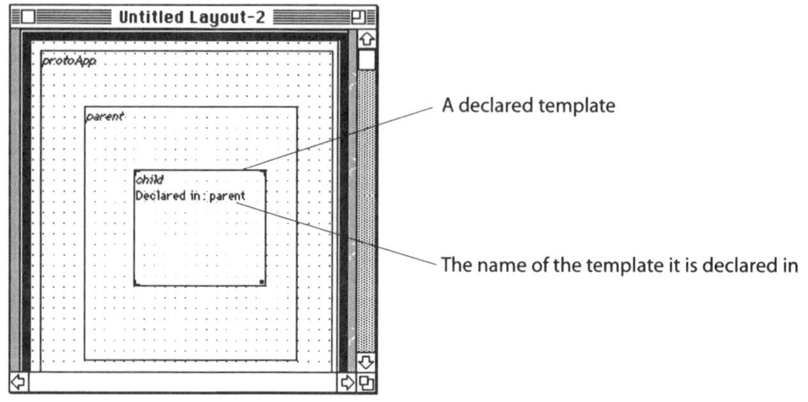

FIGURE F.35 A child template declared to its parent.

Removing Templates

If you want to get rid of a template, just select it in the layout window and select Cut from the Edit menu. The template is now gone. In cases where it is hard to select a template in the layout window, you can also do the following:

1. In the browser window, double-click on the template you want to remove.

2. Make the layout window the frontmost window and then select Cut from the Edit menu. The template will be removed from the layout.

An Introduction to the Browser

You just encountered a new term in the last section—the Browser. This is the tool provided by NTK to allow editing the slots in a template. The browser window is created by selecting New Browser in the Window menu. The browser window has three parts (see FIGURE F.36).

The left top shows a list of templates and their hierarchical relationship. The top right shows the slots in the template that is selected on the left. And the bottom of the browser shows the slot editor for the slot currently being edited. In FIGURE F.36, you see a browser window with a `protoApp` template selected on the left, its `title` slot selected on the right and the contents of its `title` slot shown below.

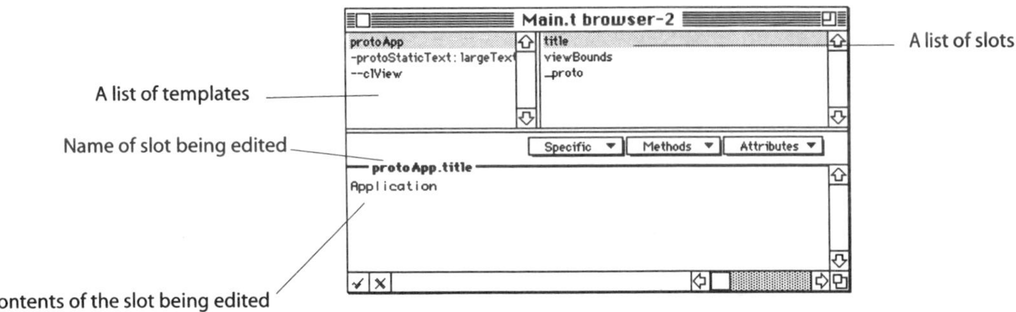

FIGURE F.36 The browser window.

New Browser window creates a browser for the selected templates in a layout. The topmost parent you have selected is the topmost template shown in the browser window (see FIGURE F.37).

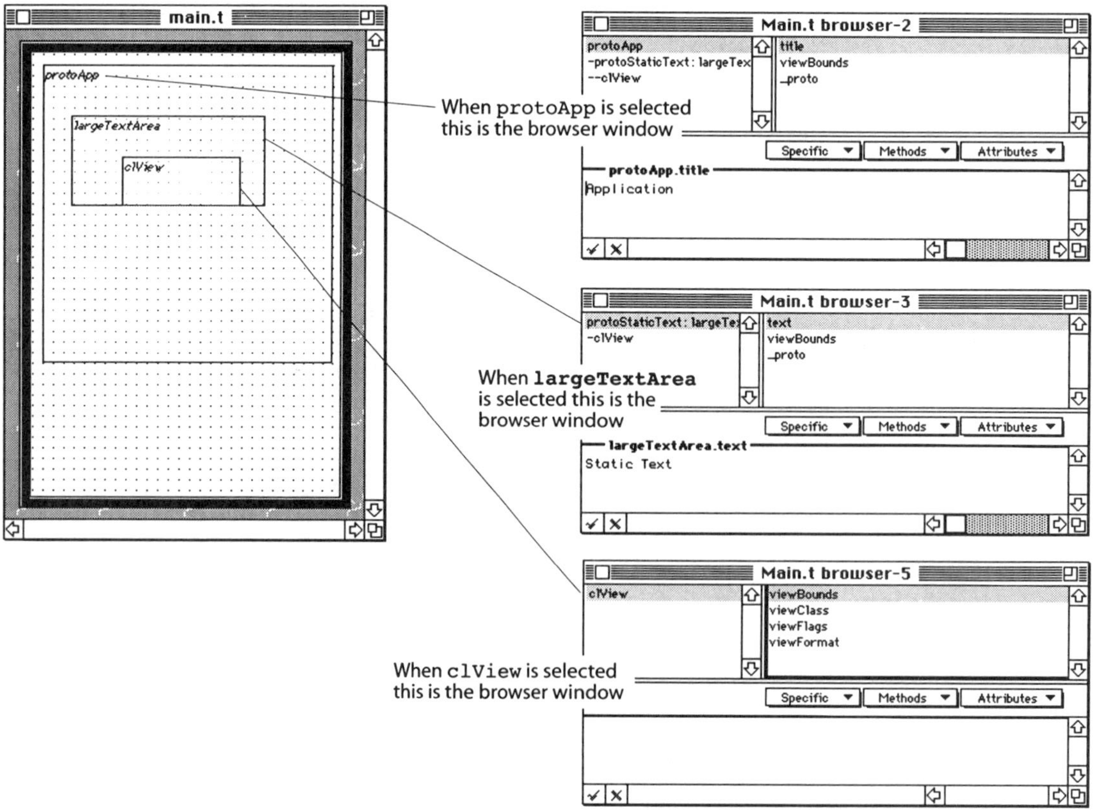

When protoApp is selected this is the browser window

When **largeTextArea** is selected this is the browser window

When clView is selected this is the browser window

FIGURE F.37 Creating browsers on various templates in a layout.

Note: Another way to open a browser window is to option–double–click on a template in a layout window. A third way is to select New Browser while a layout is selected in the project window.

You will spend a majority of your time in the browser. It is here that you will add the methods to your application and any specialization in each template's appearance.

The Template Area of the Browser

In the template area of the browser you can see the hierarchical relationship between the templates and change them as well. You can select a template and then use a combination of the option and arrow keys to move the template around in the structure. The up and down arrow keys will move the template up or down in order among its siblings. The left arrow key moves a template out to the next level (its old parent's level) and the right arrow key moves a template into the next level (makes it a child of its previous sibling).

A template's sibling is the one immediately above it in the same level of the template list (see FIGURE F.38). If the template immediately above is its parent, the template has no previous sibling.

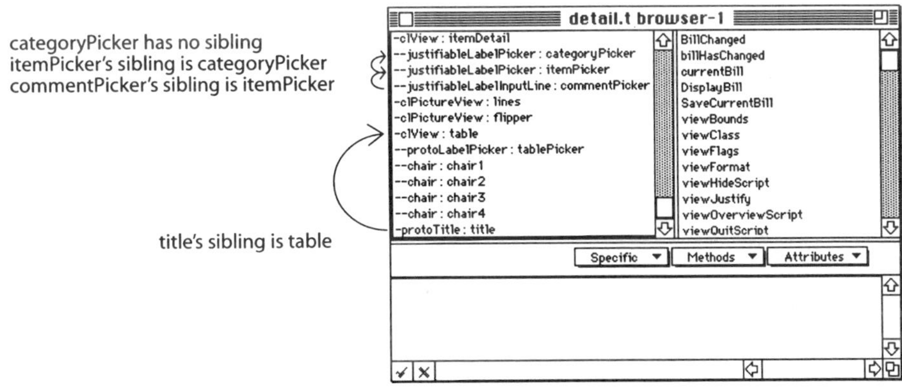

FIGURE F.38 Identifying a template's sibling.

Selecting Templates

Sometimes it can be difficult to select a template in the layout window (for example, when a child is as big as a parent template). If you need to select this type of pesky template, do the following:

1. Double-click on the template in the browser.
2. Make the layout window the frontmost window.

The Template Hierarchy

In the template list in the browser you can also see the parent–child relationship between templates. The dash symbols represent the level of ancestry:

no dash	The main template displayed in the browser.
-	Child of the main template.
--	Grandchild of the main template, or child of the last - template.

The Slot List of the Browser

As we said, the right side of the browser displays the slots of the template selected on the left. You can select the slot, edit it, remove it, or paste another slot into that template when the top right of the browser is selected.

Editing an Existing Slot

To actually edit a slot from the slot list, *you need to double-click on the slot name*. You may find this a difficult aspect of NTK to work with at times. For instance, double-clicking can be annoying when you are resizing a number of templates—imagine you are in the middle of editing the third of 10 `viewBounds` slots. It is difficult to remember which slot you are changing or whether it corresponds to the one selected in the list (see FIGURE F.39).

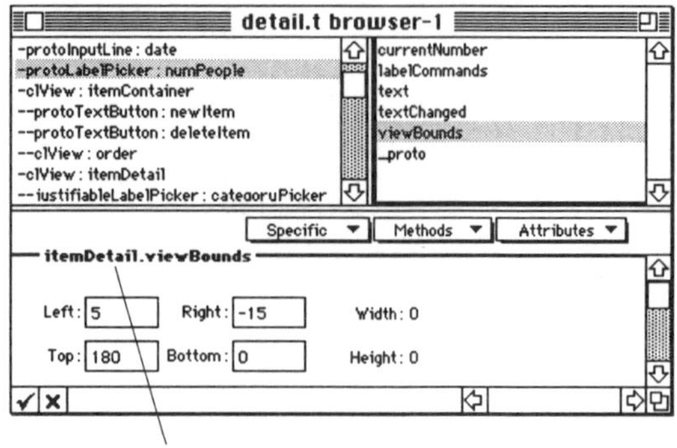

This is `itemDetail`'s `viewBounds`—not `numPeople`'s

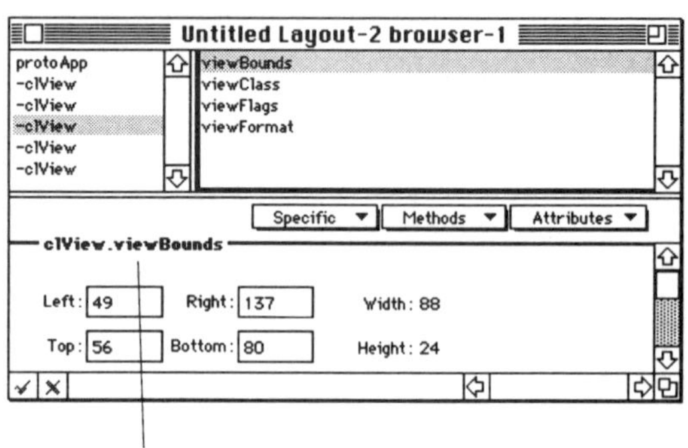

Q: Whose `viewBounds`?
A: It is actually that of the second `clView`.

FIGURE F.39 Slot editors don't always match the selected slots.

This points out another reason why it is useful to name the templates with which you are working. Without a name, you would have no way of knowing which of the five `clView` templates in FIGURE F.39 you are editing without double-clicking again in the slot editor. If templates are named, then the name of the template and its slot are always displayed in the slot editor.

Displaying a Slot's Values

You can also display the values after the slots in the list by selecting Show Slot Values from the Browser menu. This gives you an abbreviated version of the slot contents, as displayed in FIGURE F.40.

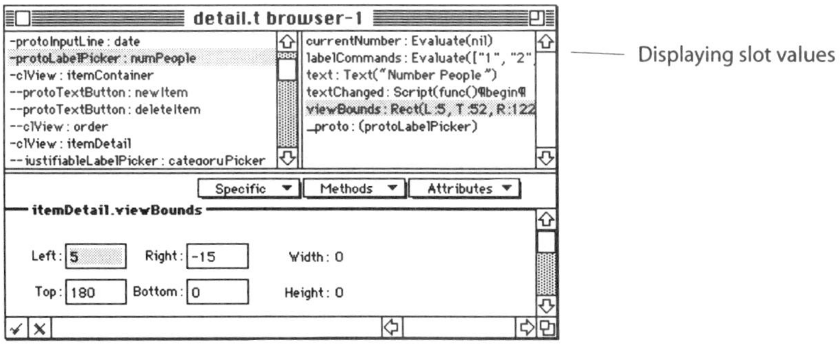

FIGURE F.40 Displaying slot values in the slot list.

Deleting and Renaming Slots

You can delete a slot by clicking on it in the slot list and selecting Cut from the Edit menu.

You can rename a slot by clicking on it in the slot list and selecting Rename Slot from the Browser menu. Make sure not to use a name of a predefined slot unless you intend to override that slot.

Adding Slots

You can add new slots to a template in the slot editor. The most convenient method for adding a slot is only available for predefined slots. For these slots, you can select the slot by its name in one of the three popup menus: Specific, Methods, or Attributes. FIGURE F.41 displays the contents of these three popups.

Note: The Specific popup menu's contents will vary, depending on the type of template. The specific items are those slots that are standard optional additions to a template.

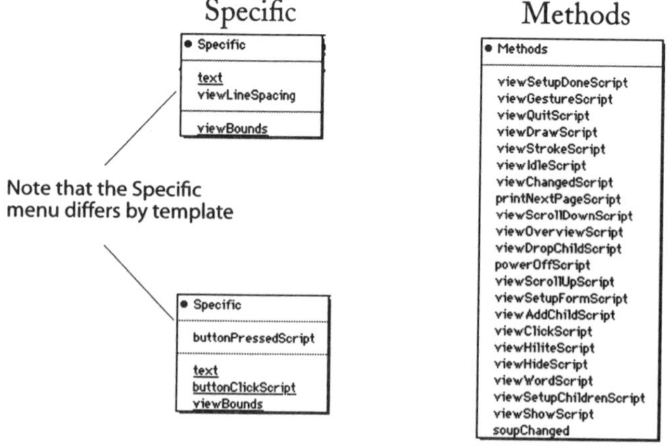

FIGURE F.41 Popup menus in the slot editor of the browser.

These slots are all the standard slots that NTK provides to you by name. This is roughly how the slots are grouped in the popup menus:

Specific	These are the slots specific to a particular type of template. They are slots in the proto that you might want to override in a particular template. For example, the `viewFont` slot in `protoStaticText` is available in the Specific menu.
Methods	These are standard methods associated with views.
Attributes	These are slots that are used with many different kinds of templates.

The New Slot Dialog

You can also add a slot using the New Slot dialog, which is found in the Browser menu. When you select New Slot in the menu, you get the dialog shown in FIGURE F.42. It is also in this dialog that you will add all of your own methods to a template. At the top of the dialog are the same popup menus available in the slot editor. Below is a text area to enter the name of your slot.

To the right of the name is another popup, where you can specify one of nine types of slot editors:

Evaluate	A generic editor, preset to `nil`.
Custom	A custom editor (currently disabled).

Script	An editor for functions, preset to hold
	`func()`
	`begin`
	`end`

Text	An editor for text, preset to surround the slot value with double quotes, " ".
Number	An editor for integers.
Boolean	An editor for `True` or `nil` values.
Rectangle	An editor to enter left, right, top, and bottom values.
Picture	An editor with a file picker and a resource display area.
Font	A font editor (currently disabled).

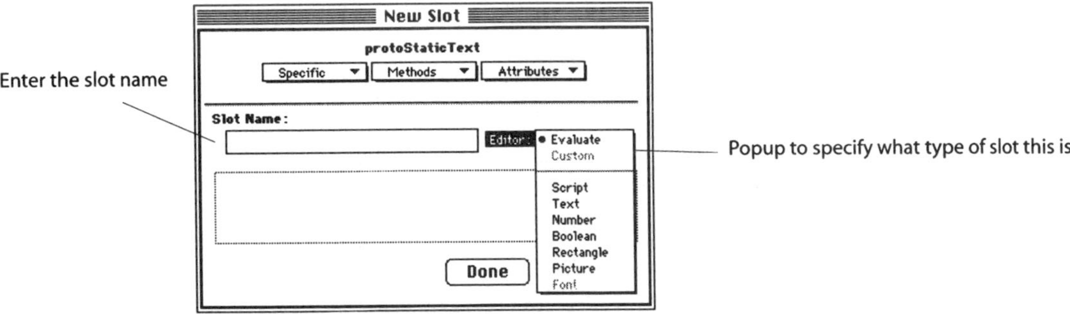

FIGURE F.42 NTK's new slot dialog.

The Slot Editor

The slot editor is located in the lower portion of the browser window (see FIGURE F.43). Its ingredients are fairly simple. The name of the slot being edited is displayed at the top left of the editor. The value found in the slot is displayed in the editor window. Below the editor window are two checkboxes, Apply and Revert.

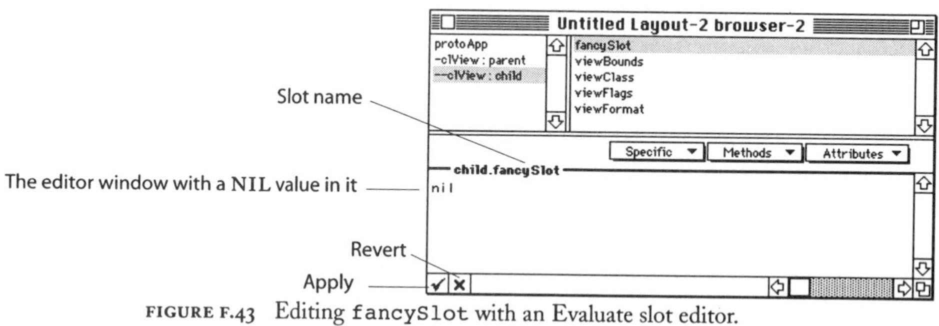

FIGURE F.43 Editing `fancySlot` with an Evaluate slot editor.

The Apply and Revert Buttons

Think of the editor window as an interim work area. Whatever you enter there is temporary until you take the further step of writing it to the slot—which you do by clicking the Apply button. You can go edit another slot or work in another part of the project; the slot you were working on will remain unchanged if you do not hit the Apply button. If you have Auto Apply selected in the Settings dialog, then changes in the editor window are automatically applied whenever you build or otherwise leave that editor window.

Hitting the Revert button lets you change your mind about a proposed change—the contents of the slot editor window are removed and the slot's original value is redisplayed. You can no longer revert once you have clicked Apply; the Apply button takes a change you have entered and writes that to the slot.

Types of Slots Editors

There are many different types of slot editors. The type of slot editor is dependent on the type of slot you are editing. In the New Slot dialog, there is a popup menu where you can specify the type of editor. Many of the predefined slots also have specific types of editors that are worth a brief explanation.

viewEffect

You can set various view effects in the effect picker in this editor (see FIGURE F.44).

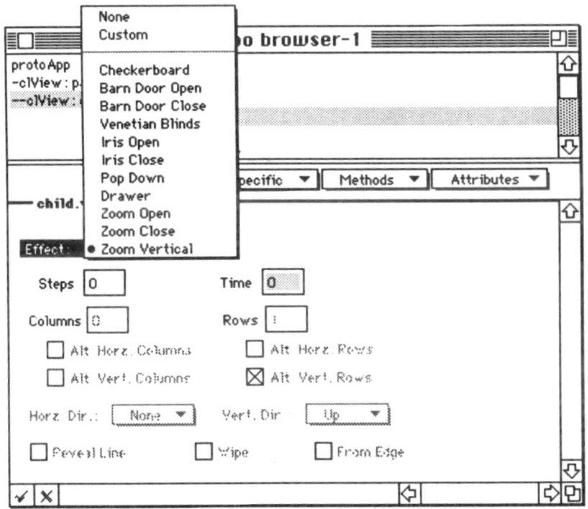

FIGURE F.44 The `viewEffect` slot.

viewFormat

You can set various aspects of a view's frame and fill in this editor (see FIGURE F.45).

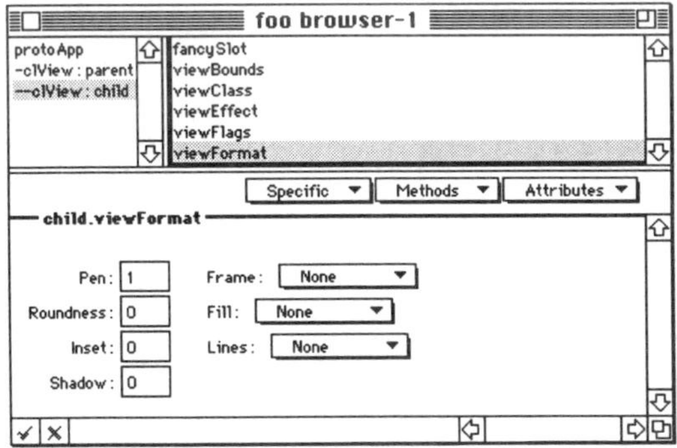

FIGURE F.45 The `viewFormat` slot.

viewBounds

You can set the left, right, top, and bottom bounds of a view template in this editor (see FIGURE F.46).

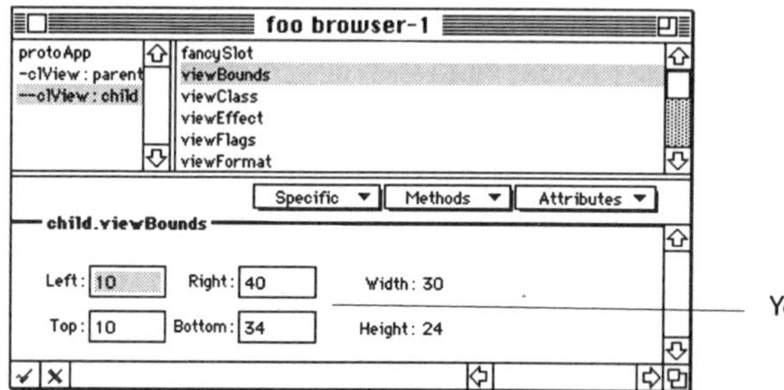

FIGURE F.46 The viewBounds slot.

viewFlags

You can set various view options in this editor (see FIGURE F.47).

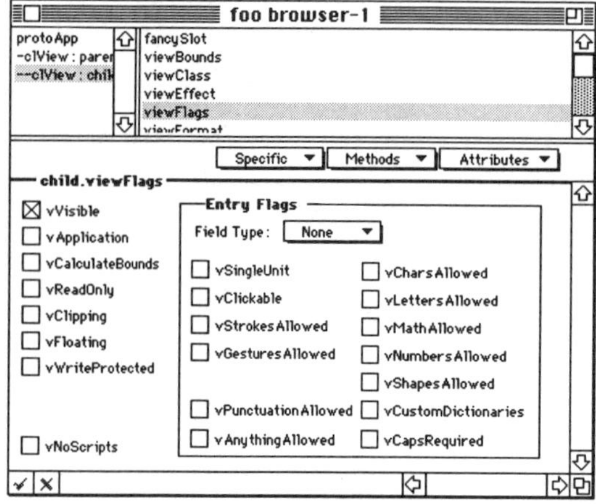

FIGURE F.47 The viewFlags slot.

viewJustify

You set the justification of one view relative to another in this editor. Note that you can also set the internal justification of graphics and text (see FIGURE F.48).

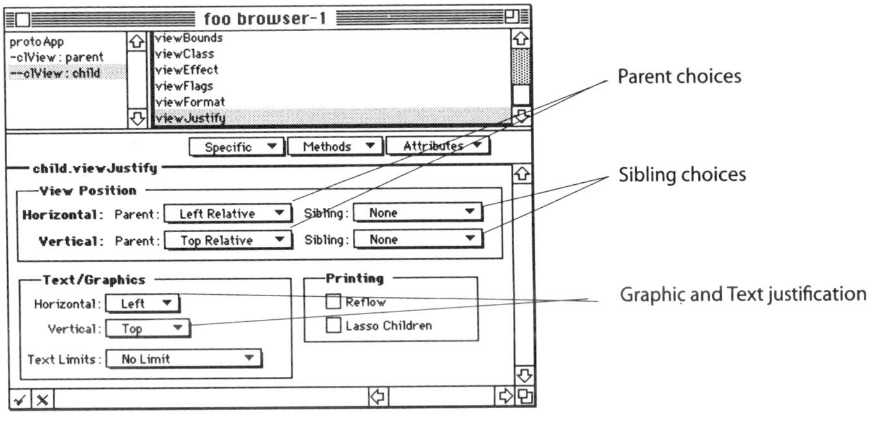

FIGURE F.48 The `viewJustify` slot.

⚠ *Caution: Bad Bug Alert!* If you change the sibling justification, the `viewJustify` slot editor automatically changes the parent justification to match. This is awful because parent justification should always be Top Relative (or Left Relative) when using sibling justification; make sure to reset the parent justification to one of those two values after modifying the sibling justification. If the parent justification isn't Top (or Left) Relative, sibling views will appear in incorrect locations on the Newton.

Additional Parts of Your Project

The Project Data File

The Project Data file is a text file in which you can define the following:

constants — These are constants available anywhere in your application. They might include the application symbol, for instance.

`InstallScript`	A function executed as your package is installed.
`RemoveScript`	A function executed when your package is deinstalled. Note that your application templates are no longer present when this function is called.

You can edit this file in any text editor, or you can edit it within NTK by selecting Project Data from the Project menu (which creates the file in your project folder if it doesn't already exist).

Project Settings

The Project Settings dialog (see FIGURE F.49) controls various aspects of the project build. Select Settings from the Project menu to bring up the dialog.

Caution: One of the limitations of Demo NTK is that the Project Settings dialog cannot be edited. See "Differences between Demo NTK and NTK" on page 396 for more details.

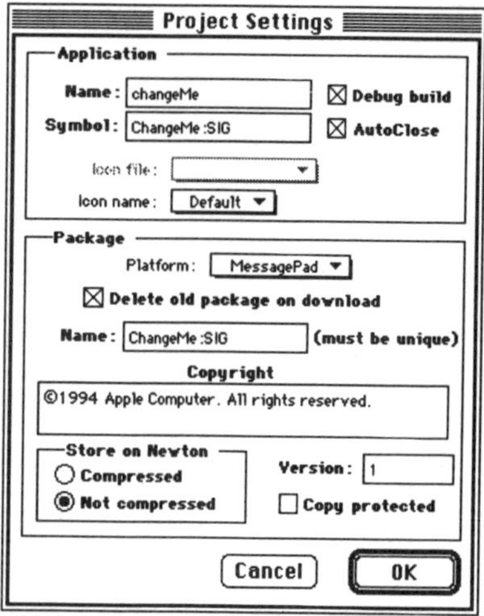

FIGURE F.49 Project Settings dialog.

Name

This is the name of the application that appears in the Extras drawer.

Symbol

This is the unique application symbol. At runtime, a slot is created in the root view with this symbol; the slot's value is the application base view.

Debug Build

If this option is set, then NTK creates a `debug` slot for all named templates. (If you have manually added such a slot to a template, NTK does not override it.) The value stored in the slot is the name of the template. The `debug` slot is created when the project is compiled, therefore you do not see it in the browser.

In addition, NTK sets the compile time constant `debugOn` to `true` if Debug Build is checked. If the option is off, the constant is set to `nil`.

Auto-Close

If this is checked, opening this application closes all other auto-close applications. You should normally keep this checked.

Icon File/Icon Name

These two popups control the icon that displays above your application name in the Extras drawer.

The Icon File popup presents a list of resource files in your project. The Icon Name popup presents a list of named PICT resources from the chosen file.

Platform

Choose a platform (one of the files in the Platforms folder). The platform determines the size of the drawing view shown in the layout window.

Package Name

This unique name appears when the user selects Remove Software.

Copyright

This string is embedded in your package.

Store on Newton

These following two options have no effect on how NTK builds your application. Rather, they control how the package is stored on the Newton:

Compressed	When selected, the package is compressed on the Newton. It will be smaller but slower.
Not Compressed	When selected, the application is not compressed.

Version

This number is used to differentiate between different versions of the same application.

Copy Protected

Setting this flag specifies that the application should not be copied. In particular, this means that Newton Connection will not copy the application when doing a backup. Other software that copies applications should respect this flag, although there is nothing that requires it.

Building and Downloading

Building a Package

To build an application from the project file, use Build Package from the Project menu. This builds a package file in the same folder as your project. It uses the name of the project followed by the suffix ".pkg".

NTK Toolkit App

The NTK Toolkit App on the Newton is used for downloading packages (including applications) you build with NTK. It is also used to debug your applications using the Inspector.

The Inspector

NTK contains an Inspector window, where you type code that is compiled into NewtonScript, downloaded to the Newton, executed, and then the result is displayed in the Inspector window.

When you press the enter key, the selected text (or the line with the insertion point) is compiled, downloaded, and evaluated.

To work, the Inspector must be connected to the Newton. Here is how to connect the Inspector:

On the Macintosh

- Select Connect Inspector from the Window menu.

On the Newton

- Tap the Toolkit icon to open the Toolkit App slip (make sure the type of connection is set to Macintosh serial).
- Tap the Connect Inspector button. The Newton displays a progress slip momentarily and then that slip disappears (see FIGURE F.50).

FIGURE F.50 Connecting the Inspector on the Newton.

Selecting Open Inspector on the Macintosh merely opens the Inspector window. You still need to select Connect Inspector to establish a connection.

Downloading a Package

You must already have installed NTK Toolkit App and have a serial cable between your Newton and your Macintosh (see "Installing Demo NTK" on page 395) in order to download a package. Choose Download Package from the Project menu. Then, on your Newton tap the Toolkit icon to open the Toolkit App slip. Make sure the type of connection is set to Macintosh serial and tap the Download Package button. The Newton will display a progress slip, while NTK will display a progress dialog. Within a few seconds, your package should appear in the Extras drawer.

Downloading a Package with the Inspector Connected

While the Inspector is connected, you can download packages automatically from the Macintosh—just select Download Application from the Project menu.

Common Problems and Questions about NTK

Here are a list of the most common problems people have with NTK.

1. When I look at my templates in my layout window, they sometimes appear to be in the wrong location. When I check the `viewBounds` and `viewJustify` slots of these templates, the values are correct. Nevertheless, both preview and nonpreview displays look wrong (see FIGURE F.51).

   A: Download your application and check it on the Newton. If it is correct on the Newton it doesn't matter what it looks like on the Macintosh screen. There is a problem in how NTK displays templates that use relative full justification. This should be fixed in future versions of NTK, but for now remember the most important rule: If it is right on the Newton, it is right (regardless of what the NTK layout window looks like).

2. I have created a template that I can't see in the NTK layout window. How do I select it in order to delete it?

A: From the browser window, double-click on the template in the template list (FIGURE F.52). This selects the template in the layout window. Now bring the layout window to the front and select Cut from the Edit menu.

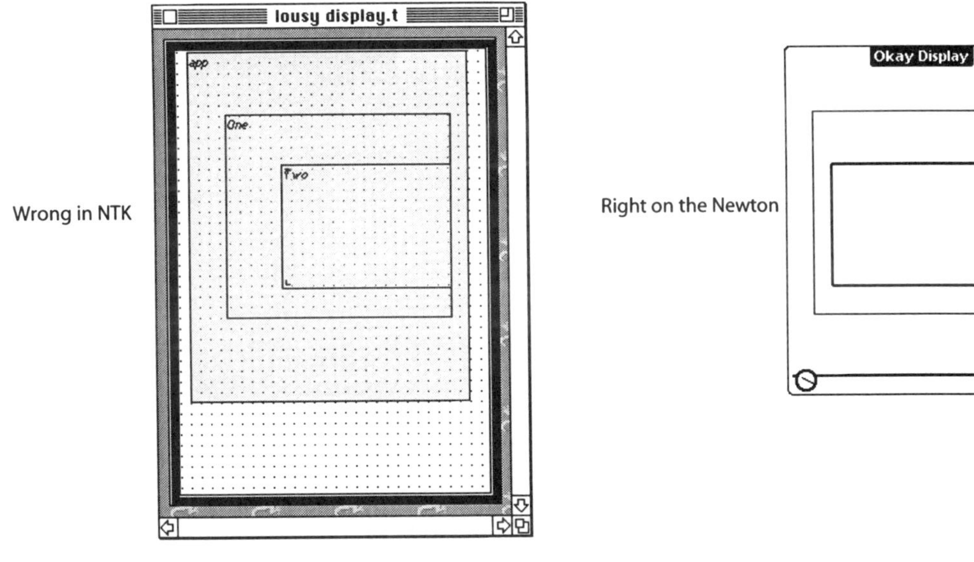

FIGURE F.51 NTK display anomaly.

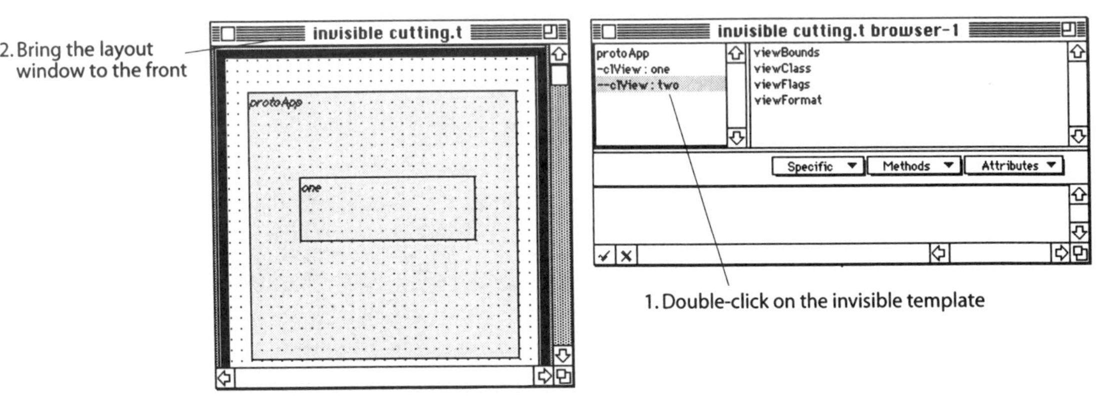

FIGURE F.52 Deleting an invisible view.

3. I can't get my PowerBook Duo to make a connection to my Newton.

   A: Make sure you have turned AppleTalk off. If this doesn't fix the problem, make sure you have the right port selected (see the note on page 400).

4. I have heard that Newton Toolkit comes with the file "AppleTalk ADSP Tool". Do I need it to use Demo NTK?

   A: No. The only additional capability this file provides is to allow downloading packages from NTK over AppleTalk rather than using a serial connection. Connecting the Inspector requires a serial connection.

Appendix G

Where To Go From Here

Life is to be lived. If you have to support yourself, you had bloody well better find some way that is going to be interesting. And you don't do that by sitting around wondering about yourself.

—*Katharine Hepburn*

There are quite a few places that you can go to learn more about Newton programming. Apple has a couple of different developer programs, one of which might be right for you. There are several different on-line services that contain both sample code and formats for asking and receiving answers to programming problems. There are also several periodicals that are good sources for articles on Newton programming. Last of all we will talk about a couple of very useful tools you should consider getting if you are going to do any real Newton programming.

Apple's Associates and Partners Program for Newton

There are two main programs for Newton developers that are offered by Apple Computer. The first program, the Associates Program, is the place that most of you will find yourselves. The second program is for individuals or companies that want more extensive interactions. For more information about either program you can call Apple Computer at (408) 974-4897.

Newton Associates Program

To be a member of the Associates Program you pay $400 per year. For this annual fee you get a pile of stuff, including the following:

- Quantities of sample code shipped to you on CD.
- Technical documentation.
- Hardware and software discounts (for both Newton and Macintosh platforms).
- A monthly developer mailing.
- Invitations to annual developers' conferences.

Newton Partners Program

Being a Newton Partner gets you more help for more money. For an annual fee of $2500 you get all the things that Associates get as well as access to Developer Technical Support engineers via email. They will respond to your questions as well as look through your source code and at problems if you so request.

Comp.Sys.Newton.Programmer

The Usenet newsgroup **comp.sys.newton.programmer** has a number of Newton programmers (as well as a surprising number of Apple engineers) discussing Newton programming.

Newton Material on eWorld

There is information for Newton programmers in **Computer Center:Apple Customer Center:Apple Products & Technologies:The World of Newton:Llaama Lounge (Developers)**. The Sample Code and Tools folder contains official DTS sample code. Developer's Info Source contains official DTS questions and answers. As well, there are on-line interviews with various Newton programming personalities (at the time this was written, these were conducted Thursdays at 6:30 PM PST).

Newton Material on America Online

Under the keyword PDA, you'll find many discussions on Newton programming.

Newton Material on CompuServe

There is a Newton Developers+ Forum on CompuServe. You can get there using the GO keyword **NEWTDE**.

Newton Material on the Internet

The ftp site **newton.sys.uea.ac.uk** contains mirrors of many Newton ftp sites around the world.

The ftp site **ftp.apple.com** contains Developer Technical Support sample code and questions and answers.

The www page **http://newton.uiowa.edu** has a great deal of information on Newton development.

The NPC Mailing List

The NPC (Newton Programmer's Collective) mailing list is a programmer's mailing list. Send mail with the subject "Subscribe" to npc-request@mit.edu to subscribe.

PDA Developers Magazine

This is a magazine devoted to providing detailed technical information (both reviews and source code) on developing PDA software for Newton, Psion, GEOS, Magic Cap and Hewlett Packard devices. It is issued 6 times (bimonthly) a year by Creative Digital Systems. Subscriptions are $60 per year, or $80 per year for overseas orders. You can send a subscription request to:

PDA *Developers* (ISSN 1075-5829)

Creative Digital Systems

293 Corbett Avenue

San Francisco, CA 94144

(415) 621-4252

(415) 621-4922 (FAX)

You can also contact Creative Digital Systems via email at **cds@netcom.com** or **74774.50@compuserve.com.**

PDA *Developers* is a boon to any Newton developer particularly to a brand new one. Each issue is filled with a reviews of Newton and other wireless software and hardware products and a wealth of Newton source code. While it is true that some of the source code is better than other bits, it is all well worth a look.

develop Magazine

develop, The Apple Technical Journal is a quarterly publication of Apple Computer's Developer Press group. It is published quarterly in March, June, September, and December. *develop* issues and code are also available on AppleLink.

This publication has a monthly Newton Q&A section and occasionally contains Newton programming articles. These articles are always of high quality and well worth reading. Subscriptions to *develop* can be had from APDA. Contact via email at **dev.subs@applelink.apple.com** or you can write to *develop* directly at:

> *develop*
> Apple Computer, Inc.
> P.O. Box 531
> Mt. Morris, IL 61054

APDA

APDA offers programmers access for a wide variety of important technical material, including development tools, resources, training products, and information. The APDA Tools Catalog contains Apple and third-party development products.

To order products or to receive a catalog call: (800) 282-2732 in the U.S., (800) 637-0029 in Canada, (716) 871-6555 internationally, or (716) 871-6511 for faxes. You can order via email at any of the following addresses:

- AppleLink: APDA
- Internet apda@applelink.apple.com
- CompuServe 76666, 2405
- America Online: APDA

Newton Programming Classes

Apple Developer University provides Newton training for both the beginning and advanced Newton programmer. Currently there are three classes offered including Newton Essentials (beginning Newton programming), Newton Extended Topics (advanced topics in Newton programming), and Newton Communications. Both of us teach these five-day beginning and advanced programming classes for DU. On-site instruction can also be arranged. For more information you can reach the DU registrar at (408)974-6215.

NewtRTFM

This is a quick reference guide of the NTK 1.0.1 technical documentation in the form of a Newton digital book. The guide was written by Sandeep Shah. If you like electronic documentation, then this is for you. It offers reasonable navigating tools on nearly 800 Newton topics grouped into about 20 categories. NewtRTFM gives you the sample programs and documentation in one.

Both Mac and Window disks are available. You can purchase this product from Creative Digital Systems for $90 [293 Corbett Avenue, San Francisco, CA 94144, (415) 621-4252]. PDA *Developers* subscribers get a $10 discount.

ViewFrame

ViewFrame is an essential debugging tool for every Newton developer. It lets you look at any application's frames, including the ability to view and edit them. ViewFrame is composed of three different elements. These are:

- Programmer's Keyboard
- View Frame Editor
- ViewFrame.

The Programmer's Keyboard has all the features of the normal onscreen keyboard (except for word entry into the user dictionary) and can be installed on your Newton. This can make entering NewtonScript a much more pleasant task. The keyboard is available from both the editor and ViewFrame via the keyboard icon. Multiple keyboard layouts are available.

The ViewFrame Editor is a simple text editor in which you can create and execute NewtonScript code. Actually any Newton code that works in a program will also work here. ViewFrame allows you to view or modify any object on the Newton. ViewFrame is especially useful for displaying debugging information about your program when the Inspector is not available (for example, with serial communications).

Both Mac and Window disks are available. You can purchase this product from Creative Digital Systems for $70 [293 Corbett Avenue, San Francisco, CA 94144, (415) 621-4252]. PDA *Developers* subscribers get a $10 discount.

Index

A

C

D

E

H

I

K

L

M

N

O

P

Q

R

S

T

U

V

W

Apple Computer, Inc. Software License

PLEASE READ THIS LICENSE CAREFULLY BEFORE OPENING THE PACKAGE CONTAINING THE SOFTWARE. BY USING THE SOFTWARE YOU ARE AGREEING TO BE BOUND BY THE TERMS OF THIS LICENSE. IF YOU DO NOT AGREE TO THE TERMS OF THIS LICENSE, PROMPTLY RETURN THE UNUSED SOFTWARE DISK AND BOOK TO THE PLACE WHERE YOU OBTAINED IT AND YOUR MONEY WILL BE REFUNDED.

1. **License.** The application, demonstration, system and certain other software accompanying this License, whether on disk, in read only memory, or on any other media (the "Apple Software"), the related documentation and fonts are licensed to you by Apple. You own the disk on which the Apple Software and fonts are recorded but Apple and/or Apple's Licensor(s) retain title to the Apple Software, related documentation and fonts. This License allows you to use the Apple Software and fonts on a single Apple computer and make one copy of the Apple Software and fonts in machine-readable form for backup purposes only. You must reproduce on such copy the Apple copyright notice and any other proprietary legends that were on the original copy of the Apple Software and fonts. You may also transfer all your license rights in the Apple Software and fonts, the backup copy of the Apple Software and fonts, the related documentation and a copy of this License to another party, provided the other party reads and agrees to accept the terms and conditions of this License.

2. **Restrictions.** The Apple Software contains copyrighted material, trade secrets and other proprietary material and in order to protect them you may not decompile, reverse engineer, disassemble or otherwise reduce the Apple Software to a human-perceivable form. You may not modify, network, rent, lease, loan, distribute or create derivative works based upon the Apple Software in whole or in part. You may not electronically transmit the Apple Software from one computer to another or over a network. The use authorized hereunder is expressly limited to use in connection with doing the exercises contained in the book with which it is distributed and does not include the right to use the Apple software for purposes of building applications for any Newton-based products.

3. **Termination.** This License is effective until terminated. You may terminate this License at any time by destroying the Apple Software, related documentation and fonts and all copies thereof. This License will terminate immediately without notice from Apple if you fail to comply with any provision of this License. Upon termination you must destroy the Apple Software, related documentation and fonts and all copies thereof.

4. **Export Law Assurances.** You agree and certify that neither the Apple Software nor any other technical data received from Apple, nor the direct product thereof, will be exported outside the United States except as authorized and as permitted by the laws and regulations of the United States. If the Apple Software has been rightfully obtained by you outside of the United States, you agree that you will not re-export the Apple Software nor any other technical data received from Apple, nor the direct product thereof, except as permitted by the laws and regulations of the United States and the laws and regulations of the jurisdiction in which you obtained the Apple Software.

5. **Government End Users.** If you are acquiring the Apple Software and fonts on behalf of any unit or agency of the United States Government, the following provisions apply. The Government agrees:

(i) if the Apple Software and fonts are supplied to the Department of Defense (DoD), the Apple Software and fonts are classified as "Commercial Computer Software" and the Government is acquiring only "restricted rights" in the Apple Software, its documentation and fonts as that term is defined in Clause 252.227-7013(c)(l) of the DFARS; and

(ii) if the Apple Software and fonts are supplied to any unit or agency of the United States Government other than DoD, the Government's rights in the Apple Software, its documentation and fonts will be as defined in Clause 52.227-19(c)(2) of the FAR or, in the case of NASA, in clause 18-52.227-86(d) of the NASA Supplement to the FAR.

6. **Limited Warranty on Media.** Apple warrants the diskettes and/or compact disc on which the Apple Software and fonts are recorded to be free from defects in material and workmanship under normal use for a period of ninety (90) days

from the date of purchase as evidenced by a copy of the receipt. Apple's entire liability and your exclusive remedy will be replacement of the diskettes and/or compact disc not meeting Apple's limited warranty and which is returned to Apple or an Apple authorized representative with a copy of the receipt. Apple will have no responsibility to replace a disk/disc damaged by accident, abuse or misapplication. ANY IMPLIED WARRANTIES ON THE DISKETTES AND/OR COMPACT DISC, INCLUDING THE IMPLIED WARRANTIES OF MERCHNATABILITY AND FITNESS FOR A PARTICULAR PURPOSE ARE LIMITED IN DURATION TO NINETY (90) DAYS FORM THE DATE OF DELIVERY. THIS WARRANTY GIVES YOU SPECIFIC LEGAL RIGHTS, AND YOU MAY ALSO HAVE OTHER RIGHTS WHICH VARY BY JURISDICTION.

7. Disclaimer of Warranty on Apple Software. You expressly acknowledge and agree that use of the Apple Software and fonts is at your sole risk. The Apple Software, related documentation and fonts are provided "AS IS" and without warranty of any kind and Apple and Apple's Licensor(s) (for the purposes of provisions 7 and 8, Apple and Apple's Licensor(s) shall be collectively referred to as "Apple") EXPRESSLY DISCLAIM ALL WARRANTIES, EXPRESS OR IMPLIED, INCLUDING, BUT NOT LIMITED TO, THE IMPLIED WARRANTIES OF MERCHANTABILITY AND FITNESS FOR A PARTICULAR PURPOSE. APPLE DOES NOT WARRANT THAT THE FUNCTIONS CONTAINED IN THE APPLE SOFTWARE WILL MEET YOUR REQUIREMENTS, OR THAT THE OPERATION OF THE APPLE SOFTWARE WILL BE UNINTERRUPTED OR ERROR-FREE, OR THAT DEFECTS IN THE APPLE SOFTWARE AND THE FONTS WILL BE CORRECTED. FURTHERMORE, APPLE DOES NOT WARRANT OR MAKE ANY REPRESENTATIONS REGARDING THE USE OR THE RESULTS OF THE USE OF THE APPLE SOFTWARE AND FONTS OR RELATED DOCUMENTATION IN TERMS OF THEIR CORRECTNESS, ACCURACY, RELIABILITY, OR OTHERWISE. NO ORAL OR WRITTEN INFORMATION OR ADVICE GIVEN BY APPLE OR AN APPLE AUTHORIZED REPRESENTATIVE SHALL CREATE A WARRANTY OR IN ANY WAY INCREASE THE SCOPE OF THIS WARRANTY. SHOULD THE APPLE SOFTWARE PROVE DEFECTIVE, YOU (AND NOT APPLE OR AN APPLE AUTHORIZED REPRESENTATIVE) ASSUME THE ENTIRE COST OF ALL NECESSARY SERVICING, REPAIR OR CORRECTION. SOME JURISDICTIONS DO NOT ALLOW THE EXCLUSION OF IMPLIED WARRANTIES, SO THE ABOVE EXCLUSION MAY NOT APPLY TO YOU.

8. Limitation of Liability. UNDER NO CIRCUMSTANCES INCLUDING NEGLIGENCE, SHALL APPLE BE LIABLE FOR ANY INCIDENTAL, SPECIAL OR CONSEQUENTIAL DAMAGES THAT RESULT FROM THE USE OR INABILITY TO USE THE APPLE SOFTWARE OR RELATED DOCUMENTATION, EVEN IF APPLE OR AN APPLE AUTHORIZED REPRESENTATIVE HAS BEEN ADVISED OF THE POSSIBILITY OF SUCH DAMAGES. SOME JURISDICTIONS DO NOT ALLOW THE LIMITATION OR EXCLUSION OF LIABILITY FOR INCIDENTAL OR CONSEQUENTIAL DAMAGES SO THE ABOVE LIMITATION OR EXCLUSION MAY NOT APPLY TO YOU.

In no event shall Apple's total liability to you for all damages, losses, and causes of action (whether in contract, tort (including negligence) or otherwise) exceed the amount paid by you for the Apple Software and fonts.

9. Controlling Law and Severability. This License shall be governed by and construed in accordance with the laws of the United States and the State of California, as applied to agreements entered into and to be performed entirely within California between California residents. If for any reason a court of competent jurisdiction finds any provision of this License, or portion thereof, to be unenforceable, that provision of the License shall be enforced to the maximum extent permissible so as to effect the intent of the parties, and the remainder of this license shall continue in full force and effect.

10. Complete Agreement. This License constitutes the entire agreement between the parties with respect to the use of the Apple Software, related documentation and fonts, and supersedes all prior or contemporaneous understandings or agreements, written or oral, regarding such subject matter. No amendment to or modification of this License will be binding unless in writing and signed by a duly authorized representative of Apple.

About the Enclosed Floppy Disk

Wireless for the Newton

· Julie McKeehan and Neil Rhodes ·

The enclosed floppy disk contains all of the sample applications from this book, as well as a fully-functional, demonstration version of Newton Toolkit™ (NTK™), Apple Computer's complete development environment for the Newton®. With this disk and the system described below, you should be able to try out NewtonScript™ and build real, working Newton® applications. In order to build the most robust Newton® applications and have the right to distribute them, you'll have to buy the Newton Toolkit™ from the Apple Developer Association.

System Requirements for the Enclosed Disk:

Macintosh or compatible
Newton device
Serial cable to hook your Newton to the Macintosh

To Order Newton Toolkit
Call APDA at 1-800-282-2732

For information on developing for Newton
call Newton Developer Relations
at 408-862-7226

For Newton MessagePad support
call 1-800-SOS-APPL

Newton Developer Relations
Apple Computer, Inc.
1 Infinite Loop
MS 305-3A
Cupertino, CA 95014

For Newton Training

Apple Developer University
20525 Mariani Avenue, M/S 305-ITU
Cupertino, CA 95014

Telephone: 408-974-4897
Facsimile: 408-974-0544
AppleLink: DEVUNIV
Internet: devuniv@applelink.apple.com